PHYSICO-CHEMICAL WASTEWATER TREATMENT AND RESOURCE RECOVERY

Edited by **Robina Farooq** and **Zaki Ahmad**

Physico-Chemical Wastewater Treatment and Resource Recovery

http://dx.doi.org/10.5772/67803
Edited by Robina Farooq and Zaki Ahmad

Contributors

Eduardo Alberto López-Maldonado, Mercedes Teresita Oropeza-Guzmán, Karla Alejandra Suárez-Meraz, Agnieszka Węgrzyn, Tao Zhang, Rongfeng Jiang, Yaxin Deng, Gonzalo Montes-Atenas, Fernando Valenzuela, Magdalena Urbaniak, Anna Wyrwicka, Madhumita Ray, Adnan Khan, Tulip Chakraborty, Jing Wan, Nillohit Mitra Ray, Sreejon Das, Dalia Saad, Pay Drechsel, Deirdre Byrne, Ainhoa Rubio Clemente, Edwin Chica, Gustavo Peñuela, Mehrab Mehrvar, Ciro Fernando Bustillo-Lecompte, Jan Derco, Katarína Šimovičová, Jozef Dudáš, Mária Valickova, Carlos Barrera-Diaz, Nelly González-Rivas, Taner Yonar, Ayse Kurt, Nihan Goral, Berna Kiril, Ozge Sivrioglu

CBS, Edition **2018**

Published by InTech

Janeza Trdine 9, 51000 Rijeka, Croatia

Publishing Process Manager Martina Usljebrka

Technical Editor SPi Global

Cover Designer InTech Design Team

Additional hard copies can be obtained from orders@intechopen.com

Physico-Chemical Wastewater Treatment and Resource Recovery, Edited by Robina Farooq and Zaki Ahmad
p. cm.
Print ISBN 978-953-51-3129-8
Online ISBN 978-953-51-3130-4

Contents

Preface

This volume covers the physico-chemical treatment technologies for wastewater treatment and recovery of materials, energy and water as a resource. It mainly consists of processes including oxidation, adsorption and management of persistent pollutants in wastewater. It consists of three sections. Oxidation of contaminants in wastewater is well placed in Section 1. The chapter 'Removal of BTX Contaminants with O_3 and O_3/UV Process' describes a promising procedure for the removal of BTX components from aquatic environment. A kinetic model for pollutant degradation by UV/H_2O_2 is developed and validated in the chapter 'Kinetic Modelling of the UV/H_2O_2 Process: Determining the Effective Hydroxyl Radical Concentration'. Fenton reagent and electrolysis process are used to generate hydroxyl radicals in the chapter 'Electrooxidation-Ozonation: A Synergistic Sustainable Treatment Process'. Main parameters to control the process are addressed with examples. Iron-containing active phase that was deposited on natural layered silicate (vermiculite) using several techniques is discussed in the chapter 'Removal of Phenol from Wastewater Using Fenton-Like Reaction over Iron Oxide–Modified Silicates'. This chapter also explains experimental methods for the treatment of model industrial effluent of phenol. The chapter 'Treatment of Antibiotics in Wastewater Using Advanced Oxidation Process' presents an overview of the literature on antibiotics and their removal from water by advanced oxidation processes. It was found that most of the investigated AOP methods for the oxidation of antibiotics in water are direct and indirect photolyses with the combinations of H_2O_2, TiO_2, ozone and Fenton's reagent.

Section 2, 'Micropollutants and Their Removal Processes', describes studies about the micropollutants entering into the environment after conventional wastewater treatment facilities. Its chapter 'Micropollutants in Wastewater: Fate and Removal Processes' explains the estimation of micropollutants by physical properties instead of using tedious and cost-intensive analysis. The section includes information concerning wastewater treatment plants functioning in respect to PCDDs/PCDFs and PCBs. Fate and impact of these pollutants are extensively elaborated in 'PCDDs/PCDFs and PCBs in Wastewater and Sewage Sludge'.

Section 3, 'Resource Recovery and Management', is designated for recovery of chemicals and management of wastewater. The study about 'phosphorus recovery by struvite crystallization from livestock wastewater and reuse as fertilizer' represents that struvite crystallization is a promising tool for recovering phosphorus from livestock wastewater.

The chapter 'Slaughterhouse Wastewater: Treatment, Management and Resource Recovery' draws the attention of readers and researchers towards the wastewater of meat processing industry. In this chapter, the regulatory framework relevant to the SWW management, environmental impacts, health effects and treatment methods is discussed. The chapter 'Wastewater Treatment via Low-Cost Adsorption Technologies' addresses the wastewater treatment of mining residues for removal of metals via adsorption methodologies, along with classic models of adsorption thermodynamics and kinetics that are well presented.

The chapter 'The Integral Use of Nejayote: Characterization, New Strategies for Physicochemical Treatment and Recovery of Valuable Products' describes physico-chemical treatment strategy. It addresses the problem of *nejayote* and its reuse using waste from shrimp shells. The chapter 'Effective Management of Wastewater for Future Water Security "Social Perspectives"' discusses how the adopting of a holistic methodology that acknowledges sociological factors including community participation and public involvement, social perception, attitudes, impression, gender roles and public acceptance would lead to improvements in wastewater management practice.

I would like to express my gratitude to Prof. Dr. Zaki Ahmad who started to work with me as coeditor. Prof. Zaki passed away during his work on this book. His efforts and contributions are highly appreciated, and his services as book editor are highly acknowledged.

I would like to thank Ms. Martina Usljebrka, Publishing Process Manager, for enabling me to publish this book. I want to thank my husband, Prof. Dr. Saleem Farooq Shaukat, my daughter Kinza Farooq, sons Abdul Basit and Faisal Farooq and grandchildren Zoha Fatima and Aarib Basit who kept me motivated to accomplish this work. I am grateful for my father Mr. Muhammad Mukhtar and my mother Mrs. Rafia Mukhtar, my sisters and my brothers who always supported and encouraged me throughout my life.

Prof. Dr. Robina Farooq
Department of Chemical Engineering
COMSATS Institute of Information Technology
Lahore, Pakistan

Prof. Dr. Zaki Ahmad
COMSATS Institute of Information Technology
Lahore, Pakistan

Note from the publisher

It is with great sadness and regret that we inform the contributing authors and future readers of this book that the editor, Prof. Zaki Ahmad, passed away during his work on publications and before having a chance to see them.

Prof. Ahmad was InTech's long-term collaborator and edited his first book with us in 2011 *Recent Trends in Processing and Degradation of Aluminium Alloys,* followed by publications 'Aluminium Alloys - New Trends in Fabrication and Applications', 'New Trends in Alloy Development, Characterization and Application' and 'High Temperature Corrosion'. This fruitful collaboration continued until his final days when he was acting as a coeditor on the books *Biological Wastewater Treatment and Resource Recovery* and *Physico-chemical Treatment of Wastewater and Resource Recovery.*

We would like to acknowledge Dr. Zaki Ahmad's contribution to open access scientific publishing, which he made during 6 years of dedicated work on edited volumes, and express our gratitude for his always pleasant cooperation with us.

InTech Book Department Team

Section 1

Oxidation of Contaminants in Wastewater

Chapter 1

Removal of BTX Contaminants with O_3 and O_3/UV Processes

Ján Derco, Katarína Šimovičová, Jozef Dudáš and Mária Valičková

Additional information is available at the end of the chapter

http://dx.doi.org/10.5772/65889

Abstract

The legal basis for the monitoring of priority and priority hazardous substances in water, sediment, and biota follows from Directive 2008/105/EC which defines the good chemical status to be achieved by all Member States together with the Water Framework Directive 2000/60/EC. The BTX compounds are considered to be the most toxic components of gasoline. Thus, organic petroleum components can induce a serious problem to public health and the aquatic environment. The effect of ozone and ozone/UV on degradation of the BTX in a model water was studied. The results indicate that the highest BTX removal rates were observed during the first 5 min of the process for all investigated pollutants. The treatment efficiencies above 90% were observed in all investigated pollutants after 40 min of ozonation. The results show a significant proportion of stripping in the removal of BTX components. Higher overall efficiency was observed by O_3/UV process after abstracting share of stripping process. Application of investigated processes appears to be a promising procedure for removal of petrol aromatic hydrocarbons from aquatic environment. However, for practical application, an improvement of process removal efficiency and investigation of impact of ozonation intermediates and products on aquatic microorganisms are required.

Keywords: BTX, jet-loop reactor, ozone, O_3/UV, wastewater treatment

1. Introduction

The adoption of the Framework Directive on water [1] provides a policy tool that enables sustainable protection of water resources. The Decision No. 2455/2001/EC of the European Parliament and the Council of November 2001 [2] established the list of 33 priority substances or group of substances, including the priority hazardous substances, presenting a significant risk to water pollution or *via* the aquatic environment including such risks to water used for the abstraction of drinking water.

Hazardous substances are defined as substances or groups of substances that are toxic, persistent, and liable to bioaccumulation, and other substances or groups of substances which give rise to an equivalent level of concern. The EC member countries have extended this list with relevant pollutants for individual countries. Thus, in the supplement of the Water Act [3] there have been identified altogether 59 relevant substances for SR.

The BTEX contaminants consist of benzene, ethyl benzene, toluene, and three isomers of xylene. These compounds are the volatile organic compounds (VOCs) found in petroleum derivatives such as petrol (gasoline). They represent one of the main groups of soluble organic compounds that is present in wastewater from refinery. They are the most toxic components of gasoline. These substances can lead to serious health problems ranging from irritation of eyes, skin, and mucous membranes and ending with weakened nervous system, decreased bone marrow function, and cancer. Benzene in particular is highly toxic. The World Health Organization classifies the substance as carcinogenic. It is also on the list of the priority substances [4].

Many oil substances have acute toxic effect on aquatic microorganisms with possible chronic consequences [5].

Commonly used wastewater treatment processes usually apply physical and physiochemical processes. Thus, the discharge of organic pollutants may create some environmental problems, particularly at microlevel. The aromatic oil fraction consists mainly of polyaromatic hydrocarbons (PAHs) and is more toxic and persistent than the aliphatic hydrocarbons [6]. Leakages including release of petroleum products, e.g., gasoline, diesel fuel, lubricating, and heating oil, from leaking oil tanks are the most frequent sources of soil and groundwater contaminations with BTEX substances. They are polar and readily soluble, and thus they are able to penetrate into soil and groundwater and cause serious environmental problems. These compounds dispose of acute and long-term toxic effects [4].

BTEX are among the most frequently detected contaminants in US public drinking-water systems that rely on groundwater sources [7]. These organic compounds make up a significant percentage of petroleum. The most contaminated locality of hazardous BTEX substances in the Slovak Republic is the airport at Sliač, Sliač-Vlkanová territory, contaminated by the Soviet Army, and the gas station in Rajecké Teplice where BTEX contaminants were identified as dominant in the groundwater. BTEX were also found in groundwater in Bratislava due to poor technical conditions of technological equipment (old stocks of aviation fuel) and the subsequent uncontrolled release of oil into the rock mass at the Airport of M. R. Štefanik [8].

Ozone is a very powerful oxidizing agent ($E^\circ = 2.07$ V). Ozone may react with organic compounds in two ways: by direct reaction as molecular ozone or by indirect reaction through formation of secondary oxidants like free radical species [9–11]. In practice, both mechanisms may occur depending on the type of chemical wastewater pollution.

At low pH, the predominant reaction mechanisms are the direct electrophilic attack by molecular ozone [12], i.e., ozonolysis.

Under such conditions, ozone is a selective oxidant and reacts with multiple bonds (C=C, C=N, N=N, etc.) , but only at low rates with single bonds (C-C, C-O, O-H). At high pH, indirect reaction occurs, i.e., organics are degraded by secondary oxidants/chain reaction involving powerful radicals including OH, which are produced by ozone decomposition. These radicals are very strong and nonselective oxidants. Hydroxyl radicals can be formatted by increasing pH or by decomposition of O_3 with homogeneous and heterogeneous catalysts.

The main goal of our research was to study the feasibility of ozone and combination of O_3/UV processes for removal of selected benzene, toluene, and xylenes (BTX) from water/wastewater. Investigation of process kinetics and stripping of volatile substances were also performed.

2. Experimental methods

2.1. Experimental equipment

Ozonation trials were performed in a laboratory ozonation reactor. A schematic of the ozonation apparatus is illustrated in **Figure 1** [13]. Ozonation jet-loop reactor was operated in batch mode with regard to wastewater and in continuous mode with regard to gas. Active volume of the reactor was 3.5 L. The treated wastewater was transferred into ozonation reactor before starting operation of the reactor. A membrane pump was used to maintain external circulation of liquid reaction mixture. Pulsation of recirculated external flow was minimized with diaphragm pulsation damper (SERA 721.1, Seybert & Rahier, Immenhausen, Germany). Ozone was generated using a Lifetech generator with maximum production of 5 g h^{-1}. Ozone was generated at 50% of the maximum ozone generator's power and maintaining continuous oxygen flow of 60 L h^{-1}. A mixture of O_3 and O_2 was injected into a wastewater sample through a Venturi ejector. At the same time, the ejector sucked the mixture of O_3 and O_2 from the reactor headspace.

This, together with external circulation, should improve the efficiency of ozone utilization in the ozonation reactor. Pen-Ray UV lamp with wavelength 254 nm was used to generate hydroxyl radicals in the reactor. The outfall of reaction-gas mixture was transported into a destruction glass column by a fine-bubble porous distributive device. The destructive reactive column contained a potassium iodide solution. The active volume of the destructive reactive column was 1.0 dm^3. An excess O_3 was destructed in this device [13].

2.2. Analytical methods

For determination BTEX compounds in model wastewater, gas chromatography was used with MS detector in connection with headspace autosampler.

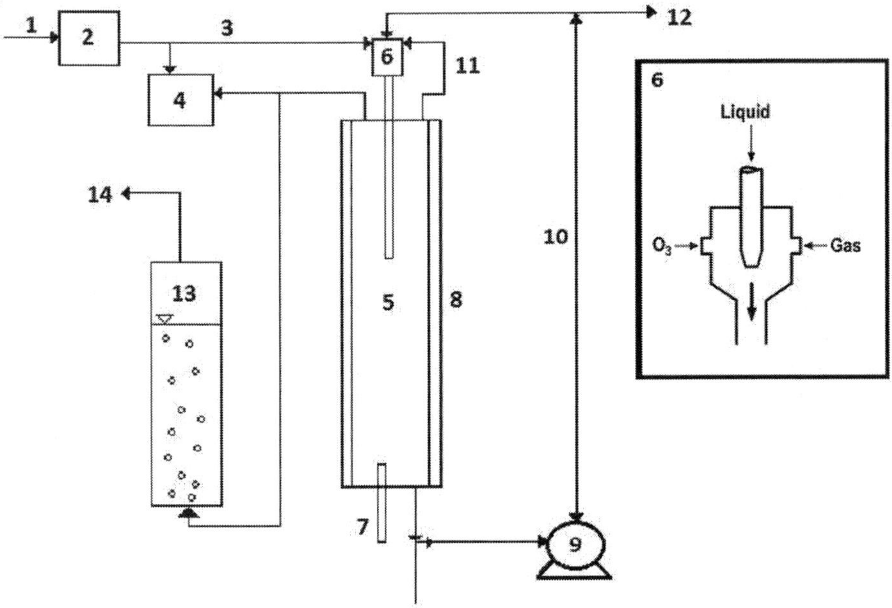

Figure 1. Schematic scheme of the ozonation equipment. 1—O_2, 2—O_3 generator, 3—O_2/O_3 gas flow, 4—UV detector of O_3, 5—O_3 reactor, 6—Venturi nozzle, 7—UV source, 8—tempering jacket, 9—recirculation pump, 10—external reaction mixture recirculation loop, 11—sucking of gas from head space, 12—sampling device, 13—outfall of gas, 14—destruction device for residual O_3.

Gas chromatographic method after static headspace extraction was used for quantification of the BTX in water. Headspace part was analyzed by gas chromatography with MS detector (Agilent Technologies 7890A GC Systems). All substances used for preparation of model wastewater and standard stock solutions were purchased from Dr. Ehrenstorfer.

2.3. Processing of experimental data

Desorption of volatile pollutants from water solution is proportional to its concentration in water solution

$$\frac{dS}{dt} = -k \cdot S \tag{1}$$

where k/h^{-1} is desorption rate.

Experimental data of BTX degradation were fitted by the zero (Eq. (2))-, first (Eq. (3)-, and second (Eq. (4))-order reaction kinetic models. For a batch reaction system, under the assumption of a constant reaction volume, the following relationships were obtained.

$$S_t = S_0 - k_0 t \tag{2}$$

$$S_t = S_0 \exp(-k_1 t) \quad (3)$$

$$S_t = \frac{S_0}{(1 + S_0 k_2 t)} \quad (4)$$

where S_t (g m^{-3}) stands for the content of BTX substances in model wastewater in time t, S_0 (g m^{-3}) is the beginning content of BTX substances in model wastewater, k_0 (g m^{-3} h^{-1}), k_1 (h^{-1}), and k_2 (g^{-1} m^3 h^{-1}) are the rate constants for the kinetics of the zero, the first, and the second reaction order, respectively [13].

The grid search optimization method was applied to calculate values of parameters of the used mathematical models. The objective function was defined as the sum of squares between the measured and calculated values of BTX components divided with the number of measurements reduced by the number of estimated parameters [13, 14].

3. Results of the work

The removals of studied compounds with the ozonation time are presented in **Figure 2**. The initial concentrations of benzene, toluene, o-xylene, and p-xylene were 800, 1600, 800, and 600 μg l^{-1}, respectively. From **Figure 2**, it is obvious that for all studied pollutants measured, the highest removal rates were observed within the first 5 min of the process. The highest affinity of ozone was measured toward p-xylene (59.6% removal efficiency). On the other hand, the lowest treatment efficiency was measured for benzene (20%) within the same ozonation time.

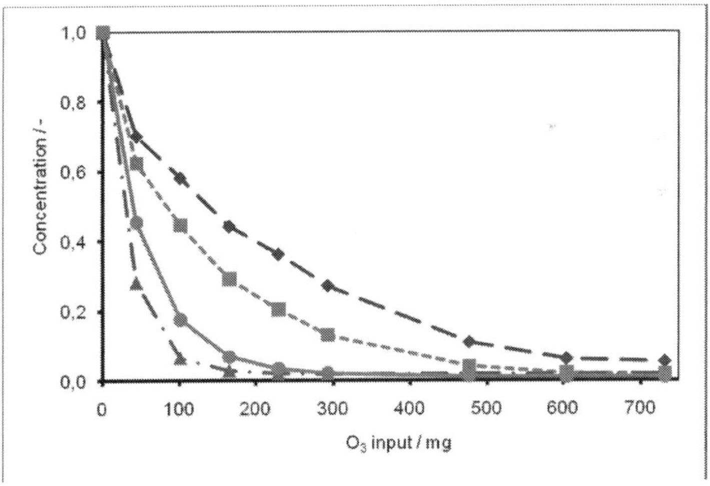

Figure 2. BTX (dimensionless values) removal during ozonation of model wastewater (◆ benzene, ■ toluene, ▲ p-xylene, ● o-xylene).

t [min]	Benzene [%]	Toluene [%]	p-Xylene [%]	o-Xylene [%]
5	20.1	26.4	59.6	44.4
20	53.9	68.3	85.7	86.4
60	84.6	89.6	90.0	89.9

Table 1. Removal efficiency values of BTX compounds with ozone.

The treatment efficiencies of BTX components increased with the increase of ozonation time [15]. The highest treatment efficiency was observed for p-xylene (81.3% during the first 20 min of ozonation). The second highest treatment efficiency (72.2%) was measured for o-xylene. Final removal efficiencies of BTX constituents were observed in the range from 86.4 to 90%. The values of removal efficiency during O_3 process are summarized in **Table 1**.

The best fit of experimental degradation data of all studied pollutants was obtained by the first-order kinetic model.

The removals of studied BTX by O_3/UV treatment time are presented in **Figure 3**. It is obvious that the removal rates of BTX by O_3/UV process are higher in comparison with the removal rates observed with ozone alone. This is confirmed also by the values of removal efficiency given in **Table 2**.

Comparisons of p-xylene and o-xylene removals using O_3 only for oxidation and O_3/UV treatment of model wastewater are presented in **Figures 4** and **5**, respectively. Slightly higher removal rates of these two pollutants were measured when treated with ozone alone.

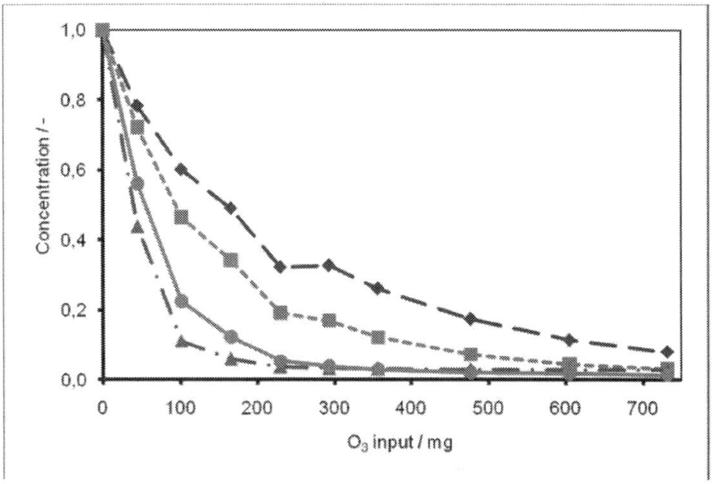

Figure 3. BTX (dimensionless values) removal by O_3/UV treatment of model wastewater (◆ benzene, ■ toluene, ▲ p-xylene, ● o-xylene).

http://dx.doi.org/10.5772/65889

t [min]	Benzene [%]	Toluene [%]	p-xylene [%]	o-xylene [%]
5	21.7	27.9	56.9	43.9
20	67.8	80.8	96.3	94.7
60	91.9	96.9	97.2	98.6

Table 2. Removal efficiency values of BTX compounds by O_3/UV process.

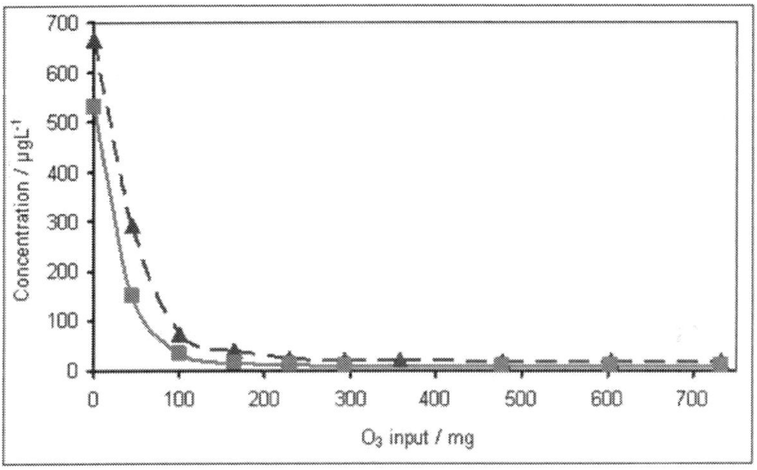

Figure 4. p-Xylene concentration profiles during O_3 and O_3/UV treatment of model wastewater (■ p-xylene-O_3, ▲ p-xylene-O_3/UV).

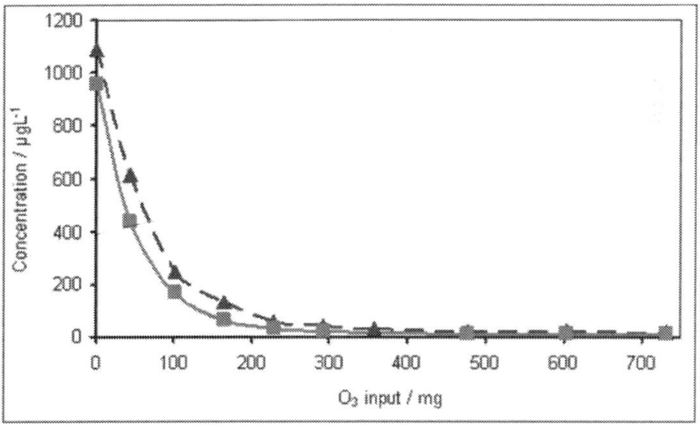

Figure 5. o-YLENE concentration profiles during O_3 and O_3/UV treatment of model wastewater (■ o-xylene-O_3, ▲ o-xylene-O_3/UV).

However, insignificant differences in the removal rates and treatment efficiencies follow from the treatment of the investigated BTX compounds using O_3 and O_3/UV. The data presented above represent overall removal of pollutants during O_3, or O_3/UV treatment, i.e., an effect of stripping of pollutants content is also included in the data.

The effect of gas striping of the investigated BTX compounds at the conditions of ozonation and O_3/UV trials was also studied. Volatility of substances depends on the size of molecules as well as on the vapor pressure [16]. With the increase of the molecular weight, the solubility of substance in water decreases. Important factor influencing solubility in water is hydrophobicity of substances. Solubility in water decreases with increase hydrophobicity of substance [17]. The information on hydrophobicity gives octanol-water partitioning coefficient value. Evaporation of substances correlates with vapor pressure [17] and is strongly influenced by temperature and the pressure of the system [18].

Volatility of substances can be quantified by the values Henry low. With the increase of the Henry's constant value, solubility of substance in water decreases. The values of basic physical-chemical properties of BTX components are given in **Table 3** [19].

Comparison of o-xylene, p-xylene, benzene, and toluene concentration profiles measured during the stripping only and ozonation treatment of model water are presented in **Figures 6–9**, respectively. As it was already mentioned, 10 min of ozonation corresponds to input of 45 mg O_3 per liter of active volume of the jet-loop ozonation reactor.

It is obvious from the presented results that stripping can significantly contribute to the removal of investigated compounds during the ozonation and O_3/UV treatments. The higher contribution of stripping to the overall removal of the component during the ozonation is for benzene. This observation correlates very well with the physical properties of the different components (**Table 3**). Similar results were obtained for toluene (36.8 $\mu g\ L^{-1}\ min^{-1}$) in comparison to benzene (32.0 $\mu g\ L^{-1}\ min^{-1}$).

	Benzene	**Toluene**	**o-Xylene**	**p-Xylene**
Molecular weight[g mol^{-1}]	78.11	92.13	106.16	106.16
Water solubility[mg L^{-1}]	1700	515	175	198
Vapor pressure(at 20°C) [kPa]	12.6923	3.7863	0.8799	0,8799
Boling point[°C]	80.1	110.6	144.4	138.4
Octanol-water partition coefficient (at 20°C)[log Kow]	2.13	2.69	2.77	3.15
Henry's law constant(at 25°C) [kPa m^3 mol^{-1}]	0.55	0.67	0.50	0.71
AA-EQS [$\mu g\ L^{-1}$]	10[a]	100[a]	10[b]	10 [b]

Notes: AA-EQS, Annual Average Value of Environmental Quality Standard (EQS).
[a] European Commission (2008) List of EQS [20].
[b] Regulation of the Slovak Government [21] List a EQS of total xylenes of 10 $\mu g\ L^{-1}$.

Table 3. Basic characteristics for studied organic components of BTX.

Concentration profiles of the studied BTX substances during ozonation, i.e., after excluding removal of the compounds by stripping process during ozonation of the model wastewater, are presented in **Figure 10**. In other words, the values plotted in **Figure 10** were obtained by subtraction of the concentrations of individual components due to stripping by oxygen flow from the total concentrations obtained during ozonation of the model wastewater.

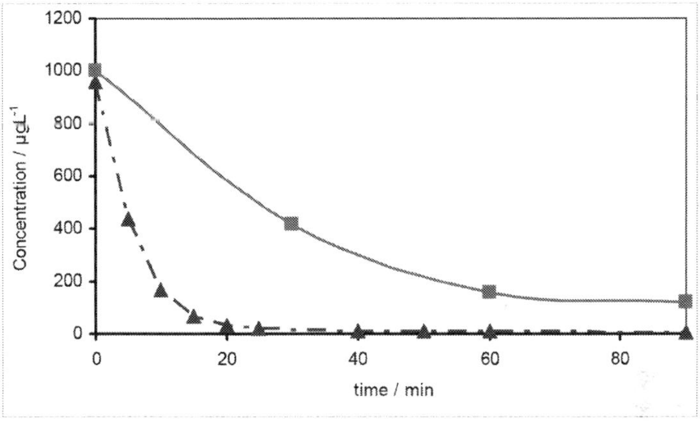

Figure 6. o-Xylene concentration profiles during stripping and ozonation of model wastewater (■ o-xylene-striping, ▲ o-xylene-ozonation).

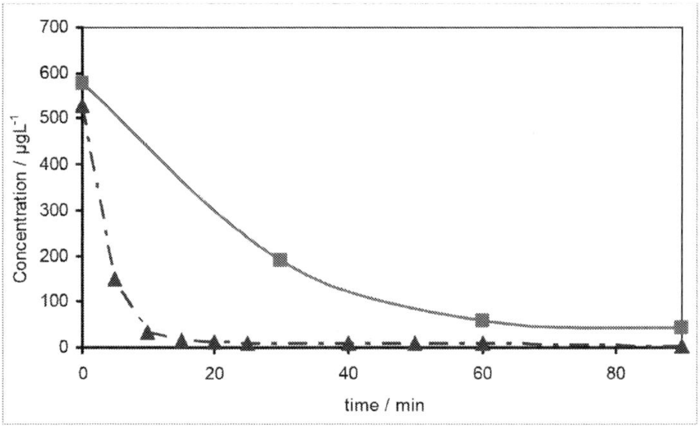

Figure 7. p-Xylene concentration profiles during stripping and ozonation of model wastewater (■ p-xylene-striping, ▲ p-xylene-ozonation).

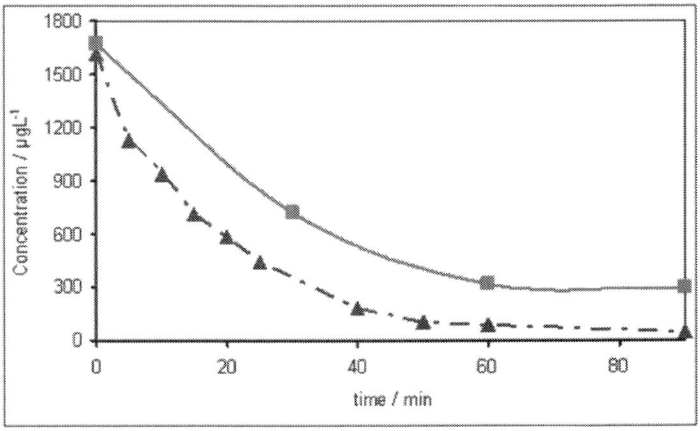

Figure 8. Benzene concentration profiles during stripping and ozonation of model wastewater (■ benzene-striping, ▲ benzene-ozonation).

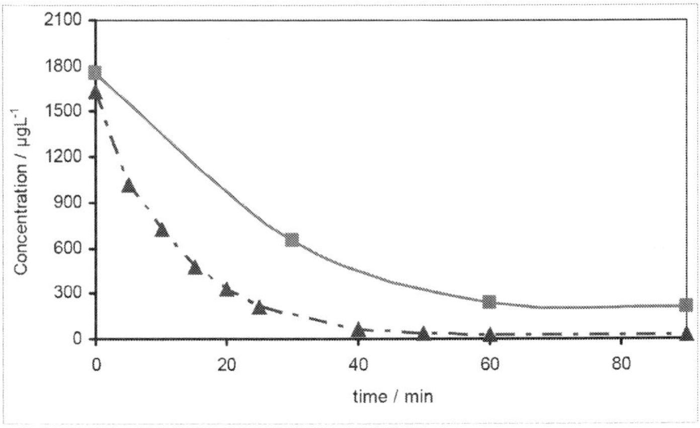

Figure 9. Toluene concentration profiles during stripping and ozonation of model wastewater (■ toluene-striping, ▲ toluene-ozonation).

Ozonation and stripping experimental trials were performed at the same operational conditions except presence of ozone in the system for stripping tests. The removal efficiencies for BTX components due to ozone oxidation of the model wastewater sample are given in **Table 4**.

Experimental data were processed using kinetic models to evaluate an order of the reaction. The calculated concentration profiles obtained by kinetic models corresponded to the best fit of experimental data (**Table 4**) for ozone oxidation, i.e., after excluding contribution of stripping to overall BTX concentrations during ozonation.

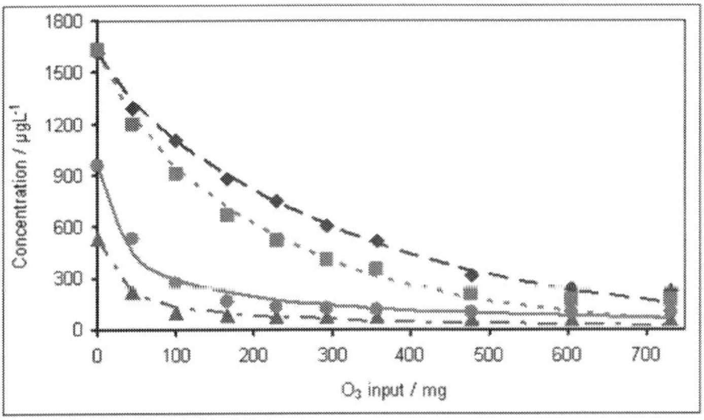

Figure 10. Experimental (◆ benzene, ■ toluene, ▲ p-xylene, ● o-xylene) and calculated (lines) BTX concentration profiles during ozonation of model wastewater.

t [min]	Benzene [%]	Toluene [%]	p-Xylene [%]	o-Xylene [%]
5	20.1	26.4	59.6	44.4
20	53.9	68.3	85.7	86.4
60	84.6	89.6	90.0	89.9

Table 4. Removal efficiency values of BTX compounds by O_3 (stripping excluded).

t [min]	Benzene [%]	Toluene [%]	p-Xylene [%]	o-Xylene [%]
5	21.7	27.9	56.9	43.9
20	67.8	80.8	96.3	94.7
60	91.9	96.9	97.2	98.6

Table 5. Removal efficiency values of BTX compounds by O_3/UV.

The highest removal rates were observed during the first 5 minutes of ozonation for all investigated pollutants. However, there is approximately 10% difference in removal efficiency (**Table 1**) and it is caused by ozonation reaction only (**Table 4**).

The best fit of experimental degradation data of benzene and toluene was obtained by the first-order kinetic model. On the other hand, the second-order kinetic model was more appropriate for description degradation of xylenes. The rate constant values and the values of the correlation coefficient r_{XY} corresponding to treated kinetic data after subtracting volatilized portions due to oxygen stripping are summarized both for O_3 and O_3/UV treatments in **Table 6**.

Pollutants		k_1 [h^{-1}]	k_2 [g m^{-3} h^{-1}]	
Benzene	O_3	3.92×10^{-2}	–	0.9964
	O_3 /UV	5.42×10^{-2}	–	0.9770
Toluene	O_3	5.55×10^{-2}	–	0.9885
	O_3 /UV	4.56×10^{-2}	–	0.9977
o-Xylene	O_3	–	2.38×10^{-4}	0.9793
	O_3/UV	–	1.78×10^{-4}	0.9636
p-Xylene	O_3	–	5.78×10^{-4}	0.9839
	O_3/UV	–	3.82×10^{-4}	0.9980

Table 6. Kinetic parameters and statistical characteristics–striping excluded.

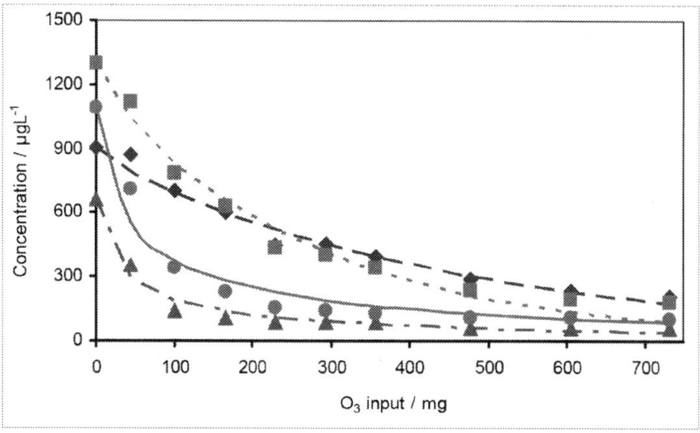

Figure 11. Experimental (◆ benzene, ■ toluene, ▲ p-xylene, ● o-xylene) and calculated (lines) BTX concentration profiles during O_3/UV treatment of model wastewater.

Similar concentration profiles for the studied BTX compounds during the O_3/UV treatment of the model wastewater are presented in **Figure 11**. The values presented in **Figure 11** were also obtained by subtraction of concentrations of individual components caused by gas stripping. The calculated concentration profiles were also obtained by kinetic models corresponding to the best fit of experimental values (**Table 6**) for O_3/UV treatment, i.e., after excluding contributions of stripping of individual compounds. The removal efficiencies values are given in **Table 5**.

Removal efficiencies are very close to those given in **Table 2**. In the case of o-xylene and p-xylene, the best fit was obtained by the second-order kinetic model (**Table 6**). The first-order kinetic models for benzene and toluene may indicate significant influence of gas stripping on total removal of these compounds from solution and the process is probably determined by physical phenomena rather than chemical.

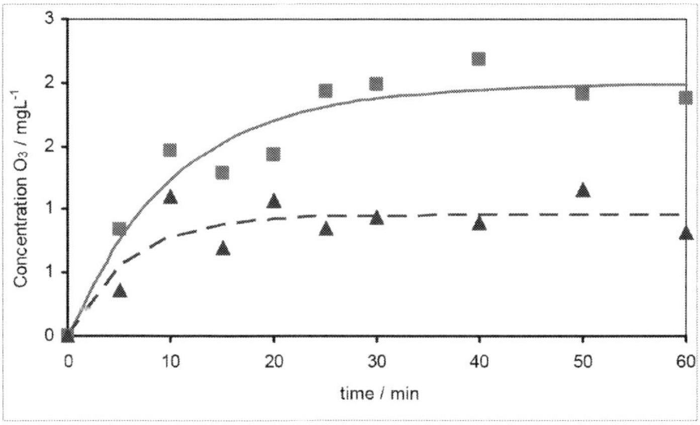

Figure 12. Ozone concentration profiles in the jet-loop reactor during treatment of model wastewater with O_3 (■) and O_3/UV (▲).

Figure 12 illustrates ozone concentration profiles in the jet-loop reactor during treatment of model wastewater with O_3 and O_3/UV.

Lower ozone concentration in water or higher ozone consumption in the system is obvious for O_3/UV treatment. Lower experimental ozone concentration for O_3/UV reaction system can be explained by decomposition and hydroxyl radicals formation. However, insignificant increase of removal rates was observed as a result of radical reaction mechanism. On the other hand, the values of kinetic constant for all compounds (**Table 6**) are slightly higher when ozone alone was applied in comparison to O_3/UV treatment. Thus, the higher removal rates for ozone treatment of BTX were observed in comparison to the O_3/UV treatment process.

4. Conclusion

Effect of ozone and O_3/UV treatments on BTX components were investigated in this study. Investigation of stripping of volatile substances was also performed.

The highest removal rates for all investigated BTX components were observed during the first 5 min of processing for both ozonation and O_3/UV treatment processes.

Ozone showed the highest affinity to p-xylene. The lowest removal efficiency was measured for benzene. Treatment efficiencies above 90% were observed for all investigated pollutants after 40 min of ozonation. Longer ozonation time resulted in very low enhancements of removal efficiencies of both ozonation and O_3/UV treatment processes.

Application of O_3/UV treatment had no significant effect in comparison with ozonation only, particularly for benzene. In case of o-xylene and p-xylene removal efficiencies, over 90% were observed after 20 min of the process. Forty minutes of the process were needed for more than 90% removal efficiency of ethylbenzene.

Due to high volatility of BTX components their removal from liquid phase can be significantly influenced by stripping. According to physical characteristics, the highest stripping can be expected for benzene and toluene [15].

From the processing of experimental data by simple kinetic models [13, 14], removal of o-xylene and p-xylene was best achieved by the second-order kinetic model. On the other hand, best fit of experimental data for benzene and toluene was obtained using the first-order kinetic model.

From the result of the study one can conclude that ozonation is a prospective process and a promising procedure for the removal of BTX components from aquatic environment. However, further research should be performed to enhance process efficiency and to study the impact of reaction intermediates and products on aquatic ecosystem.

Acknowledgements

This work was supported by the Slovak Research and Development Agency under the contract No. APVV-0656-12. The authors would like to thank also for the support from the VEGA Grant 1/0859/14.

Author details

Ján Derco[1]*, Katarína Šimovičová[2], Jozef Dudáš[1] and Mária Valičková[1]

*Address all correspondence to: jan.derco@stuba.sk

1 Institute of Chemical and Environmental Engineering, Faculty of Chemical and Food Technology, Slovak University of Technology, Bratislava, Slovak Republic

2 Water Research Institute, Bratislava, Slovak Republic

References

[1] European Commission (2000). Directive 2000/60/EC of the European Parliament and of the Council of 23 October 2000 establishing a framework for community action in the field of water policy. Off. J. Eur. Communities, L 327/1.

[2] European Commission (2001). Decision No. 2455/2001/EC of the European Parliament and of the Council establishing the list of priority substances in the field of water policy and amending Directive 2000/60/EC, 2001. Off. J. Eur. Union, OJ L 44, pp. 1–5.

[3] Water Act (2004). Act No. 364/2004 on water (Water Act) (in Slovak).

[4] Mathur AK, Majumder CB, Chatterjee S. Combined removal of BTEX in air stream by using mixture of sugar cane bagasse, compost and GAC as biofilter media. *J. Hazard. Mater.* 2007; **148**: 64–74. DOI: 10.1016/j.jhazmat.2007.02.030

[5] Wiszniowski J, Ziembinska A, Ciesielski S. Effect of petroleum organic compounds in synthetic wastewater on treatment performance and microbial community structures in membrane biological reactor (MBR). *Special Abstracts/J. Biotechnol.* 2010; **150 S**: 277–278. DOI: 10.1016/j.jbiotec.2010.09.198.

[6] Reddy CM, Quinn JG. GC-MS analysis of total petroleum hydrocarbons and polycyclic aromatic hydrocarbons in. *Mar. Pollut. Bull.* 1999**; 38(2)**: 126–35. DOI: 10.1016/S0025-326X(98)00106-4.

[7] USEPA, *Occurrence Estimation Methodology and Occurrence Findings Report for the Six-Year Review of Existing National Primary Drinking Water Regulations.* Washington, DC: EPA-815-R-03-006. Office of Water. 2003.

[8] Frankovská J, Kordík J, Slaninka I, Jurkovič L, Greif V, Šottník P, Dananaj I, Mikita S, Dercová K, Jánová V. The Atlas of Remediation Methods for Contaminated Sites. Štátny geologický ústav Dionýza Štúra. Bratislava. 2010. ISBN 978-80-8934338-6 (in Slovak).

[9] Hoigné J, Bader H. Rate constants of reactions of ozone with organic and inorganic compounds in water. I. Non-dissociating organic compounds. *Water Res.* 1983; **17**: 173–183. DOI: 10.1016/0043-1354(83)90098-2.

[10] Arslan-Alaton I. The effect of pre-ozonation on the biocompatibility of reactive dye hydrolysates. *Chemosphere.* 2003; **51(9)**: 825–833. DOI: 10.1016/S0045-6535(03)00231-5.

[11] Sánchez-Polo M, Rivera-Utrilla J, Prados-Joya G, Ferro-García MA, Bautista-Toledo I. Removal of pharmaceutical compounds, nitroimidazoles, from waters by using the ozone/carbon system. *Water Res.* 2008; **42(7)**: 4163–4171. DOI: 10.1016/j.watres.2008.05.034.

[12] Alvares ABC, Diaper C, Parsons SA. Partial oxidation by ozone to remove recalcitrance from wastewaters–a review. *Environ. Technol.* 2001; **22**: 409–427. DOI: 10.1080/09593332208618273.

[13] Derco J, Melicher M, Kassai A, Dudáš J, Valičková M. Removal of benzothiazols by ozone pretreatment. Environ. Eng. Sci. 2011; **28(11):** 781–785. DOI: 10.1089/ees.2010.0399.

[14] Derco J, Valičková M, Šilhárová K, Dudáš J, Luptaková A. Removal of selected chlorinated micropollutants by ozonation. Chem Pap. 2013; **67**: 1585–1593. DOI: 10.2478/s11696-013-0324-x.

[15] Šimovičová K, Derco J, Valičková M, Melicher M, Sumegová L. Reducing of component of gasoline in water by O_3/H_2O_2 process. *46th Internatiol Conference on Petroleum Processing,* Bratislava, 2013, ISBN 978-80-969792-4-0.

[16] Bedient PB. Groundwater contamination, transport and remediation. Englewood Cliffs, New Jersey: Prentice Hall PTR. 1994.

[17] Srijata M, Pranab R. BTEX : a serious ground-water contaminant. *Res. J. Environ. Sci.* 2011; **5:** 394–398. DOI: 10.3923/rjes.2011.394.398.

[18] Ullmann F. *Ullmann's Encyclopedia of Industrial Chemistry* (6th, completely rev. ed.) Electronic Release. Weinheim, Germany: Wiley – VCH. 2011. ISBN: 9783527303878.

[19] Mateju V. ed. The Compendium of Clean-up Technologies. 2006; Vodní zdroje Ekomonitor spol. s.r.o., Chrudim. ISBN: 80-86832-15-5 (in Czech).

[20] European Commission (2008). Directive 2008/105/EC of the European Parliament and of 10 the Council of 16 December 2008 on environmental quality standards in the field of water policy, amending and subsequently repealing Council Directives 82/176/EEC, 83/513/EEC, 84/156/EEC, 84/491/EEC, 86/280/EEC and amending Directive 2000/60/EC.

[21] Regulation of the government of the Slovak Republic N° 269/2010: Quality objectives of surface waters and on limit values for pollution indicators of waste waters and special waters (in Slovak).

Chapter 2

Kinetic Modeling of the UV/H_2O_2 Process: Determining the Effective Hydroxyl Radical Concentration

Ainhoa Rubio-Clemente, E. Chica and
Gustavo A. Penuela

Additional information is available at the end of the chapter

http://dx.doi.org/10.5772/65096

Abstract

A kinetic model for pollutant degradation by the UV/H_2O_2 system was developed. The model includes the background matrix effect, the reaction intermediate action, and the pH change during time. It was validated for water containing phenol and three different ways of calculating HO° level time-evolution were assumed (non-pseudo-steady, pseudo-steady and simplified pseudo-steady state; denoted as kinetic models A, B and C, respectively). It was found that the kind of assumption considered was not significant for phenol degradation. On the other hand, taking into account the high levels of $HO_2°$ formed in the reaction solution compared to HO° concentration (~10^{-7} M >>>> ~10^{-14} M), $HO_2°$ action in transforming phenol was considered. For this purpose, phenol-$HO_2°$ reaction rate constant was calculated and estimated to be $1.6x10^3$ M^{-1} s^{-1}, resulting in the range of data reported from literature. It was observed that, although including $HO_2°$ action allowed slightly improving the kinetic model degree of fit, HO° developed the major role in phenol conversion, due to their high oxidation potential. In this sense, an effective level of HO° can be determined in order to be maintained throughout the UV/ H_2O_2 system reaction time for achieving an efficient pollutant degradation.

Keywords: UV/H_2O_2 process, matrix background, kinetic model, reaction rate constant

1. Introduction

Nowadays, one of the major problems associated with the presence of toxic and persistent pollutants in the aquatic environment is the unfeasibility of conventional treatments for the effective removal of those substances [1-3]. Hence the application of alternative technologies, such as advanced oxidation processes, is needed [4]. Among these techniques, the UV/H_2O_2 system is included. It consists of the photolysis of hydrogen peroxide (H_2O_2) by applying ultraviolet (UV) radiation resulting in the generation of hydroxyl radicals (HO°) [5, 6]. This process may be performed at room temperature and pressure, it has no mass transfer problems, it is easy to maintain and operate, no sludge requiring a subsequent treatment and disposal is produced, and it may achieve a complete pollutant mineralization [5]. Therefore, the UV/H_2O_2 system seems to be a promising alternative for the treatment of water containing toxic and recalcitrant substances. However, this kind of technology can be expensive due to the associated electrical and oxidant costs [7].

In order to reduce costs and make the process more feasible for industrial applications, the UV/H_2O_2 system optimization is required [8] and kinetic models can be considered as functional tools for this purpose. Up to date, several kinetic models have been proposed for describing the UV/H_2O_2 process and predicting different pollutant removal rates [6–15]. In some of these models [13, 15], the proposed set of ordinary differential equations (ODE) defining the studied pollutant degradation rate can be simplified into a pseudo-first–order kinetic expression, whose solution is an exponential one. In such as models, experimental results are fit to that solution. Subsequently, model predictions agree well with laboratory data. In that kind of models the calculated reaction rate constants for the tested pollutant degradation are apparent reaction rate constants (K_{app}), which include pollutant removal reaction rate constants and the values of parameters such as the quantum yield for the oxidant, the conjugate base (HO_2^-), and the contaminant photolysis, the initial level of the chemical species involved in pollutant oxidation, the optical path length of the system, the UV-light intensity, and the molar extinction coefficients of H_2O_2/HO_2^- and pollutant, among others. Therefore, knowing those parameters is not required.

On the other hand, there are dynamic kinetic models that try to solve the considered ODE set, for which the values of the mentioned variables are required, increasing the complexity of the kinetic model. In order to solve the proposed ODE set, the pseudo-steady state approximation assumption for reactive intermediates, such as HO°, is invoked by arguing that these chemical species are as transient ones as their concentration can be presumed to be at a pseudo-steady state [10, 12, 14]. In other models [6–15], on the contrary, the non-pseudo-steady state premise in free radical rate expressions for predicting the degradation of the probe compound in a more accurate way is applied. However, although hydroperoxyl radicals ($HO_2°$) are involved in those models, none of them, excluding Huang and Shu [11] and Liao and Gurol [12] models, includes $HO_2°$ in the target pollutant oxidation final expression. Additionally, some of these models cannot be reproduced unless a conversion factor is included, as demonstrated by Audenaert et al. [8].

In this sense, the aim of this work was to develop a kinetic model based on the main chemical and photochemical reactions for pollutant degradation in water systems by the UV/H_2O_2 process taking into account the decomposition of the pollutant through direct photolysis, HO° oxidation, and $HO_2°$ and superoxide radical ($O_2^{°-}$) transformation. Furthermore, HO° scavenging effects of carbonate (CO_3^{2-}), bicarbonate (HCO_3^-), sulfate (SO_4^{2-}), and chloride (Cl^-) ions were considered. pH changes in the bulk and the detrimen-

http://dx.doi.org/10.5772/65096

tal action of the organic matter (OM) and the reaction intermediates in shielding UV and quenching HO° were studied. The influence of the pseudo-steady and non-pseudo-steady state hypothesis for determining HO° concentration-evolution with time was also examined and the second-order HO_2° reaction rate constant for the studied pollutant was determined. MATLAB software was used to solve the ODE set that characterizes the current model and the results were validated by using experimental data obtained from the literature for phenol (PHE) degradation by the UV/H_2O_2 process in a completely mixed batch photoreactor.

2. Experimental model approach

A mathematical model for predicting pollutant degradation and the concentrations of the main species involved in a UV/H_2O_2 system was developed. The developed model describes radical chain reactions occurring during the UV/H_2O_2 process in the presence of HO° scavengers and UV-radiation absorbers, such as dissolved organic matter (DOM), anions and reaction intermediate products. Pollutant degradation mechanisms, direct UV photolysis and radical attack by HO°, HO_2°, $O_2^{\circ-}$ and other anion radicals ($CO_3^{\circ-}$, $SO_4^{\circ-}$, H_2ClO°, $HClO^{\circ}$, Cl°, and $Cl_2^{\circ-}$) were included. Additionally, the model incorporates the competitive UV-radiation absorption by H_2O_2, the parent compound and the DOM in terms of dissolved organic carbon (DOC), as well as the formation and disappearance of intermediate products, also considered as DOC. Moreover, it accounts for the solution pH change due to the mineralization of organic compounds and the formation of acids.

2.1. UV/H_2O_2 system fundamentals

The UV/H_2O_2 system initiates with the primary photolysis of H_2O_2 or HO_2^-, producing HO° according to Eqs. (1) and (2) [6, 13]. Based on the Beer-Lambert law and quantum yield definition, the reaction rates for H_2O_2/HO_2^- direct photodegradation and HO° generation are obtained through Eqs. (3)—(5), respectively.

$$H_2O_2 \xrightarrow{h\upsilon} 2HO^{\circ} \tag{1}$$

$$HO_2^- + H_2O \xrightarrow{h\upsilon} 2HO^{\circ} + OH^- \tag{2}$$

$$\frac{d[H_2O_2]}{dt} = -\phi_{H_2O_2} I_{a,\,H_2O_2} \tag{3}$$

$$\frac{d\left[HO_2^-\right]}{dt} = -\phi_{HO_2^-} I_{a,\ HO_2^-} \tag{4}$$

$$\frac{d\left[HO^\circ\right]}{dt} = 2\phi_{H_2O_2} I_{a,\ H_2O_2} + 2\phi_{HO_2^-} I_{a,\ HO_2^-} \tag{5}$$

where $\phi_{H_2O_2}$ and $\phi_{HO^-_2}$ (mol Ein^{-1}) are the quantum yields of the photochemical reactions of H_2O_2 and HO_2^-. $I_{a,\ H_2O_2}$ and $I_{a,\ HO^-_2}$ (Ein L^{-1} s^{-1}) refer to the UV-radiation intensities absorbed by H_2O_2 and HO_2^-, respectively, calculated according to Eqs. (6) and (7), where $f_{H_2O_2}$ and $f_{HO^-_2}$ are the fractions of the UV-radiation absorbed by H_2O_2 and HO_2^-, respectively (Eqs. (8) and (9)) [7, 8].

$$I_{a,H_2O_2} = I_0 f_{H_2O_2}\{1 - exp[-2.3l(\varepsilon_{H_2O_2}[H_2O_2] + \varepsilon_{HO_2^-}[HO_2^-] + \varepsilon_C[C] + \varepsilon_{DOC}[DOC])]\} \tag{6}$$

$$I_{a,HO_2^-} = I_0 f_{HO_2^-}\left\{1 - exp\left[-2.3l\left(\varepsilon_{H_2O_2}\left[H_2O_2\right] + \varepsilon_{HO_2^-}\left[HO_2^-\right] + \varepsilon_C\left[C\right] + \varepsilon_{DOC}\left[DOC\right]\right)\right]\right\} \tag{7}$$

$$f_{H_2O_2} = \frac{\varepsilon_{H_2O_2}\left[H_2O_2\right]}{\varepsilon_{H_2O_2}\left[H_2O_2\right] + \varepsilon_{HO_2^-}\left[HO_2^-\right] + \varepsilon_C\left[C\right] + \varepsilon_{DOC}\left[DOC\right]} \tag{8}$$

$$f_{HO_2^-} = \frac{\varepsilon_{HO_2^-}\left[HO_2^-\right]}{\varepsilon_{H_2O_2}\left[H_2O_2\right] + \varepsilon_{HO_2^-}\left[HO_2^-\right] + \varepsilon_C\left[C\right] + \varepsilon_{DOC}\left[DOC\right]} \tag{9}$$

in which $[H_2O_2]$, $[HO_2^-]$, [C], and [DOC] are H_2O_2, HO_2^-, contaminant and DOM, in terms of DOC, concentrations, respectively. $\varepsilon_{H_2O_2}$, $\varepsilon_{HO_2^-}$, ε_C and ε_{DOC} (M^{-1} m^{-1}) are the molar extinction coefficients of H_2O_2, HO_2^-, the pollutant and the DOC, respectively. In turn, l (mm) is the photoreactor path length and I_0 (Ein s^{-1}), the incident UV-light intensity.

In addition to the oxidant photolysis, the target pollutant (C) may interact with the UV-radiation, undergoing degradation and producing reaction intermediates. A fraction of those by-products can be dissolved in the solution [16]. This fraction is denoted as DOC (Eq. (10)) [6]. The reaction rate for the contaminant direct photolysis is obtained through Eq. (11).

$$C \xrightarrow{h\upsilon} DOC + ? \tag{10}$$

$$\frac{d[\mathrm{C}]}{dt} = -\phi_c I_{a,\,\mathrm{C}} \tag{11}$$

where ϕ_{c} is the quantum yield of pollutant photolysis and $I_{a,\mathrm{C}}$ is the amount of UV-light absorbed by the contaminant, calculated through Eqs. (12) and (13).

$$I_{a,\,\mathrm{C}} = I_0 f_{\mathrm{C}}\left\{1 - \exp\left[-2.3l\left(\varepsilon_{\mathrm{H_2O_2}}\left[\mathrm{H_2O_2}\right] + \varepsilon_{\mathrm{HO_2^-}}\left[\mathrm{HO_2^-}\right] + \varepsilon_{\mathrm{C}}\left[\mathrm{C}\right] + \varepsilon_{\mathrm{DOC}}\left[\mathrm{DOC}\right]\right)\right]\right\} \tag{12}$$

$$f_{\mathrm{C}} = \frac{\varepsilon_{\mathrm{C}}\left[\mathrm{C}\right]}{\varepsilon_{\mathrm{H_2O_2}}\left[\mathrm{H_2O_2}\right] + \varepsilon_{\mathrm{HO_2^-}}\left[\mathrm{HO_2^-}\right] + \varepsilon_{\mathrm{C}}\left[\mathrm{C}\right] + \varepsilon_{\mathrm{DOC}}\left[\mathrm{DOC}\right]} \tag{13}$$

Once HO° are produced, they rapidly react with the pollutant of interest, degrading it to form reaction intermediates (Eq. (14)), which subsequently can be attacked by HO° and undergo further degradation to produce final products, such as CO_2, H_2O, and mineral acids (Eq. (15)) [7]. In this model, intermediate substances were considered as HO° scavengers as well as UV-light absorbers. Additionally, it was assumed that the pH of the solution decreased due to the conversion of the target pollutant, and consequently the reaction intermediates, into carbon dioxide (i.e., $H_2CO_3^*$ in the aqueous phase); although it must be highlighted that not all the DOC is mineralized, since carboxylic acids are also formed during the oxidation process, making the pH of the bulk decreases as well [8]. Under this presumption, Eq. (15) is simplified as Eq. (16). The mass balances for the evolution of the pollutant, the dissolved organic fraction of the formed by-products, HO° and the $H_2CO_3^*$ concentrations with time are shown by Eqs. (17)–(20), respectively.

$$\mathrm{C} + \mathrm{HO}^\circ \xrightarrow{k_{\mathrm{C,HO^\circ}}} \mathrm{DOC} + ? \tag{14}$$

$$\mathrm{DOC} + \mathrm{HO}^\circ \xrightarrow{k_{\mathrm{DOC,\,HO^\circ}}} \mathrm{CO_2} + \mathrm{H_2O} + \text{mineral acids} \tag{15}$$

$$\mathrm{DOC} + \mathrm{HO}^\circ \xrightarrow{k_{\mathrm{DOC,\,HO^\circ}}} \mathrm{H_2CO_3^*} + \mathrm{H_2O} \tag{16}$$

$$\frac{d[\mathrm{C}]}{dt} = -k_{\mathrm{C,HO^\circ}}[\mathrm{HO^\circ}][\mathrm{C}] \quad (17)$$

$$\frac{d[\mathrm{DOC}]}{dt} = k_{\mathrm{C,HO^\circ}}[\mathrm{HO^\circ}][\mathrm{C}] - k_{\mathrm{DOC,HO^\circ}}[\mathrm{HO^\circ}][\mathrm{DOC}] \quad (18)$$

$$\frac{d[HO^\circ]}{dt} = -k_{\mathrm{C,\,HO^\circ}}[\mathrm{HO^\circ}][\mathrm{C}] - k_{\mathrm{DOC,\,HO^\circ}}[\mathrm{HO^\circ}][\mathrm{DOC}] \quad (19)$$

$$\frac{d[\mathrm{H_2CO_3^*}]}{dt} = k_{\mathrm{DOC,HO^\circ}}[\mathrm{HO^\circ}][\mathrm{DOC}] \quad (20)$$

where [C], [DOC], [HO°], and [$H_2CO_3^*$] correspond to pollutant, dissolved by-products, HO° and $H_2CO_3^*$ levels, respectively. $k_{\mathrm{C,HO^\circ}}$ and $k_{\mathrm{DOC,HO^\circ}}$ are the rate constants of Eq. (14) and (15) or (16).

In the UV/H_2O_2 system, recombination of HO° can occur to produce H_2O_2. However, these free radicals can also react with H_2O_2 and HO_2^-, particularly when the oxidant is in excess, to produce HO_2°. Although HO_2° are less reactive than HO° (E° = 0.98 and 2.8 V, respectively) [8], they can also be involved in pollutant degradation, especially if these radicals are produced in high amounts in the system. Furthermore, HO_2° can produce $O_2^{\circ-}$, which subsequently can participate in pollutant degradation and mineralization [17]. Therefore, the role of these reactive oxygen species was included in the proposed kinetic model.

It is important to note that as the oxidation process develops, the pH of the solution generally goes down, and consequently, some chemical species appear while other species vanish. In order to consider the change of chemical species inside the bulk according to the pH of the solution in the kinetic model, a correction factor (δ_{R_i}) was introduced for the photolysis of H_2O_2, HO_2^-, and for the reaction between H_2O_2 and HO°. This correction factor can adopt two values (0 and 1). When δ_{R_i}=1, reaction R_i is promoted (i.e., the time-varying concentrations of the chemical species involved in reaction R_i are taken into account in the model). When δ_{R_i}=0, reaction R_i is not considered in the model and, subsequently, the chemical species taking part in reaction R_i are neglected.

On the other hand, species commonly present in water, such as DOM and inorganic anions (e.g., CO_3^{2-}, HCO_3^-, SO_4^{2-}, and Cl^-, among others) may also have a significant effect because of their ability to absorb UV-light and/or to scavenge HO°. The HO° scavenging effect of matrix constituents drastically limits the oxidation action of HO°, leading to a decrease in the performance of the system [18].

Taking into account all the mentioned processes and in order to give a more realistic view of what happens in the UV/H_2O_2 process, the kinetic equation describing pollutant degradation can be expressed as Eq. (21).

$$-\frac{dC}{dt} = \phi_c I_{a,\,C} + k_{14}[\mathrm{C}][\mathrm{HO}^\circ] + k_{15}[\mathrm{C}]\left[\mathrm{HO}_2^\circ\right] + k_{16}[\mathrm{C}]\left[\mathrm{O}_2^{\circ-}\right] + \sum_{i=35}^{40} k_i[\mathrm{C}][\mathrm{AR}_i] \tag{21}$$

where the terms $\phi_c I_{a,\,C}$, $k_{14}[C][HO^\circ]$, $k_{15}[C]\left[HO_2^\circ\right]$, $k_{16}[C]\left[O_2^{\circ-}\right]$ and $\sum_{i=35}^{40} k_i[C]\left[AR_i\right]$ represent the specific contributions of UV-radiation, the oxidation of HO°, HO_2°, $O_2^{\circ-}$ and the formed anion radicals (AR) (including $CO_3^{\circ-}$, $SO_4^{\circ-}$, H_2ClO°, $HClO^\circ$, Cl°, and $Cl_2^{\circ-}$) to the overall pollutant degradation, respectively.

In order to quantitatively evaluate the contribution of the cited terms in the contaminant removal, parameters *d*, *f*, *g*, and *h* were introduced in Eq. (21), as described by Eq. (22).

$$-\frac{dC}{dt} = \phi_c I_{a,\,C} + k_{14}[\mathrm{C}][\mathrm{HO}^\circ]d + k_{15}[\mathrm{C}]\left[\mathrm{HO}_2^\circ\right]f + k_{16}[\mathrm{C}]\left[\mathrm{O}_2^{\circ-}\right]g + \sum_{i=35}^{40} k_i[\mathrm{C}][\mathrm{AR}_i]h \tag{22}$$

When $d = f = g = h = 0$ (i.e., when the initial concentrations of oxidant radical species are equal to zero), the kinetic model only describes the degradation of the pollutant by photolysis. If $d = h = 1$ and $f = g = 0$, in addition to the photolysis conversion, in a deionized water, the model can predict pollutant transformation through HO° and $CO_3^{\circ-}$ (the latter from the reaction between HO° and HCO_3^- or CO_3^{2-}). When $d = f = h = 1$ and $g = 0$, contaminant removal by photolysis and the action of HO° and HO_2°, as well as $CO_3^{\circ-}$, is described. When all the parameters are equal to 1 (i.e., $d = f = g = h = 1$) and there are no inorganic anions different from HCO_3^- and CO_3^{2-} in the studied water, the kinetic model accounts for the degradation of the pollutant by direct photolysis and oxidation through HO°, HO_2°, $O_2^{\circ-}$, and $CO_3^{\circ-}$. As the model includes the reactions where Cl^-, SO_4^{2-}, CO_3^{2-}, and HCO_3^- are involved, it can be used for the treatment of different types of water.

Based on the reactions illustrated in **Figures 1** and **2**, and the different involved parameters, the mass balances and the corresponding ODE of the species of interest (C, DOC, H_2O_2, HO_2^-, HO°, HO_2°, $O_2^{\circ-}$, $CO_3^{\circ-}$, $SO_4^{\circ-}$, $HClO^{\circ-}$, H_2ClO°, $HClO^\circ$, Cl°, $Cl_2^{\circ-}$, OH^-, H^+, $H_2CO_3^*$, CO_3^{2-}, HCO_3^-, HSO_4^-, SO_4^{2-}, and Cl^-) are summarized in **Table 1**.

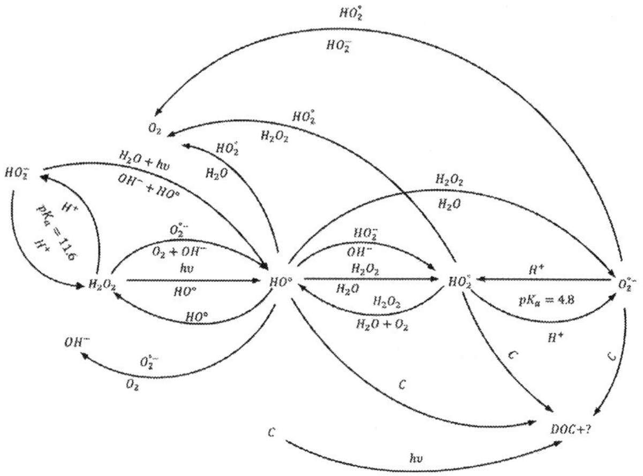

Figure 1. Reaction network of the UV/H_2O_2 process for the degradation of the parent compound without considering the effect of matrix constituents. The species above and below the arrows are the co-reactants and the co-products, respectively.

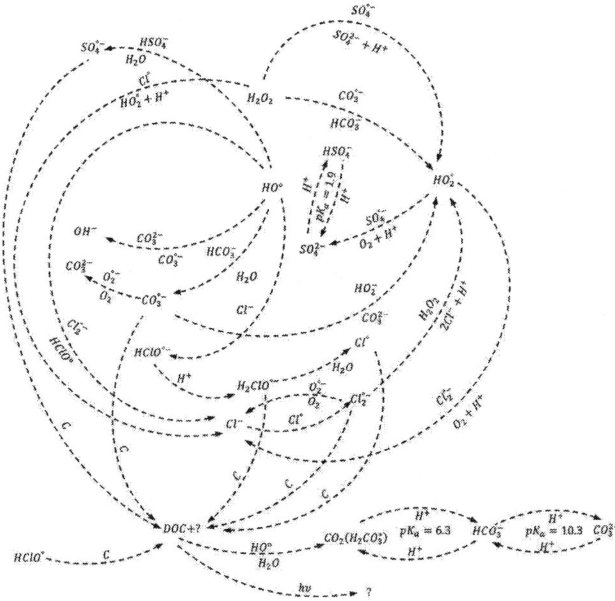

Figure 2. Reaction network describing the effect of the matrix background in the UV/H_2O_2 system with co-reactants and co-products above and below the arrows, respectively.

http://dx.doi.org/10.5772/65096

No.	Species	Kinetic expressions
ODE_1	C	$\frac{d[C]}{dt}$ $= -\phi_c I_{a,C} - k_{14}[C][HO^{\circ}]d - k_{15}[C][HO_2^{\circ}]f$ $- k_{16}[C][O_2^{\circ-}]g$ $- (k_{35}[C][CO_3^{\circ-}] + k_{36}[C][SO_4^{\circ-}] + k_{37}[C][H_2ClO^{\circ}] + k_{38}[C][HClO^{\circ}] + k_{39}[C][Cl^{\circ}] + k_{40}[C][Cl_2^{\circ-}])h$
ODE_2	H_2O_2	$\frac{d[H_2O_2]}{dt} = -\phi_{H_2O_2} I_{a,H_2O_2}\delta_{R_3} - k_1[H_2O_2] + k_2[HO_2^-][H^+] - k_5[H_2O_2][HO^{\circ}]\delta_{R_7} - k_5[H_2O_2][HO^{\circ}]\delta_{R_8}$ $+ k_7[{}^{\circ}OH][HO^{\circ}] + k_{10}[HO_2^{\circ}][HO_2^{\circ}] - k_{11}[H_2O_2][HO_2^{\circ}] - k_{12}[H_2O_2][O_2^{\circ-}]$ $- k_{20}[H_2O_2][CO_3^{\circ-}] - k_{24}[SO_4^{\circ-}][H_2O_2] - k_{31}[Cl_2^{\circ-}][H_2O_2] - k_{32}[Cl^{\circ}][H_2O_2]$
ODE_3	HO_2^-	$\frac{d[HO_2^-]}{dt} = -\phi_{HO_2^-} I_{a,HO_2^-}\delta_{R_4} + k_1[H_2O_2] - k_2[HO_2^-][H^+] - k_6[HO^{\circ}][HO_2^-] + k_{13}[HO_2^{\circ}][O_2^{\circ-}]$ $- k_{21}[HO_2^-][CO_3^{\circ-}]$
ODE_4	${}^{\circ}OH$	$\frac{d[HO^{\circ}]}{dt} = 2\phi_{H_2O_2} I_{a,H_2O_2}\delta_{R_3} + 2\phi_{HO_2^-} I_{a,HO_2^-}\delta_{R_4} - k_5[H_2O_2][HO^{\circ}]\delta_{R_7}$ $- k_5[H_2O_2][HO^{\circ}]\delta_{R_8} - k_6[HO^{\circ}][HO_2^-] - k_7[HO^{\circ}][HO^{\circ}] - k_8[HO^{\circ}][HO_2^{\circ}]$ $- k_9[HO^{\circ}][O_2^{\circ-}] + k_{11}[H_2O_2][HO_2^{\circ}] + k_{12}[H_2O_2][O_2^{\circ-}] - k_{14}[C][HO^{\circ}]d$ $- k_{17}[DOC][HO^{\circ}] - k_{18}[CO_3^{2-}][HO^{\circ}] - k_{19}[HCO_3^-][HO^{\circ}] - k_{23}[HSO_4^-][HO^{\circ}]$ $- k_{26}[Cl^-][HO^{\circ}] - k_{30}[Cl_2^{\circ-}][HO^{\circ}]$
ODE_5	HO_2°	$\frac{d[HO_2^{\circ}]}{dt}$ $= -k_3[HO_2^{\circ}] + k_4[O_2^{\circ-}][H^+] + k_5[H_2O_2][HO^{\circ}]\delta_{R_8}$ $+ k_6[HO^{\circ}][HO_2^-] - k_8[HO^{\circ}][HO_2^{\circ}] - k_{10}[HO_2^{\circ}][HO_2^{\circ}] - k_{11}[H_2O_2][HO_2^{\circ}] - k_{13}[HO_2^{\circ}][O_2^{\circ-}]$ $- k_{15}[C][HO_2^{\circ}]f + k_{20}[H_2O_2][CO_3^{\circ-}] + k_{21}[HO_2^-][CO_3^{\circ-}] + k_{24}[SO_4^{\circ-}][H_2O_2] - k_{25}[SO_4^{\circ-}][HO_2^{\circ}]$ $+ k_{31}[Cl_2^{\circ-}][H_2O_2] + k_{32}[Cl^{\circ}][H_2O_2] - k_{33}[Cl_2^{\circ-}][HO_2^{\circ}]$
ODE_6	$O_2^{\circ-}$	$\frac{d[O_2^{\circ-}]}{dt} = k_3[HO_2^{\circ}] - k_4[O_2^{\circ-}][H^+] + k_5[H_2O_2][HO^{\circ}]\delta_{R_7}$ $- k_9[HO^{\circ}][O_2^{\circ-}] - k_{12}[H_2O_2][O_2^{\circ-}] - k_{13}[HO_2^{\circ}][O_2^{\circ-}] - k_{16}[C][O_2^{\circ-}]g$ $- k_{22}[O_2^{\circ-}][CO_3^{\circ-}] - k_{34}[Cl_2^{\circ-}][O_2^{\circ-}]$
ODE_7	OH^-	$\frac{d[OH^-]}{dt} = \phi_{HO_2^-} I_{a,HO_2^-}\delta_{R_4} + k_6[HO^{\circ}][HO_2^-] + k_9[HO^{\circ}][O_2^{\circ-}] + k_{12}[H_2O_2][O_2^{\circ-}] + k_{18}[CO_3^{2-}][HO^{\circ}]$
ODE_8	H^+	$\frac{d[H^+]}{dt} = k_1[H_2O_2] - k_2[HO_2^-][H^+] + k_3[HO_2^{\circ}] - k_4[O_2^{\circ-}][H^+] + k_5[H_2O_2][HO^{\circ}]\delta_{R_7}$ $+ k_{24}[SO_4^{\circ-}][H_2O_2] + k_{25}[SO_4^{\circ-}][HO_2^{\circ}] - k_{27}[HClO^{\circ-}][H^+] + k_{31}[Cl_2^{\circ-}][H_2O_2]$ $+ k_{32}[Cl^{\circ}][H_2O_2] + k_{33}[Cl_2^{\circ-}][HO_2^{\circ}] + k_{41}[H_2CO_3^{*}] - k_{42}[HCO_3^-][H^+]$ $+ k_{43}[HCO_3^-] - k_{44}[CO_3^{2-}][H^+] + k_{45}[HSO_4^-] - k_{46}[SO_4^{2-}][H^+]$
ODE_9	DOC	$\frac{d[DOC]}{dt} = \phi_c I_{a,C} + k_{14}[C][HO^{\circ}]d + k_{15}[C][HO_2^{\circ}]f + k_{16}[C][O_2^{\circ-}]g - k_{17}[DOC][HO^{\circ}] - \phi_{DOC} I_{a,DOC}$ $+ (k_{35}[C][CO_3^{\circ-}] + k_{36}[C][SO_4^{\circ-}] + k_{37}[C][H_2ClO^{\circ}] + k_{38}[C][HClO^{\circ}] + k_{39}[C][Cl^{\circ}]$ $+ k_{40}[C][Cl_2^{\circ-}])h$
ODE_{10}	CO_3^{2-}	$\frac{d[CO_3^{2-}]}{dt} = -k_{18}[CO_3^{2-}][HO^{\circ}] + k_{21}[HO_2^-][CO_3^{\circ-}] + k_{22}[O_2^{\circ-}][CO_3^{\circ-}] + k_{43}[HCO_3^-] - k_{44}[CO_3^{2-}][H^+]$
ODE_{11}	$CO_3^{\circ-}$	$\frac{d[CO_3^{\circ-}]}{dt} = k_{18}[CO_3^{2-}][HO^{\circ}] + k_{19}[HCO_3^-][HO^{\circ}] - k_{20}[H_2O_2][CO_3^{\circ-}] - k_{21}[HO_2^-][CO_3^{\circ-}]$ $- k_{22}[O_2^{\circ-}][CO_3^{\circ-}] - k_{35}[C][CO_3^{\circ-}]h$

ODE_{12}	HCO_3^-	$\frac{d[HCO_3^-]}{dt} = -k_{19}[HCO_3^-][HO^{\circ}] + k_{20}[H_2O_2][CO_3^{\circ-}] + k_{41}[H_2CO_3^*] - k_{42}[HCO_3^-][H^+] - k_{43}[HCO_3^-] + k_{44}[CO_3^{2-}][H^+]$
ODE_{13}	$H_2CO_3^*$	$\frac{d[H_2CO_3^*]}{dt} = k_{17}[DOC][HO^{\circ}] - k_{41}[H_2CO_3^*] + k_{42}[HCO_3^-][H^+]$
ODE_{14}	HSO_4^-	$\frac{d[HSO_4^-]}{dt} = -k_{23}[HSO_4^-][HO^{\circ}] - k_{45}[HSO_4^-] + k_{46}[SO_4^{2-}][H^+]$
ODE_{15}	$SO_4^{\circ-}$	$\frac{d[SO_4^{\circ-}]}{dt} = k_{23}[HSO_4^-][HO^{\circ}] - k_{24}[SO_4^{\circ-}][H_2O_2] - k_{25}[SO_4^{\circ-}][HO_2^{\circ}] - k_{36}[C][SO_4^{\circ-}]h$
ODE_{16}	SO_4^{2-}	$\frac{d[SO_4^{2-}]}{dt} = k_{24}[SO_4^{\circ-}][H_2O_2] + k_{25}[SO_4^{\circ-}][HO_2^{\circ}] + k_{45}[HSO_4^-] - k_{46}[SO_4^{2-}][H^+]$
ODE_{17}	Cl^-	$\frac{d[Cl^-]}{dt} = -k_{26}[Cl^-][HO^{\circ}] - k_{29}[Cl^{\circ}][Cl^-] + k_{30}[Cl_2^{\circ-}][^{\circ}OH] + 2k_{31}[Cl_2^{\circ-}][H_2O_2] + k_{32}[Cl^{\circ}][H_2O_2] + 2k_{33}[Cl_2^{\circ-}][HO_2^{\circ}] + 2k_{34}[Cl_2^{\circ-}][O_2^{\circ-}]$
ODE_{18}	$HClO^{\circ-}$	$\frac{d[HClO^{\circ-}]}{dt} = k_{26}[Cl^-][HO^{\circ}] - k_{27}[HClO^{\circ-}][H^+]$
ODE_{19}	H_2ClO°	$\frac{d[H_2ClO^{\circ}]}{dt} = k_{27}[HClO^{\circ-}][H^+] - k_{28}[H_2ClO^{\circ}] - k_{37}[C][H_2ClO^{\circ}]h$
ODE_{20}	$HClO^{\circ}$	$\frac{d[HClO^{\circ}]}{dt} = k_{30}[Cl_2^{\circ-}][HO^{\circ}] - k_{38}[C][HClO^{\circ}]h$
ODE_{21}	Cl°	$\frac{d[Cl^{\circ}]}{dt} = k_{28}[H_2ClO^{\circ}] - k_{29}[Cl^{\circ}][Cl^-] - k_{32}[Cl^{\circ}][H_2O_2] - k_{39}[C][Cl^{\circ}]h$
ODE_{22}	$Cl_2^{\circ-}$	$\frac{d[Cl_2^{\circ-}]}{dt} = k_{29}[Cl^{\circ}][Cl^-] - k_{30}[Cl_2^{\circ-}][HO^{\circ}] - k_{31}[Cl_2^{\circ-}][H_2O_2] - k_{33}[Cl_2^{\circ-}][HO_2^{\circ}] - k_{34}[Cl_2^{\circ-}][O_2^{\circ-}] - k_{40}[C][Cl_2^{\circ-}]h$
k_i		$k_1 = 3.7\times10^{-2}\ s^{-1}$; $k_2 = 2.6\times10^{10}\ M^{-1}s^{-1}$; $\varepsilon_{H_2O_2}(254\ nm) = 1\ 800\ M^{-1}m^{-1}$; $\varepsilon_{HO_2^-}(254\ nm) = 22\ 800\ M^{-1}m^{-1}$; $\Phi_{H_2O_2}(254\ nm) = 0.5\ molEin^{-1}$; $\Phi_{HO_2^-}(254\ nm) = 0.5\ molEin^{-1}$; $pH > 11.6 \Rightarrow \delta_{R_3} = 0, \delta_{R_4} = 1$; $pH < 11.6 \Rightarrow \delta_{R_3} = 1, \delta_{R_4} = 0$; $k_3 = 1.58\times10^5\ s^{-1}$; $k_4 = 1.0\times10^{10}\ M^{-1}s^{-1}$; $k_5 = 2.7\times10^7\ M^{-1}s^{-1}$; $pH > 4.8 \Rightarrow \delta_{R_7} = 1, \delta_{R_8} = 0$; $pH < 4.8 \Rightarrow \delta_{R_7} = 0, \delta_{R_8} = 1$; $k_6 = 7.5\times10^9\ M^{-1}s^{-1}$; $k_7 = 5.5\times10^9\ M^{-1}s^{-1}$; $k_8 = 6.6\times10^9\ M^{-1}s^{-1}$; $k_9 = 7.0\times10^9\ M^{-1}s^{-1}$; $k_{10} = 8.3\times10^5\ M^{-1}s^{-1}$; $k_{11} = 3\ M^{-1}s^{-1}$; $k_{12} = 0.13\ M^{-1}s^{-1}$; $k_{13} = 9.7\times10^7\ M^{-1}s^{-1}$; $k_{C,HO^{\circ}} = k_{14}$; $k_{C,HO_2^{\circ}} = k_{15}$; $k_{C,O_2^{\circ-}} = k_{16}$; $k_{DOC,HO^{\circ}} = k_{17} = 2.0\times10^8\ M^{-1}s^{-1}$; $k_{18} = 3.9\times10^8\ M^{-1}s^{-1}$; $k_{19} = 8.5\times10^6\ M^{-1}s^{-1}$; $k_{20} = 4.3\times10^5\ M^{-1}s^{-1}$; $k_{21} = 3.0\times10^7\ M^{-1}s^{-1}$; $k_{22} = 6.5\times10^8\ M^{-1}s^{-1}$; $k_{23} = 3.5\times10^5\ M^{-1}s^{-1}$; $k_{24} = 1.2\times10^7\ M^{-1}s^{-1}$; $k_{25} = 3.5\times10^9\ M^{-1}s^{-1}$; $k_{26} = 4.3\times10^9\ M^{-1}s^{-1}$; $k_{27} = 3.0\times10^{10}\ M^{-1}s^{-1}$; $k_{28} = 5.0\times10^4\ s^{-1}$; $k_{29} = 8.5\times10^9\ M^{-1}s^{-1}$; $k_{30} = 1.0\times10^9\ M^{-1}s^{-1}$; $k_{31} = 4.1\times10^4\ M^{-1}s^{-1}$; $k_{32} = 1.1\times10^9\ M^{-1}s^{-1}$; $k_{33} = 3.0\times10^9\ M^{-1}s^{-1}$; $k_{34} = 2.0\times10^9\ M^{-1}s^{-1}$; $k_{35} =?$; $k_{36} =?$; $k_{37} =?$; $k_{38} =?$; $k_{39} =?$; $k_{40} =?$; $k_{41} = 1\times10^{10}\ s^{-1}$; $k_{42} = 4.5\times10^3\ M^{-1}s^{-1}$; $k_{43} = 1\times10^{10}\ s^{-1}$; $k_{44} = 4.5\times10^{-1}\ M^{-1}s^{-1}$; $k_{45} = 1\times10^{10}\ s^{-1}$; $k_{46} = 4.5\times10^{11}\ M^{-1}s^{-1}$

Table 1. Set of the ODE used in the kinetic model [9, 18, 27, 30–39].

http://dx.doi.org/10.5772/65096

2.2. Proposed kinetic model

In the developed kinetic model, three different ways of calculating the evolution of HO° concentration during time were presumed: (a) a non-pseudo-steady or transient state (i.e., the net formation rate of HO° is different from zero); (b) a pseudo-steady state; and (c) a simplified pseudo-steady state; correspondingly denoted as kinetic model A, B, and C.

In the prediction model A the concentration of HO° can be calculated with Eq. (23). From Eq. (23), the HO° concentration can be written as Eq. (24) (prediction model B).

$$\begin{aligned}\frac{d[HO^\circ]}{dt} &= 2\phi_{H_2O_2} I_{a,\,H_2O_2}\delta_{R_3} + 2\phi_{HO_2^-} I_{a,HO_2^-}\delta_{R_4} - k_5[H_2O_2][HO^\circ]\delta_{R_7} - \\ & k_5[H_2O_2][HO^\circ]\delta_{R_8} - k_6[HO^\circ][HO_2^-] - k_7[HO^\circ][HO^\circ] - k_8[HO^\circ][HO_2^\circ] - \\ & k_9[HO^\circ][O_2^{\circ-}] + k_{11}[H_2O_2][HO_2^\circ] + k_{12}[H_2O_2][O_2^{\circ-}] - k_{14}[C][HO^\circ]d - \\ & k_{17}[DOC][HO^\circ] - k_{18}[CO_3^{2-}][HO^\circ] - k_{19}[HCO_3^-][HO^\circ] - k_{23}[HSO_4^-][HO^\circ] - \\ & k_{26}[Cl^-][HO^\circ] - k_{30}[Cl_2^{\circ-}][HO^\circ]\end{aligned} \tag{23}$$

$$\begin{aligned}[HO^\circ] = &(2\phi_{H_2O_2} I_{a,\,H_2O_2}\delta_{R_3} + 2\phi_{HO_2^-} I_{a,\,HO_2^-}\delta_{R_4} + k_{11}[H_2O_2][HO_2^\circ] + k_{12}[H_2O_2][O_2^{\circ-}])/ \\ &(k_5[H_2O_2]\delta_{R_7} + k_5[H_2O_2]\delta_{R_8} + k_6[HO_2^-] + k_8[HO_2^\circ] + k_9[O_2^{\circ-}] + k_{14}[C]d + \\ & k_{17}[DOC] + k_{18}[CO_3^{2-}] + k_{19}[HCO_3^-] + k_{23}[HSO_4^-] + k_{26}[Cl^-] + k_{30}[Cl_2^{\circ-}])\end{aligned} \tag{24}$$

Considering that the oxidant is in a high level (i.e., $k_5[H_2O_2]\delta_{R_7} + k_5[H_2O_2]\delta_{R_8} \gg \Sigma k_i[X_i]$, where $\Sigma k_i[X_i] = k_6[HO_2^-] + k_8[HO_2^\circ]k_9[O_2^{\circ-}] + k_{14}[C]d + k_{17}[DOC] + k_{18}[CO_3^{2-}] + k_{19}[HCO_3^-] + k_{23}[HSO_4^-] + k_{26}[Cl^-] + k_{30}[Cl_2^{\circ-}]$) and $2\phi_{H_2O_2} I_{a,H_2O_2}\delta_{R_3} + 2\phi_{HO_2^-} I_{a,HO_2^-}\delta_{R_4} \gg \Sigma(k_{11}[H_2O_2][HO_2^\circ] + k_{12}[H_2O_2][O_2^{\circ-}]$, Eq. (24) can be simplified to Eq. (25) (prediction model C). k_i and $[X_i]$ are the reaction rate constants between HO° and species i, and the concentration of species i, respectively.

$$[HO^\circ] = \frac{2\phi_{H_2O_2} I_{a,\,H_2O_2}\delta_{R_3} + 2\phi_{HO_2^-} I_{a,\,HO_2^-}\delta_{R_4}}{k_5[H_2O_2]\delta_{R_7} + k_5[H_2O_2]\delta_{R_8}} \tag{25}$$

The initial values of DOC and inorganic anionic species (CO_3^{2-}, HCO_3^-, SO_4^{2-}, Cl^-, etc.) are set according to the conditions of the water to be treated.

2.3. Numerical solution of the proposed kinetic models

The ODE system compiled in **Table 1** was solved applying MATLAB software and ODE15S function. For simultaneously solving the ODE set of the proposed kinetic models, it was necessary to define several photochemical parameters such as $\phi_{H_2O_2}$, $\phi_{pollutant}$, ϕ_{DOC}, $\varepsilon_{H_2O_2}$, $\varepsilon_{HO_2^-}$, $\varepsilon_{pollutant}$, ε_{DOC}, I_0 and l. Additionally, the initial concentrations of all the species involved in the system and the rate constants of the chemical reactions between those species were required (**Table 1**). During setting up the model, differential rate equations describing the time dependence of the concentration of the variety of the considered species were defined and plotted.

3. Results and discussion

In order to validate the proposed kinetic models, experimental data for PHE degradation by the UV/H_2O_2 process were used from Alnaizy and Akgerman [19] study. These authors conducted a set of experiments in a completely mixed batch cylindrical photoreactor made on Pyrex glass. The used photochemical parameters and the kinetic reaction rate constants of PHE with HO°, $O_2^{\circ-}$, and $CO_3^{\circ-}$ are presented in **Table 2**.

Parameters	Notation	Numerical values	References
Quantum yield	$\phi_{phenol} = \phi_C$	0.07 mol Ein^{-1}	[23]
Molar extinction coefficient	$\varepsilon_{phenol} = \varepsilon_C$	51 600 M^{-1} m^{-1}	[19]
Path length	l	63.5 mm	
Incident UV-light intensity (radiation of 254 nm > 90% and power = 15 W)	I_0	1.516×10^{-6} Ein L^{-1} s^{-1}	
Kinetic rate constant phenol-HO°	$k_{phenol, HO^\circ} = k_{14}$	6.6×10^9 M^{-1} s^{-1}	[40]
Kinetic rate constant phenol-$O_2^{\circ-}$	$k_{phenol, O_2^{\circ-}} = k_{16}$	5.8×10^3 M^{-1} s^{-1}	[41]
Kinetic rate constant phenol-$CO_3^{\circ-}$	$k_{phenol, CO_3^{\circ-}} = k_{35}$	2.2×10^7 $M^{-1}s^{-1}$	[42]

Table 2. Values of the parameters used in the kinetic model validation for phenol degradation.

3.1. Assumptions taken into consideration

As stated previously, the developed kinetic models A, B, and C employed the non-pseudo-steady, the pseudo-steady, and the simplified pseudo-steady state assumption, respectively, to estimate HO° concentration. In the proposed models, the impact of UV radiation individually and/or the combined action of H_2O_2 and UV light (including the effect of HO°, HO_2°, $O_2^{°-}$, and $CO_3^{°-}$) on PHE degradation was studied.

A PHE concentration of 2.23 × 10^{-3} M and a H_2O_2/PHE ratio of 495 were selected for validating the model. The used PHE solution was prepared by adding the appropriate amount of pollutant solution to deionized water [19]. Therefore, the effect of inorganic anions, excluding HCO_3^- and CO_3^{2-} was not taken into account. In this sense, in the PHE degradation rate expression (ODE_1) the contribution of inorganic anion radicals, such as $SO_4^{°-}$, H_2ClO°, HClO°, Cl°, and $Cl_2^{°-}$ was not studied. It was assumed that the other terms included in ODE_1 contributed to pollutant degradation.

As the treated water was deionized, the presence of OM different from the parent compound in the initial solution was neglected ($[DOC]_0 = 0$ M). Hence, the DOC in the solution came from PHE photolysis and free radical (HO°, HO_2°, $O_2^{°-}$, and $CO_3^{°-}$) oxidation.

On the other hand, several authors agree that OM reduction by direct photolysis in a UV/ H_2O_2 oxidation process can be neglected [7–9]. That is the reason why this was not included in the proposed kinetic model. However, it is highlighted that the OM is able to absorb UV-light, preventing UV-penetration into the bulk and avoiding H_2O_2/HO_2^- and pollutant direct photolysis. Therefore, the detrimental effect of UV-shielding in PHE degradation was taken into consideration. For including this effect in the kinetic model, OM molar extinction coefficient, referred as DOC molar extinction coefficient (ε_{DOC}), must be previously known. Although this parameter has already been measured [7, 8], its value is not a universal one, since DOM is a complex group of aromatic and aliphatic hydrocarbon structures with attached functional groups [20, 21]. As Alnaizy and Akgerman [19] did not measure this variable, a mean value from Peuravuori and Pihlaja [22] study was presumed. This value corresponded to $\varepsilon_{DOC(280\,nm)} = 35\,967\ M^{-1}\ m^{-1}$, which was in the same order of magnitude than ε_{PHE}. This number could be acceptable due to the formed aromatic intermediates during PHE conversion [19] conserve structural similarities with the parent compound, like the aromatic ring, responsible for the molecule excitation.

Additionally, it is widely known that the quantum yield of a compound is dependent on the excitation wavelength and the pH of the solution. For 254 nm, PHE quantum yield in an aqueous solution was found to be in the range of 0.02-0.12 mol Ein^{-1} at pH 1.6-3.2 [23]. Therefore, an average value (0.07 mol Ein^{-1}) was taken as PHE quantum yield.

Moreover, it is widely recognized that the solution pH decreases as the process proceeds. This variation in the pH can cause difficulties in modeling studies, since the presence of radical species such as HO_2°, and $O_2^{°-}$, among other chemical species involved in the oxidation system, is significantly dependent on the pH of the medium. Therefore, to give a more realistic view of what happens inside the reaction medium, H_2O_2 and HO_2^- photolysis reactions (Eqs. (1) and (2), correspondingly), as well as reactions expressed in Eqs. (26) and (27) were discriminated in the model according to the solution pH time evolution, as it is simplified in **Figure 3**. For selecting the suitable reactions with regard to the pH changes over time, previous information about the evolution of the pH during the performance of the process is required. However, in some occasions this is not provided. In this case, the initial pH of the solution was 6.8 [19]. At this pH, one of the predominant species into the bulk was H_2O_2, since the pKa of the H_2O_2/HO_2^- equilibrium is 11.6. Therefore, the photolysis of HO_2^- was neglected in the model performance (i.e., δ_{R_3} and δ_{R_4}). Additionally, the initial concentrations of HO° and other considered species were assumed to be zero, with the exception of the pollutant and H_2O_2. On the other hand, it was found that the pH of the medium rapidly dropped from 6.8 to 4.7–4.2 (4.5 as a mean value) within the first 30 min of radiation [19]. Therefore, and according to **Figure 3**, Eq. (26) was considered during the first 30 min of the reaction while Eq. (27), after that time (i.e., $\delta_{R_7} = 1$, $\delta_{R_8} = 0$ and $\delta_{R_7} = 0$, $\delta_{R_8} = 1$, before and after the first 30 min of the process, respectively).

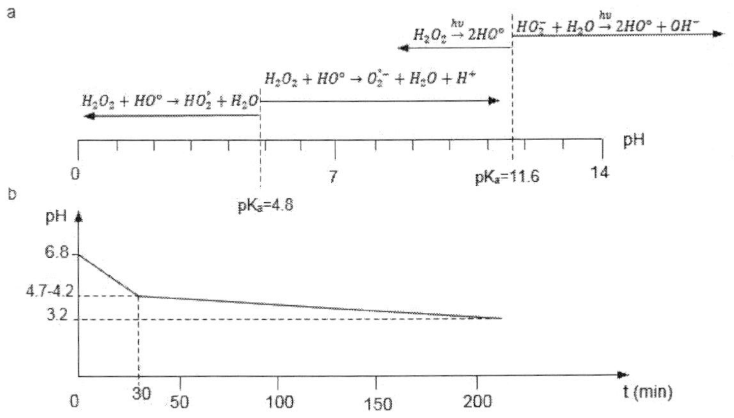

Figure 3. (a) Reaction discrimination diagram as a function of the solution pH and (b) pH evolution of the medium during the UV/H_2O_2 process in a completely mixed batch photoreactor. Operating conditions: H_2O_2/PHE = 495; $[C]_0 = 2.23 \times 10^{-3}$ M; $t = 220$ min.

$$O_2^{°-} + H^+ \rightarrow HO_2^° \tag{26}$$

$$H_2O_2 + HO^° \rightarrow O_2^{°-} + H_2O + H^+ \tag{27}$$

3.2. Kinetic model validation

Initially, the associated ODE sets with model A, B, and C were solved. Parameters $f = g = 0$ and $d = h = 1$ were considered in order to solely investigate the influence of the direct photolysis, HO°, and $CO_3^{°-}$ on PHE degradation.

Figure 4 compares the simulation results of the proposed kinetic models A, B, and C with the experimental data for 2.23 × 10^{-3} M PHE with a H_2O_2/PHE ratio of 495 and a reaction time of 220 min. The measured and simulated results for PHE direct photolysis alone are also presented. It is observed that more than 90% of the initial PHE concentration was removed after 220 min due to both direct photolysis and indirect degradation (primarily due to HO° attack). The effect of $CO_3^{\circ-}$ could be seen as marginal because of the reduced number of those radicals, in the range of 10^{-15} M, and the low reaction rate constant with PHE compared to HO°. In addition, it is shown that PHE was not completely removed by direct UV photolysis under the tested photochemical conditions (**Table 2**), since approximately 50% of the total PHE degradation was attributed to UV photolysis, as it was experimentally determined by Alnaizy and Akgerman [19]. On the other hand, the figure demonstrates that the prediction kinetic model C was in good agreement with the available experimental data with a relative high correlation factor (R^2=99.34%). For prediction models A and B, R^2 were 97.41 and 97.37%, respectively. As a result of the subtle differences between kinetic models A and B, the depicted line describing model A overlaps model B line. An acceptable agreement between model predictions considering only the UV radiation and experimental data was also verified (R^2 = 98.25%).

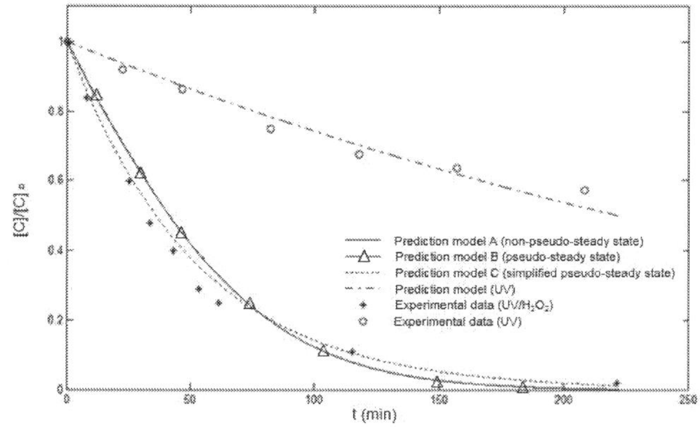

Figure 4. Comparison of the kinetic model predictions (lines) versus experimental data (o) and (*). Operating conditions: H_2O_2/PHE = 495; $[C]_0$ = 2.23 × 10^{-3} M; t = 220 min.

Furthermore, **Figure 4** clearly presents that the removal rate of the target pollutant was not significantly dependent on the hypothesis assumed to estimate the HO° level (non-pseudo-steady, pseudo-steady and simplified pseudo-steady state assumptions). This could be explained from the relatively low concentration of those reactive species in the solution (with a magnitude order of 10^{-14} M) when compared to the level of other species involved in the system, such as HO_2° and $O_2^{\circ-}$, whose concentrations were in the range of 10^{-7} M. The number of HO° remaining in the solution is in concordance with the low final HO° levels found in the literature [24, 25] and even higher than those reported by Ray and Tarr [26].

In **Figure 5**, the evolution of the HO°, HO_2°, O_2°⁻, DOC and $H_2CO_3^*$ normalized estimated concentrations using the prediction model A is depicted. It is observed that HO_2° number increased as the oxidation system proceeded, while O_2°⁻ level decreased. Typically, the effect of HO_2° and O_2°⁻ radicals are neglected in the UV/H_2O_2 system [6–8, 15, 17, 27, 28] since they are found to be less reactive than HO°, as stated previously. However, when they are produced in a high level, they could also participate in the contaminant oxidation. That is the current case of HO_2°, as [HO_2°] >>>> [HO°]. Therefore, the contribution of HO_2° to PHE degradation should be studied. On the other hand, in this work the action of O_2°- in PHE conversion can be omitted since O_2°- level decreased with the reaction time, as expected, because of the drop in the pH solution.

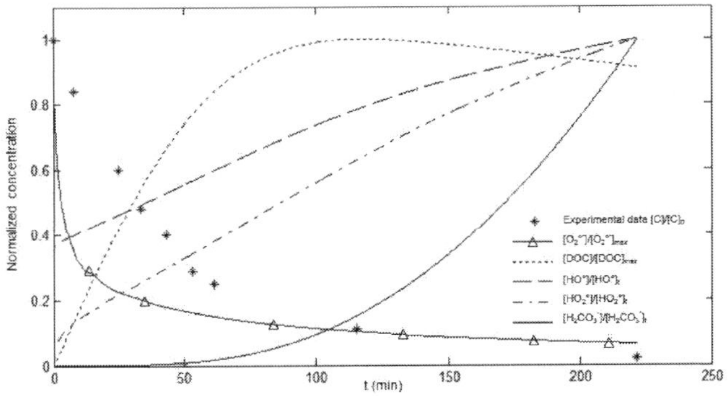

Figure 5. Concentration-time profiles of HO° (---), HO_2° (-.-.-), O_2°– (-Δ-), DOC (···) and $H_2CO_3^*$ (—) using the prediction kinetic model A. (*) represents PHE experimental evolution. Operating conditions: H_2O_2/PHE = 495; $[C]_0 = 2.23 \times 10^{-3}$ M; $t = 220$ min.

Furthermore, generally, there is a drop in the pH of the medium as the system progresses. This is probably due to acidic compound formation, such as carboxylic acids and $H_2CO_3^*$ resulting from pollutant degradation and mineralization. In this study the decrease of the pH in the solution was predicted via DOC conversion and the sole generation of $H_2CO_3^*$ by model A. As presented in **Figure 5**, DOC generation was progressively increasing as PHE was being degraded up to a certain point (117 min, corresponding to $[DOC]_{max} = 2.061 \times 10^{-3}$ M, and equivalent to ca. 93% of PHE degradation). From this point, DOC started to decrease until $[DOC]_f = 1.879 \times 10^{-3}$ M. That breakpoint represented the moment at which PHE was almost completely transformed into by-products. In addition, at this point, PHE mineralization began to be more evident, since $H_2CO_3^*$ level rose approximately in a linear way, up to a final level of 6.493×10^{-14} M, with the subsequent pH decrease. Similarities between the pattern of this DOC profile and that of the formed intermediate curves reported in Alnaizy and Akgerman [19] research are highlighted.

On the other hand, it is worth noting that HO° level evolution with the reaction time was different when comparing the kinetic prediction model A or B with C. Obviating HO_2° contribution to PHE degradation, **Figure 6** shows that the highest final HO° level was achieved in the prediction model A, with a maximum HO° concentration equal to 9.435 × 10^{-14} M. This value was similar to HO° final level in the prediction model B (9.400 × 10^{-14} M) and different to that of the prediction model C, where HO° final concentration was 4.486 × 10^{-14} M (ca. 48% lower than the obtained in the kinetic model A or B).

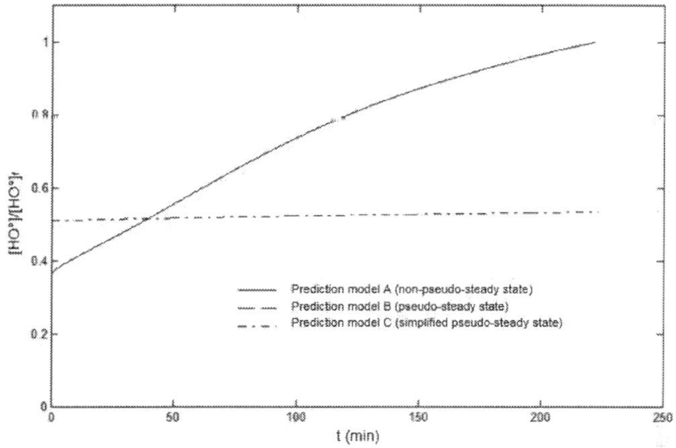

Figure 6. Estimated HO° concentration using the prediction models A (—), B (----), and C (-.-.-). Operating conditions: H_2O_2/PHE = 495; $[C]_0$ = 2.23 × 10^{-3} M; t = 220 min.

Approximately, in the first 40 min of the process a larger number of HO° in the aqueous medium with the developed kinetic model C was evidenced. Apparently, this amount of HO° was sufficient to degrade about 60% of PHE initial level under the studied experimental conditions. In contrast, models A and B, whose lines are overlapped, produced a lower number of HO° and the theoretical conversion of PHE remained above the experimental data. One possible reason for this discrepancy can be ascribed to DOM and dissolved oxygen positive effects in producing reactive oxygen species (ROS), as HO° [26], which were not considered. After 40 min of reaction, the HO° level was higher in the kinetic models A and B than in model C. This larger amount of HO° might lead to a faster conversion of the pollutant in comparison with the predicted model C, since the hypothetical depletion curve of PHE was below the measured data. Nevertheless, this rapid pollutant degradation did not occur actually. Therefore, there was an amount of HO° produced in excess that was not reacting with the contaminant. This surplus of HO° could be involved in free radical scavenging reactions. As model A and B consider the detrimental effect of HO° consuming reactions, it is suggested that their kinetic rate constants are higher than those ones used in this paper for these reactions to have a larger weight in the system. Additionally, the contradictory outcome between the actual situation and the theoretical one in the first and second stage of the process can also be attributed to the fact that just a fraction of the concentration of the species involved in the whole kinetic equations of the predicted models was actually reacting. Consequently, the real level of the species implicated in each kinetic reaction should be considered. However, it is rather difficult to determine which amount of the chemical species is exactly involving in each reaction for each time step, especially due to the high reactivity of radicals as the oxidation system progresses. In this context, further studies are required to overcome this limitation.

From these findings, it is suggested that there was an effective level of the formed HO°. Below that level, there was a lack of HO° for an efficient pollutant conversion; and above it, an excessive number of HO° was generated. That HO° effective level could be of relevance for industrial applications in order to be maintained throughout the reaction time, allowing an efficient pollutant degradation.

3.3. Estimation of PHE-HO_2° reaction rate constant

In order to study the action of HO_2° for pollutant degradation in the UV/H_2O_2 system, PHE-HO_2° rate constant was calculated. For this purpose, the kinetic model A was used and it was estimated through a non-linear least-square objective function. The objective function for minimizing the error between the predicted and the measured data was defined as Eq. (28) [17].

$$\text{Minimize}: f = \sum\left(\left[C\right]_{\text{predicted}} - \left[C\right]_{\text{measured}}\right)^2 \tag{28}$$

where $[C]_{measured}$ and $[C]_{predicted}$ correspond to the evolution of experimental and calculated pollutant concentration, respectively. This expression is a function of PHE-HO_2° rate constant. The optimum value for PHE-HO_2° second-order rate constant was found to be 1.6 × 10^3 M^{-1} s^{-1}, which is consistent with the range of the reported values by Kozmér et al. ((2.7±1.2) × 10^3 M^{-1} s^{-1}) [29]. The results of running the new prediction kinetic model A (with and without the contribution of HO_2° to pollutant conversion) and the experimental data are presented in **Figure 7**. The figure shows that the prediction model A with the estimation

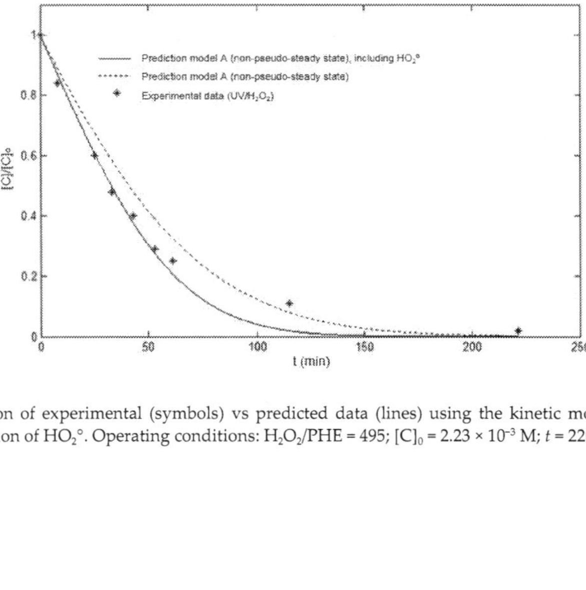

Figure 7. Comparison of experimental (symbols) vs predicted data (lines) using the kinetic model A with (—) and without (···) the action of HO_2°. Operating conditions: H_2O_2/PHE = 495; $[C]_0$ = 2.23 × 10^{-3} M; t = 220 min.

http://dx.doi.org/10.5772/65096

of PHE-$HO_2°$ rate constant was in a stronger agreement with the experimental data (R^2 = 99.0%) than the previous predicted kinetic model A studied in section 3.2 (R^2 = 97.41%). The same conclusion can be drawn for kinetic model B and C, with a R^2 of 98.78 and 99.57% using the calculated PHE-$HO_2°$ rate constant, in comparison with 97.37 and 99.34%, respectively. Therefore, in order to achieve a better fit between experimental and predicted data, $HO_2°$ action in pollutant degradation should be considered. Nonetheless, $HO_2°$ contribution to PHE degradation is non-significant in comparison with the role developed by HO° attack.

4. Conclusions

A kinetic model for studying pollutant degradation by the UV/H_2O_2 system was developed, including the background matrix effect in scavenging free radicals and shielding UV-light and the reaction intermediate action, as well as the change of the pH as the UV/H_2O_2 process proceeds. Three different ways for calculating HO° level time evolution were assumed (non-pseudo-steady, pseudo-steady and simplified pseudo-steady state; denoted as kinetic models A, B, and C, respectively). It was found that the assumption of pseudo-steady (simplified or not) or transient state for determining the HO° level evolution with time was not significant in PHE degradation rate due to the relatively low HO° level present into the bulk (~10^{-14} M). On the other hand, taking into account the high levels of $HO_2°$ formed in the reaction solution compared to HO° concentration (~10^{-7} M >>>> ~10^{-14} M), $HO_2°$ action in transforming PHE was considered. For this purpose, PHE-$HO_2°$ reaction rate constant was calculated and estimated to be 1.6×10^3 M^{-1} s^{-1}, resulting in the range of data reported from literature. It was observed that, although including $HO_2°$ action allowed slightly improving the kinetic model degree of fit, HO° developed the major role in PHE conversion, due to their high oxidation potential.

Additionally, it was found that there was an effective level of the HO° formed in solution. Below that level, there was a lack of HO° for an efficient pollutant conversion; and above it, an excessive number of HO° was generated. That HO° effective level calculated from kinetic model C could be of relevance for industrial applications in order to be maintained throughout the reaction time, allowing an efficient pollutant degradation.

In this study, there was an attempt to contemplate a wide range of the chemical reactions involved in the UV/H_2O_2 process and although high correlation factors were obtained, it is suggested to include the positive effect of the OM and the dissolved oxygen in generating ROS, as well as the effect of other anions naturally present in water bodies, as phosphate and nitrate, for the model to be a more accurate approximation of reality.

Acknowledgements

This work was financially supported by the Colombian Administrative Department of Science, Technology and Innovation (COLCIENCIAS).

Author details

Ainhoa Rubio-Clemente[1,2*], E. Chica[3] and Gustavo A. Peñuela[1]

*Address all correspondence to: ainhoarubioclem@gmail.com

1 Grupo GDCON, Facultad de Ingeniería, Sede de Investigaciones Universitarias (SIU), Universidad de Antioquia UdeA, Colombia

2 Departamento de Ciencia y Tecnología de los Alimentos, Facultad de Ciencias de la Salud, Universidad Católica San Antonio de Murcia UCAM, Spain

3 Departamento de Ingeniería Mecánica, Facultad de Ingeniería, Universidad de Antioquia UdeA, Colombia

References

[1] Rubio-Clemente, A., Torres-Palma, R.A., & Peñuela, G.A. Removal of polycyclic aromatic hydrocarbons in aqueous environment by chemical treatments: A review. Sci. Total Environ. 2014;408:201–225.

[2] Rubio-Clemente, A., Chica, E., & Peñuela, G.A. Application of Fenton process for treating petrochemical wastewater. Ingeniería y Competitividad. 2014;16(2): 211–223.

[3] Rubio-Clemente, A., Chica, E., & Peñuela, G.A. Petrochemical wastewater treatment by photo-Fenton process. Water, Air, Soil Pollut. 2015;226(3):61–78.

[4] Dopar, M., Kušić, H, & Koprivanac, N. Treatment of simulated industrial wastewater by photo-Fenton process: Part I: The optimization of process parameters using design of experiments (DOE). Chem. Eng. J. 2011;173:267–279.

[5] Litter, M., & Quici, N. Photochemical advanced oxidation processes for water and wastewater treatment. Recent Pat. Eng. 2010;4:217–241.

[6] Wols, B.A., & Hofman-Caris, C.H.M. Review of photochemical reaction constants of organic micropollutants required for UV advanced oxidation processes in water. Water Res. 2012;46:2815–2827.

[7] Song, W., Ravindran, V., & Pirbazari, M. Process optimization using a kinetic model for the ultraviolet radiation-hydrogen peroxide decomposition of natural and synthetic organic compounds in groundwater. Chem. Eng. Sci. 2008;63:3249–3270.

[8] Audenaert, W.T.M., Vermeersch, Y., Van Hulle, S.W.H., Dejans, P., Dumoulin, A., & Nopens, I. Application of a mechanistic UV/hydrogen peroxide model at full-scale:

Sensitivity analysis, calibration and performance evaluation. Chem. Eng. J. 2011;171:113–126.

[9] Crittenden, J.C., Hu, S., Hand, D.W., & Green, S.A. A kinetic model for H2O2/UV process in a completely mixed batch reactor. Water Res. 1999;33(10):2315–2328.

[10] Hong, A., Zappi, M.E., & Hill, D. Modeling kinetics of illuminated and dark advanced oxidation processes. J. Environ. Eng. 1996;122(1):58–62.

[11] Huang, C.R., & Shu, H.Y. The reaction kinetics, decomposition pathways and intermediate formation of phenol in ozonation, UV/O_3 and UV/H_2O_2 processes. J. Hazard. Mater. 1995;41:47–64.

[12] Liao, C.H., & Gurol, M.D. Chemical oxidation by photolytic decomposition of hydrogen peroxide. Environ. Sci. Technol. 1995;29(12):3007–3014.

[13] Primo, O., Rivero, M.J., Ortiz, I., & Irabien, A. Mathematical modelling of phenol photooxidation: Kinetics of the process toxicity. Chem. Eng. J. 2007;134(1–3):23–28.

[14] Rosenfeldt, E.J., & Linden, K.G. The ROH,UV concept to characterize and the model UV/H2O2 process in natural waters. Environ. Sci. Technol. 2007;41(7): 2548–2553.

[15] Yao, H., Sun, P., Minakata, D., Crittenden, J.C., & Huang, C.H. Kinetics and modeling of degradation of ionophore antiobiotics by UV and UV/H2O2. Environ. Sci. Technol. 2013;47(9):4581–4589.

[16] Zeng, Y., Hong, P.K.A., & Wavrek, D.A. Integrated chemical-biological treatment of benzo(a)pyrene. Environ. Sci. Technol. 2000;34(5):854–862.

[17] Edalatmanesh, M., Dhib, R., & Mehrvar, M. Kinetic modeling of aqueous phenol degradation by UV/H_2O_2 process. Int. J. Chem. Kinet. 2007;40(1):34–43.

[18] De Laat, J., Le, G.T., & Legube, B. A comparative study of the effects of chloride, sulfate and nitrate ions on the rates of decomposition of H_2O_2 and organic compounds by Fe(II)/H_2O_2 and Fe(III)/H_2O_2. Chemosphere. 2004;55(5):715–723.

[19] Alnaizy, R., & Akgerman, A. Advanced oxidation of phenolic compounds. Adv. Environ. Res. 2000;4(3):233–244.

[20] Leenheer, J.A., & Croué, J.P. Characterizing aquatic organic matter: Understanding the unknown structures is key to better treatment of drinking water. Environ. Sci. Technol. 2003;37(1):18A–26A.

[21] Wang, Z., Wu, Z., & Tang, S. Characterization of dissolved organic matter in a submerged membrane bioreactor by using three-dimensional excitation and emission matrix fluorescence spectroscopy. Water Res. 2009;43(6):1533–1540.

[22] Peuravuori, J., & Pihlaja, K. Molecular size distribution and spectroscopic properties of aquatic humic substances. Anal. Chim. Acta. 1997;337(2):133–149.

[23] Alapi, T., & Dombi, A. Comparative study of the UV and UV/VUV-induced photolysis of phenol in aqueous solution. J. Photochem. Photobiol. A. 2007;188((2–3)):409–418.

[24] Gallard, H., & De Laat, J. Kinetic modelling of Fe(III)/H_2O_2 oxidation reactions in dilute aqueous solution using atrazine as a model organic compound. Water Res. 2000;34(12): 3107–3116.

[25] Rosenfeldt, E.J., Linden, K.G., Canonica, S., & von Gunten, U. Comparison of the efficiency of °OH radical formation during ozonation and the advanced oxidation processes O_3/H_2O_2 and UV/H_2O_2. Water Res. 2006;40(20):3695–3704.

[26] Ray, P.Z., & Tarr, M.A. Petroleum films exposed to sunlight produce hydroxyl radical. Chemosphere. 2014;103:220–227.

[27] Kralik, P., Kušić, H., Koprivanac, N., & Božić, A.L. Degradation of chlorinated hydrocarbons by UV/H_2O_2: The application of experimental design and kinetic modeling approach. Chem. Eng. J. 2010;158(2):154–166.

[28] Kušić, H., Koprivanac, N., Božić, A.L., Papić, S, Peternel, I., & Vujević, D. Reactive dye degradation by AOPs: Development of a kinetic model for UV/H_2O_2 process. Chem. Biochem. Eng. Q. 2006;20(3):293–300.

[29] Kozmér, Z., Arany, E., Alapi, T., Takács, E., Wojnárovits, L., & Dombi, A. Determination of the rate constant of hydroperoxyl radical reaction with phenol. Radiat. Phys. Chem. 2014;102:135–138.

[30] Bielski, B.H.J., Cabelli, D.E., Arudi, L.R., & Ross, A.B. Reactivity of HO_2/O_2^- radicals in aqueous solution. J. Phys. Chem. Ref. Data. 1985;14(4):1041–1100.

[31] Buxton, G.V., Greenstock, C.L., Helman, W.P., & Ross, A.B. Critical review of data constants for reactions of hydrated electrons, hydrogen atoms and hydroxyl radicals in aqueous solutions. J. Phys. Chem. Ref. Data. 1988;17(2): 513–886.

[32] Christensen, H.S., Sehested, K., & Corftizan, H. Reaction of hydroxyl radicals with hydrogen peroxide at ambient temperatures. J. Phys. Chem. 1982;86:15–88.

[33] Schested, K., Rasmussen, O.L., & Fricke, H. Rate constants of OH with HO_2, O_2^- and H_2O_2+ from hydrogen peroxide formation in pulse-irradiated oxygenated water. J. Phys. Chem. 1968;72:626–631.

[34] Weinstein, J., Benon, H.J., & Bielski, H.J. Kinetics of the interaction of HO_2 and O_2-radicals with hydrogen peroxide. The Haber-Weiss reaction. J. Amer. Chem. Soc. 1979;101:58–62.

[35] Wols, B.A., Harmsen, D.J.H., Beerendonk, E.F., & Hofman-Caris, C.H.M. Predicting pharmaceutical degradation by UV (LP)/H_2O_2 processes: A kinetic model. Chem. Eng. J. 2014;255:334–343.

[36] Draganic, Z.D., Negron-Mendoza, A., Sehested, K., Vujosevic, S.I., Navarro-Gonzales, R., Albarran-Sanchez, M.G. Radiolysis of aqueous solutions of ammonium bicarbonate over a large dose range. Radiat. Phys. Chem. 1991;38(3):317–321.

[37] Neta, P., Huie, R.E., & Ross, A.B. Rate constants for reactions of inorganic radicals in aqueous solution. J. Phys. Chem. Ref. Data. 1988;17(3):1027–1284.

[38] Fang, G.D., Dionysiou, D.D., Wang, Y., Al-Abed, S.R., & Zhou, D.M. Sulfate radical-based degradation of polychlorinated biphenyls: Effects of chloride ion and reaction kinetics. J. Hazard. Mater. 2012;227–228:394–401.

[39] Mazellier, P., Leroy, É., De Laat, J., & Legube, B. Transformation of carbendazim induced by the H_2O_2/UV system in the presence of hydrogenocarbonate ions: Involvement of the carbonate radical. New J. Chem. 2002;26(12):1784–1790.

[40] Dorfman, L.M., & Adams, G.E. Reactivity of the hydroxyl radical in aqueous solutions. Standard Reference Data System. National Bureau of Standards and Technology. 1973;46.

[41] Yasuhisa, T., Hideki, H., & Muneyoshi, Y. Superoxide radical scavenging activity of phenolic compounds. Int. J. Biochem. 1993;25(4):491–494.

[42] Chen, S.N., & Hoffman M.Z. Rate constants for the reaction of the carbonate radical with compounds of biochemical interest in neutral aqueous solution. Radiat. Res. 1973;56(1):40–47.

Chapter 3

Electrooxidation-Ozonation: A Synergistic Sustainable Wastewater Treatment Process

Carlos E. Barrera-Diaz and Nelly González-Rivas

Additional information is available at the end of the chapter

http://dx.doi.org/10.5772/65887

Abstract

Advanced oxidation processes (AOPs) have shown to be very useful technologies for application in different wastewater treatment areas. These processes use the very strong oxidizing power of hydroxyl radicals to oxidize organic compounds to carbon dioxide and water. These procedures usually involve the use of O_3, H_2O_2, Fenton's reagent and electrolysis to generate the hydroxyl radicals. However, some recent investigations have found that the use of a coupled processes using O_3/electrooxidation increases the effectiveness of the process and also could reduce the operating costs associated to the application of AOPs. In this chapter, there is a description of our work in the treatment of wastewater using an ozonation-electrooxidation combined process. The main parameters to control for having a successful application of such method are discussed. Several examples for different kinds of polluted water are addressed.

Keywords: organic pollutants, degradation, mass transfer, mineralization, removal

1. Introduction

Traditional wastewater treatments involve the addition of chemicals or the use of microorganisms to treat polluted water. However, in both processes, there is always a residue known as sludge. The sludge management and final disposal could represent up to 50% of the total wastewater treatment plant cost. Therefore, novel ways to deal with this issue should be developed. In this way, the use of advanced oxidation processes (AOPs), in which the HO• radical production is favored, could represent an interesting option for treat wastewater with less or without sludge production.

The final goal of AOPS is the complete degradation of the pollutants present in wastewater, aiming its final mineralization, yielding as final products: carbon dioxide, water and inorganic compounds. These methodologies solve the problem of the final disposal of sludge; because when they are well developed, there is no production of sludge. Obviously, not always is possible the complete mineralization of contaminates. Nevertheless, most of the times, the final products of the destruction of contaminants are harmless compared to the original ones.

Electrochemical techniques use one of the cleanest reagents: "the electron." Thus, since the main reactive used in the oxidation process is green, this becomes sustainable. Oxidation of the organic compounds could occur at the interface of the anode/aqueous solution or in solution via intermediates. Electrochemical oxidation consists in the application of an external source of energy into an electrochemical cell that contains one or more pairs of electrodes. At the cathode, a reduction reaction occurs and the oxidation reactions takes places at the anode. The use of boron diamond doped anodes (BDD) allows the generation of HO• radicals, which reacts with organic compounds.

Electrochemical oxidation is considered a robust technology and is easy to use, for those reasons, it has been used for a diversity of wastewater treatments. The main advantages of this technology over other conventional treatments are as follows: the main reagent is the electron; many processes occur in the electrochemical cell; the addition of chemicals is not required; and the process is carried out at room temperature and atmospheric pressure.

Ozone is a powerful oxidant produced in gas phase, and by means of a diffuser, a mass transfer occurs to aqueous solution. A main advantage of ozone is that it oxidizes organic compounds without producing residual sludge. However, it was found that some compounds are ozone-resistant such as iopamidol, sucralose and atrazine-desethyl.

Combination of the ozonation processes with others, such as ozone/hydrogen peroxide, ozone with sand filtration and activated carbon filtration, have been used to remove ozone resistant contaminants; this allows enhancing the removal efficiency and reducing ozone dosage. Recent reports indicates that there are two reaction mechanisms for ozone oxidation: direct ozonation that takes place at acidic solutions and indirect HO• radical ozonation at basic solutions.

One of the major limitations for the use of ozone is the mass transfer from the gas phase to the liquid phase, the same behavior is observed at direct electrooxidation in which the HO• generation takes place at the anode surface. Thus, when both processes take place at the same time a synergy occurs, the process reaction time is decreased, this implies that the ozone and electricity consumption is also reduced.

In this chapter, there is a description of our work in the treatment of wastewater using an ozonation-electrooxidation combined process. The main parameters to control for having a successful application of such method are discussed. Several examples for different kinds of polluted water are addressed.

2. Characteristics of the hydroxyl radical

Advanced oxidation processes rely on the hydroxyl radical formation. The hydroxyl radical, OH• is a highly reactive radical, able to react unselectively and rapidly with organic

http://dx.doi.org/10.5772/65887

pollutants, including recalcitrant organic compounds, such as aromatic, chlorinated and phenolic compounds [1].

There are different technologies to produce hydroxyl radicals. Nevertheless, the most "greener" one is its electrochemical production direct from the treated water. Among all the electrodes used in the production of hydroxyl radicals, the BDD anodes have shown to be perhaps the most efficient ones. They also have some other useful characteristics that allowed his use in the wastewater treatment, such as the radicals are loosely retained in the surface of the electrode allowing its oxidative action close to the surface area.

No matter from which source the hydroxyl radicals are generated, they have some special characteristics that make the extremely useful in the treatment of wastewater polluted with recalcitrant organic compounds. Some of the characteristics of the hydroxyl radical are as follows:

- Powerful oxidant
- Highly reactive
- Easily generated
- Not selective
- Short reaction time
- Harmless

The hydroxyl radical has a high oxidation potential as shown in **Table 1**, it can be generated, chemically, electrochemically or by UV radiation combined with the presence of suitable catalyst. The major failure of the electrochemical oxidation is the high-energy consumption during the process of mineralization of pollutants.

Oxidant	Potential (V)
Fluorine	3.06
Hydroxyl radical	2.80
Ozone	2.08
Hydrogen peroxide	1.78
Hypochlorite	1.49
Chlorine	1.36

Table 1. Oxidation potential for some common oxidants [2].

3. Ozone

Ozone is a pale blue gas with a pungent odor. It is generated from oxygen. The electric discharge method is the most common process for the preparation of ozone on laboratory and industrial scale. The electrical discharge generates ionized oxygen atoms that react with oxygen molecules to producing ozone.

Ozone is a powerful oxidant and very highly unstable. For this reason, the gas should be produce *in situ* prior to its use on the wastewater treatment. Once the ozone is produced, a diffuser is used to transport the gas to aqueous solution trough a mass transfer process. A major advantage of ozone is that it fully degrades organic materials, leaving no residual sludge.

Ozone can oxidize and destroy the organics through two different pathways, the direct and the indirect ones. In the first, one of the molecules react directly with the ozone molecules. In the second case, the ozone reacts to generate oxidant species such as hydroxyl radicals that carried out the oxidation process. The oxidation pathway that operates in a particular oxidation depends on the reaction rate of ozone and the organic. Sometimes the product generated in the reaction could promote or inhibit the ozone decomposition modifying the initial oxidation mechanism.

It has been found that some compounds are resistant to the oxidation by ozone, such as iopamidol, sucralose and atrazine-desethyl. In order to overcome this limitations, the use of combined ozone processes, such as the UV light, metal oxides catalyst and hydrogen peroxides, have been proposed. In many cases, a remarkable procedure improvement was found [3–5]

Figure 1 shows an ozone bubble column reactor in which ozone is feed in the bottom part of the reactor, O_3 passes through a diffuser that allows the generation of small bubbles which reaches the wastewater contained inside the reactor.

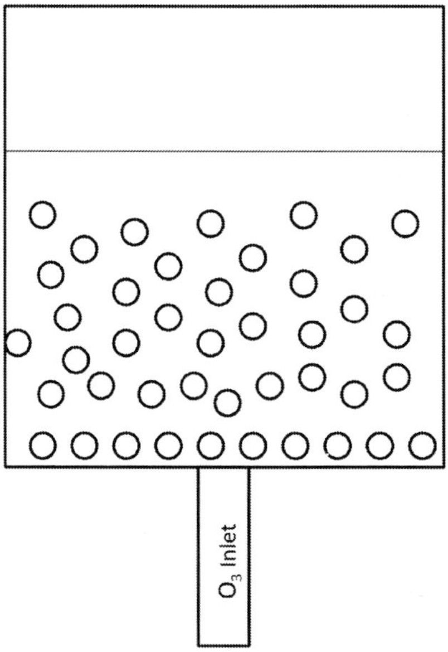

Figure 1. Schematic representation of a column bubble reactor in which ozone is feed in the bottom part.

The decomposition of ozone in water to form hydroxyl radicals, which occurs as are shown in Eqs. (**1**)–(**6**) [6]:

$$O_3 + OH^- \rightarrow O_2 + HO_2^- \tag{1}$$

$$O_3 + HO_2^- \rightarrow HO_2^{\cdot} + O_3^{\cdot -} \tag{2}$$

$$HO_2^{\cdot} \rightarrow H^+ + O_2^{\cdot -} \tag{3}$$

$$O_2^{\cdot -} + O_3 \rightarrow O_2 + O_3^{\cdot -} \tag{4}$$

$$O_3^{\cdot -} + H^+ \rightarrow HO_3^{\cdot} \tag{5}$$

$$HO_3^{\cdot} \rightarrow OH^{\cdot} + O_2 \tag{6}$$

As observed, it takes six reactions to form one hydroxyl radical; now, a mass transfer from the gas phase to the aqueous phase should take place in order to have available hydroxyl radicals in aqueous solution. The process is often limited since only a part of ozone is effectively converted to hydroxyl radicals [7].

4. Electrooxidation

Electrochemical oxidation is considered a robust technology and easy to use, for that reasons, it has been used for a diversity of wastewater treatment areas. The main advantages of this technology over other conventional treatments are as follows:

- Electron is the main reagent.
- A simple electrochemical cell is required in the process.
- Addition of chemicals is not required.
- The process is carried out at room temperature and atmospheric pressure.

In 2003, Marselli et al. demonstrated that the production of hydroxyl radicals during conductive-diamond electrolysis of aqueous wastes is possible. Consequently, a very new class of oxidation processes, the electrochemical advanced oxidation processes (EAOP) were discovered [8].

In the direct electrooxidation, pollutants in the bulk of the wastewater must reach the electrode surface and the oxidation reaction takes places once they are adsorbed onto this surface. Thus, the electrode materials influence the selectivity and efficiency of the oxidation process and mass transfer becomes a very important process. **Table 2** shows some anodic materials that have been investigated for the oxidation of organic compounds.

Material	Oxygen evolution potential
Pt	1.60
Graphite	1.70
SnO_2	1.90
PbO_2	1.90
Boron doped diamond	2.30

Table 2. Oxygen evolution potential of some electrodic materials [9].

Figure 2 shows the general scheme of an electrochemical reactor. It contains an anode made of boron diamond doped in which hydroxyl radicals are produced. The cathode is made of stainless steel and allows water reduction.

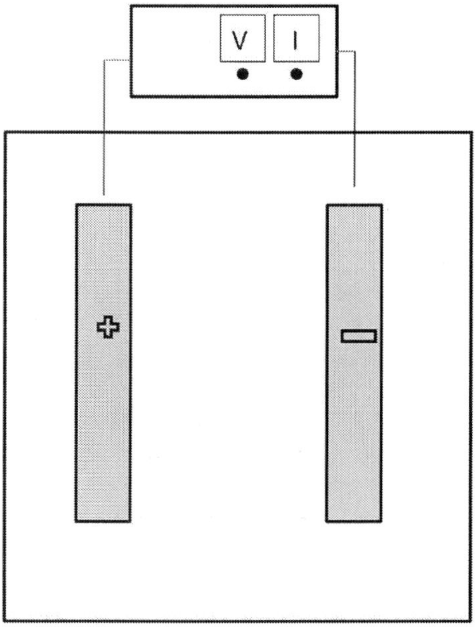

Figure 2. An electrooxidation reactor in which hydroxyl radicals are produced in the anode and water reduction takes place in the cathode.

The main reactions involved in the hydroxyl radicals production are shown in Eqs. (7)–(9)

$$DBB + H_2O \rightarrow DBB(\bullet OH) + H^+ + e^- \quad (7)$$

$$ODBB(\bullet\mathrm{OH}) + R \rightarrow DBB + mineralization\,products + \mathrm{H}^+ + \mathrm{e}^- \tag{8}$$

$$ODBB(\bullet\mathrm{OH}) \rightarrow DBB + \frac{1}{2}\mathrm{O}_2 + \mathrm{H}^+ + \mathrm{e}^- \tag{9}$$

Diamond anodes exhibits three outstanding properties as compared with other advanced oxidation technologies and with electrolysis with other anodes [10]:

- Robustness, because results found in this latter years demonstrate that it can attain the complete mineralization of almost any type of organic without producing refractory final products.
- Efficiency, because when it is operated under the no diffusion control, current efficiencies are close to 100%.
- Integration capability, because it can be easily coupled with other treatment technologies and it can be fed with green energy sources such as wind mills and photovoltaic panels.

However, as can be observed in **Figure 2**, the hydroxyl formation is limited to the anodic surface and also by the mass transfer from the liquid to the electrode.

5. Integrated ozonization-electrooxidation reactor

In order to have a synergistic effect of the two previously described processes, a couple treatment consisting in introducing electrodes inside the ozone reactor has been proposed.

Figure 3 shows an ozone bubble column reactor in which two electrodes are introduced. As can be observed, the bubbles generated by the addition of ozone in the bottom part of the reactor allowing a complete mixing of the solution.

In this reactor, the ozone and the electrooxidation reaction takes place at the same time, thus the hydroxyl radical concentration is enhanced, the mass transfer is limited and a large amount of bubbles provides an excellent mixing inside the reactor. There are several variables to control in the integrated process aiming to obtain a complete degradation of pollutants:

- Initial pollutants concentration
- Initial pH
- Current density
- Interelectrode distance
- Salt concentration (in case it is required to improve conductivity)
- Ozone flow rate
- Electrodes type

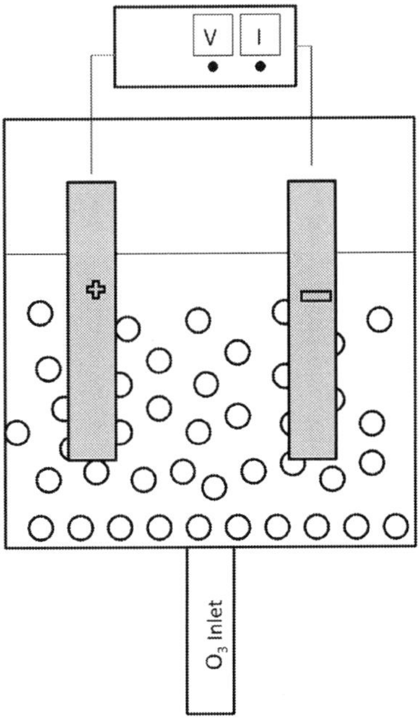

Figure 3. An electrooxidation reactor in which hydroxyl radicals are produced in the anode and water reduction takes place in the cathode.

With a set of well-optimized parameters, there is always in improvement in results compared with the two separated techniques. In **Table 3**, some samples are gathered in which the integrated ozone-electrooxidation process has been applied to different kinds of wastewater. As it is possible to observe, there is a significant improvement in the quality of the treated wastewater. In all the cases, the chemical oxygen demand (cod) is almost eliminated, and some other parameters are also decreased in an important amount. The most used parameters to control the quality of treated wastewater are as follows: conductivity, total organic carbon (TOC), color, turbidity and biochemical oxygen demand (BOD_5), the values of these parameters obtained before and after of the coupled treatment demonstrate the suitability of the proposed procedure to treat wastewater from different sources.

	Results	**References**
Industrial wastewater	Integration of the two processes at pH 7 and 20 mA cm^{-2} of current density greatly improved the reduction in COD (84%), BOD_5 (79%), color (95%), turbidity (96%) and total coliforms (99%)	[11]
Industrial wastewater	In the integrated electrochemical-ozone process with energy pulses, the COD reduction was observed to be 80% after 44 min of treatment. Initial pH was 7.5 at all experiments	[12]

	Results	References
Dye removal in denim effluents	Using the integrated process, 65% color removal, 76% turbidity removal and 37% COD reduction could be attained	[13]
Offset printing dyes	Optimal conditions are found when adding 20 mg L^{-1} AHC, followed by electrocoagulation at 4 A for 50 min, and finally, alkaline ozonation for 15 min, resulting in an overall color removal of 99.99% color and 99.35% COD	[14]
Industrial wastewater	In only 15 min, the integrated process reduced the COD by 83%, TOC by 78%, color by 93%, turbidity by 77% and conductivity by 27% at relatively low current density (12.5 mA cm^{-2})	[15]
p-Nitrophenol solutions	Up to 91%, TOC was removed after 60 min of the electrolysis-O_3 process	[16]
Industrial wastewater	COD is reduced by 99.9% along with most color and turbidity in about an hour. The coupled process practically eliminates the COD, color and turbidity without the addition of chemical and does not generate any sludge	[17]

Table 3. Examples of the electrooxidation-ozonation process applied to wastewater treatment.

6. Conclusions

Electrooxidation-ozonation is an efficient process for the treatment of different kinds of wastewater, since there is always a large reduction in COD, color, and turbidity, conductivity and BOD_5. The coupled process always has a superior performance compared with the application of separated processes. It is also noteworthy to mention that the coupled process is green, as it does not produce residual sludge. This coupled process has the potential to be used in wastewater in which other processes do not work well, including those with recalcitrant pollutants.

Acknowledgements

The authors wish to acknowledge financial support from the UAEM, and the financial support from the CONACYT through Sistema Nacional de Investigadores, which is greatly appreciated.

Nomenclature

BOD_5 Biochemical oxygen demand

COD Chemical oxygen demand

TOC Total organic carbon

Author details

Carlos E. Barrera-Diaz* and Nelly González-Rivas

*Address all correspondence to: cbd0044@gmail.com

Joint Center for Research in Sustainable Chemistry, UAEM-UNAM, Toluca Estado de México, México

References

[1] Barrera-Díaz C, Cañizares P, Fernández FJ, Natividad R, Rodrigo MA. Electrochemical advanced oxidation processes: an overview of the current applications to actual industrial effluents. Journal of the Mexican Chemical Society. 2014;**58**:256–275.

[2] Parsons A, Williams M. Advanced oxidation processes for water and wastewater treatment. IWA Publishing, London, UK, 2004.

[3] Kan C W, Cheung HF, Chan Q. A study of plasma-induced ozone treatment on the colour fading of dyed cotton. Journal of Cleaner Production. 2016;**112**:3514–3524. doi:10.1016/j.jclepro.2015.10.100

[4] Khuntia S, Majumder SK, Ghosh P. Quantitative prediction of generation of hydroxyl radicals from ozone microbubbles. Chemical Engineering Research and Design, 2015;**98**:231–239. doi:10.1016/j.cherd.2015.04.003

[5] Alsheyab MA, Muñoz AH. Reducing the formation of trihalomethanes (THMs) by ozone combined with hydrogen peroxide (H_2O_2/O_3). Desalination. 2006;**194**:121–126. doi:10.1016/j.desal.2005.10.028

[6] Andreozzi R, Caprio V, Insola A, Marotta R. Advanced oxidation processes (AOP) for water purification and recovery. Catalysis Today. 1999;**53**:51–59. doi:10.1016/S0920-5861(99)00102-9

[7] Barrera Díaz CE, González-Rivas N. The use of Al, Cu, and Fe in an integrated electrocoagulation-ozonation process. Journal of Chemistry. 2015; 6 (Article ID 158675). doi:10.1155/2015/158675

[8] Marselli B, Garcia-Gomez J, Michaud PA, Rodrigo MA, Comninellis C. Electrogeneration of hydroxyl radicals on boron-doped diamond electrodes. Journal of the Electrochemical Society. 2003;**150**: D79–D83. doi:10.1149/1.1553790

[9] Panizza M, Cerisola G. Direct and mediated anodic oxidation of organic pollutants. Chemical Reviews. 2009;**109**:6541–6569. doi:10.1021/cr9001319

[10] Rodrigo MA, Cañizares P, Sánchez-Carretero A, Sáez C. Use of conductive-diamond electrochemical oxidation for wastewater treatment. Catalysis Today. 2010;**15**:173–177. doi:10.1016/j.cattod.2010.01.058

[11] Bernal-Martínez LA, Barrera-Díaz C, Solís-Morelos C, Natividad R. Synergy of electrochemical and ozonation processes in industrial wastewater treatment. Chemical Engineering Journal. 2010;**165**:71–77. doi:10.1016/j.cej.2010.08.062

[12] Bernal-Martínez LA, Barrera-Díaz C, Natividad R, Rodrigo MA. Effect of the continuous and pulse in situ iron addition onto the performance of an integrated electrochemical–ozone reactor for wastewater treatment. Fuel. 2013;**110**:133–140. doi:10.1016/j.fuel.2012.11.067

[13] García-Morales MA, Roa-Morales G, Barrera-Díaz C, Martínez Miranda V, Balderas Hernández P, Pavón Silva TB. Integrated advanced oxidation process (ozonation) and electrocoagulation treatments for dye removal in denim effluents. International Journal of Electrochemical Science. 2013;**8**:8752–8763.

[14] Roa-Morales G, Barrera-Díaz C, Balderas-Hernández P, Zaldumbide-Ortiz F, Reyes Perez H, Bilyeu B. Removal of color and chemical oxygen demand using a coupled coagulation-electrocoagulation-ozone treatment of industrial wastewater that contains offset printing dyes. Journal of the Mexican Chemical Society. 2014;**58**:362–368.

[15] Carbajal C, Barrera-Díaz C, Roa-Morales G, Balderas-Hernández P, Natividad R, Bilyeu B. Enhancing the ozonation of industrial wastewater with electrochemically generated copper (II) ions. Separation Science and Technology. 2016;**51**:542–549. doi:10.1080/01496395.2015.1086800

[16] Qiu C, Yuan, S, Li X, Wang H, Bakheet B, Komarneni S, Wang, Y. Investigation of the synergistic effects for p-nitrophenol mineralization by a combined process of ozonation and electrolysis using a boron-doped diamond anode. Journal of Hazardous Materials. 2014;**280**:644–653. doi:10.1016/j.jhazmat.2014.09.001

[17] García-Morales MA, Roa-Morales G, Barrera-Díaz C, Bilyeu B, Rodrigo MA. Synergy of electrochemical oxidation using boron-doped diamond (BDD) electrodes and ozone (O_3) in industrial wastewater treatment. Electrochemistry Communications. 2013;**27**:34–37. doi:10.1016/j.elecom.2012.10.028

Chapter 4

Removal of Phenol from Wastewater Using Fenton-Like Reaction over Iron Oxide–Modified Silicates

Agnieszka Węgrzyn

Additional information is available at the end of the chapter

http://dx.doi.org/10.5772/65097

Abstract

Iron-containing active phase was deposited on natural layered silicate (vermiculite) using several techniques such as ion exchange, precipitation, and forced hydrolysis during hydrothermal digestion. Tuning of the synthesis conditions resulted in preparation of the catalysts with different loading of active phase and physicochemical properties. The composite materials were characterized with respect to their structure (X-ray diffraction), agglomeration state of Fe (diffuse reflectance UV-vis spectroscopy), and chemical composition. Catalytic tests were performed in semi-batch reactor under atmospheric pressure. Aqueous solution of phenol was used as a model industrial effluent, and hydrogen peroxide was added as an oxidant. Spectral techniques were used for identification of intermediate oxidation products. Spent catalysts were also characterized, and structural and chemical changes were determined, e.g., leaching degree of active phase.

Keywords: Fenton-like process, advanced oxidation processes, catalysis, silicate, vermiculite, nanocrystalline iron oxide, phenol

1. Introduction

Refractory organic compounds, such as dyes, phenols, or endocrine disrupting compounds (EDC), are characterized with high toxicity, carcinogenic properties, and this poses a serious hazard to aquatic living organisms. Difficulty of contaminations' removal is caused by their resistance to aerobic digestion, stability to light, heat, and oxidizing agents. Technologies used currently for wastewater treatment, however, used widely, suffer from design shortcomings or are very expensive. Emerging technologies, so-called advanced oxidation processes (AOP), is a large group of methods based on oxidation using strong oxidants, such as ozone or hydrogen peroxide. In AOP methods, higher conversion levels may be obtained at atmospheric pressure and temperatures lower compared to other oxidation processes [1–6]. Moreover, chemical oxidants may be accompanied by catalysts or physical agents such as sunlight, UV or γ radiation, ultrasounds, microwave, or cavitation, increasing efficiency of the reaction [7, 8].

The catalysts used in the Fenton-like system are, among others, natural iron-bearing earth materials, such as goethite, hematite, magnetite, or ferrihydrite [9–11]. Modification of iron oxides to improve their performance in organic pollutant degradation can be achieved by substitution with other transition metals [12]; however, introduction of heavy metals may be questionable from the point of view of the secondary contamination with catalytic leachates. It is also known that nanoscale materials are characterized with different properties compared to their bulk phase [8, 13, 14]. Nanocatalysts offer higher specific surface areas and few or no mass-transfer limitations. It is expected that reaction rate will be higher for nanomaterials. Also, diffusion of large organic molecules (organic dyes, pharmaceuticals) will be no longer problematic as it is observed in microporous materials. On the other hand, the separation and recycling of nanocatalysts at a technical scale still present a challenge.

To circumvent the costly catalyst separation process, magnetic properties of some iron oxides may be exploited [15, 16]. The other possibility is the immobilization on solid support. The most popular materials in this group are activated carbon, silica, and aluminum oxide [17–21]; however, more advanced technologies are also studied employing graphene oxide [22]. The encapsulation of iron oxide nanoparticles in polymer matrix or carbonized sewer sludge was reported as another possibility to stabilize oxide nanoparticles [23, 24].

Facing much more stringent environmental regulations, new waste-free technologies must be developed, based on cheaper, non-toxic materials. Clays proposed as starting materials fulfill all requirements for low-cost, ecological precursors for industrial technologies or large-scale applications. Such materials could be used as catalysts in a large group of emerging technologies consisting on oxidation processes, such as wet oxidation, catalytic wet air oxidation, and advanced oxidation processes. Natural clay minerals provide with excellent support for Fe-containing nanocrystalline active phase of the Fenton-like reaction. Vermiculite, which was used in presented work as a catalytic support, is natural clay mineral belonging to phyllosilicates. It is characterized with high thermal and mechanical stability. Moreover, its properties may be easily modified to obtain efficient adsorbents or catalysts [25, 26].

2. Materials and methods

Commercial expanded vermiculite (South Africa), fraction size 0.5–2 mm, was provided by Romico Polska Sp. z o.o. The silicate was pulverized in electrical blender, and fraction below 180 μm was separated (sample S0). Such prepared vermiculite was used as a support for deposition of nanocrystalline iron oxides.

Two standard procedures [27, 28] were applied to obtain well-defined oxide structures. Pure 2-line ferrihydrite was prepared by precipitation from 0.1 M $Fe(NO_3)_3{\cdot}9H_2O$ (p.a., POCh) solution using 1 M KOH (p.a., POCh). Potassium hydroxide solution was added dropwise at RT and constant stirring until pH was equal to 7. Product was centrifuged, washed with water, and freeze dried. Similar procedure was used to obtain vermiculite-supported ferrihydrite. Suspension of 5 g of vermiculite (S0) was prepared in 150 mL of distilled water, then 100 mL of $Fe(NO_3)_3{\cdot}9H_2O$ solution was added dropwise. Suspension was stirred for the next 2 h to allow ion exchange. In the next step, 1 M KOH solution was added to raise pH up to 7. Crystallization was continued for the next 30 min, product was centrifuged, washed, and dried. Sample codes, depending on Fe/vermiculite ratio, were S2, S3, and S4 (**Table 1**).

Sample name (precipitation)	Sample name (forced hydrolysis)	Fe/vermiculite ratio (mg/g)
S0	HS0	–
–	HS1	16.8
S2	HS2	33.6
S3	HS3	67.2
S4	–	134.4

Table 1. Intended Fe/vermiculite ratio in vermiculite-supported Fe oxide catalysts.

Pure hematite with crystal size of 4 nm was prepared by forced hydrolysis. 3.32 g of Fe(NO3)3·9H2O was dissolved in preheated HCl (p.a., POCh) solution (0.002 M, 400 mL) to obtain Fe concentration of 0.02 M.. Solution was transferred into polypropylene bottle fitted in autoclave and heated at 98°C for 7 days. Product was centrifuged, washed with water, and freeze dried. Synthesis of vermiculite-supported nano-hematite was performed using acidified iron nitrate solutions with the addition of 20 g of vermiculite. Sample codes, depending on Fe/vermiculite ratio, were HS0 (no Fe salt was added), HS1, HS2, and HS3 (**Table 1**).

Phenol removal was studied as a test reaction, and semi-batch reactor was used to minimize formation of side products [26]. Round-bottom flask equipped with reflux condenser was heated to 70°C on magnetic stirrer. Each time reactor was charged with 340 mL of phenol solution (pH = 5.4) and 600 mg of catalyst. Hydrogen peroxide (30%, p.a., POCh) was added into the reaction mixture in 13-min intervals (10 min of non-disturbed reaction and 3 min for sample withdrawal and next injection). Phenol concentration was studied spectrophotometrically (Thermo SCIENTIFIC EVOLUTION 220) as a complex with 4-aminoantipyrine. H_2O_2 concentration (using VO_3^- in 8 M H_2SO_4), Fe dissolved in reaction mixture (SCN^- complex), and colored intermediate products (sample quenched with methanol) were also determined spectrophotometrically. In each interval, pH was measured. Reaction conditions were summarized in **Table 2**.

Reaction code	Phenol concentration (g/L)	H_2O_2 volume added in one injection (mL)
Catalyst R1	1	2
Catalyst R01	0.1	2
Catalyst R01m	0.1	0.2

Table 2. Reaction conditions of phenol removal over vermiculite-supported iron oxide catalysts.

The conversion, X (%), of model pollutant (phenol) was calculated according to Eq. (1):

$$X = \frac{C_0 - C}{C_0} \cdot 100\% \tag{1}$$

where C_0 is the starting concentration and C is the concentration at a given reaction time.

Fresh and spent catalysts were characterized by X-ray diffraction method (XRD) using a powder diffractometer (Bruker, D2 PHASER) equipped with CuKα radiation source. The Sherrer equation (2) was used for determination of nano-hematite crystal size:

$$D = \frac{0.89\lambda}{\beta cos\theta} \tag{2}$$

where λ is the X-ray wavelength, β is the line broadening at half the maximum intensity, and θ is the Bragg angle.

The coordination and aggregation of iron present in the catalysts were studied by diffuse reflectance-UV-vis spectroscopy (DRS-UV-vis). The measurements were performed in the range of 190–900 nm with a resolution of 2 nm using an Evolution 600 (Thermo) spectrophotometer. Content of iron was measured using spectrophotometric technique at wavelength λ = 510 nm (Thermo SCIENTIFIC EVOLUTION 220) as a complex with 1,10-phenanthroline after leaching of metal cations in 6 M HCl.

3. Results and discussion

3.1. Catalytic tests

Three types of catalytic tests were carried out in semi-batch reactor: concentrated (R1—1 g/L) and diluted (R01—0.1 g/L) phenol solutions with the addition of significant excess of oxidant (six times 2 mL) and diluted phenol solution with the minimum amount of oxidant added (six times 0.2 mL). In each series of catalytic tests, it was observed that initiation phase is the first step, as in the case of free-radical reactions, especially for experiments carried out in concentrated phenol solution (**Figure 1A**). Initial 10–20 min are characterized with slow increase in pollutant conversion. After 30–50 min of the reaction over iron oxide-containing catalysts, conversion rapidly increased reaching values above 95%. Non-modified silicates, on the other hand, presented very low activity. Sample submitted to hydrothermal treatment (HS0) slightly increased phenol oxidation compared to non-catalytic process; however, in the latter case, conversion was not higher than 8% after 75 min. On the contrary, the addition of starting vermiculite (S0) to reaction mixture resulted in slow increase in phenol conversion up to 42%. Reduction in particle size was the only preparation step in this case; therefore, contaminations present in the starting materials, such as interlayer and adsorbed transition metal cations as well as naturally occurring iron oxides and carbonates, may be responsible for the observed catalytic effect.

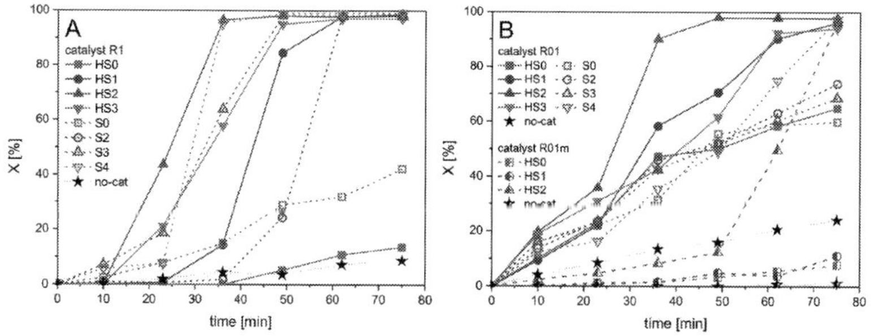

Figure 1. Conversion of phenol in oxidation reaction over vermiculite-supported iron oxide catalysts; A—initial concentration of phenol 1 g/L, volume of H_2O_2 injection = 2 mL; B—initial concentration of phenol 0.1 g/L, volume of H_2O_2 injection = 2 or 0.2 mL.

When diluted solution of phenol was used (**Figure 1B**) and accompanied by small excess of oxidant (R01m—0.2 mL), almost no effect was observed within assigned experimental time. Only one sample, doped with nano-hematite, HS2, showed catalytical properties after 50 min of reaction. The non-catalytical reaction performed with large excess of oxidant (R01—2 mL) resulted in quite significant conversion equal to 48% within 75 min. Slightly higher activity was observed when ferrihydrite-doped samples, S2 and S3, and non-modified silicates were added as catalysts. After constant gradual increase of conversion, it reached 60–75% within 75 minutes. Only one sample with the highest loading of ferrihydrite and samples doped with nano-hematite allowed to reach the conversion level above 95%. Nevertheless, it should be stressed that after initial increase in conversion, it was inhibited and much slower at longer reaction times in the case of removal of concentrated pollutant. Similar effect of the reaction stagnation, due to accumulation of the reaction products, was also observed in homogeneous Fenton reaction [29].

It may be observed that conversion of phenol was more effective with higher doping with ferrihydrite. On the contrary, regardless reaction conditions, in the series of nano-hematite-containing catalysts, an optimum amount of iron oxide results in higher efficiency of the reaction. The best sample in this case was HS2 doped with 3.36 wt.% of iron in the form of nano-hematite.

According to results of the catalytic tests described above, 2-line ferrihydrite supported on vermiculite is less active than analogous materials containing hematite. Additional experiment, performed in concentrated phenol solution (1 g/L) and using active sample HS3 as catalyst, provided information about reaction path and intermediate products. UV-vis spectra for samples withdrawn during experiment, quenched with methanol or mixed additionally with VO_3^-/H_2SO_4 solution, allowed to distinguish between transition products formed in the course of the reaction. After 29 min of the reaction, which corresponds to 43% of phenol conversion, colored products were formed. In the UV-vis spectrum recorded in methanol (**Figure 2A**) bands assigned to phenol (278 and 284 nm), hydroquinone (300 nm) and benzoquinone/quinhydrone (245, 255, and 300 nm) were identified. However, bands assigned to catechin were strongly overlapped by other strong peaks, and it cannot be excluded that this product was also formed. After 62 min of the experiment phenol conversion reached 97%, no colored products were recorded, and reaction was completed.

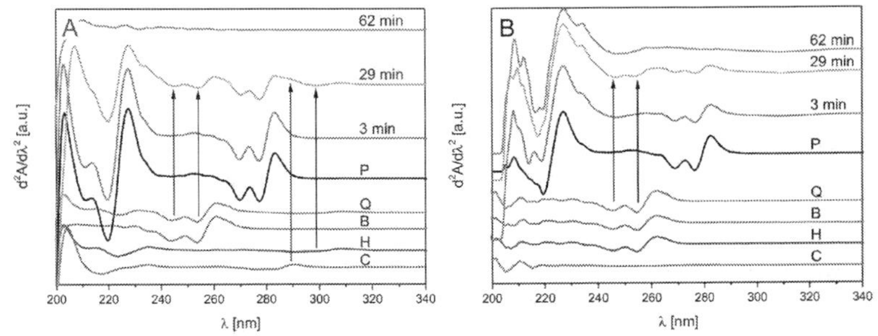

Figure 2. Identification of transition products of phenol oxidation over HS3 catalyst (reaction conditions: PhOH = 1 g/L, volume of H_2O_2 injection = 2 mL): A—derivative UV-vis spectra recorded in MeOH, B—derivative UV-vis spectra recorded in VO_3^-/H_2SO_4; P—phenol; Q—qiunhydrone; B—benzoquinone; H—hydroquinone; C—catechin.

However, in the spectra measured after the reaction of the sample of effluent with VO_3^-/H_2SO_4 mixture (**Figure 2B**), peaks below 230 nm, assigned to unidentified organic compounds, were recorded at the end of the test. Evolution of pH followed opposite trend as phenol conversion, and at 29 and 62 min, it was equal to 2.87 and 2.48, respectively. Those observations confirm that final products are not only H_2O and CO_2 but also organic acids.

3.2. Characterization of as received and spent catalysts

Iron oxide-bearing catalysts were obtained by direct deposition of formed oxide on the silicate support. Vermiculite was selected due to its mechanical and thermal stability. On the contrary to montmorillonite, it is not exfoliating rapidly in contact with water, and swelling is limited to changes of number of water molecules in the interlayer space. Moreover, mineral itself contains significant amount of iron.

As it was shown in **Figure 3**, both expected oxide structures were formed [27, 28]. XRD pattern of 2-line ferrihydrite consists of two broad reflections, while nano-hematite is characterized by the presence of several sharp but not intense peaks. Ferrihydrite structure was not observed after deposition on the support due to inherent poor ordering of the structure and low content in the composite material. On the other hand, using the Sherrer equation, it was confirmed that crystal size of pure nano-hematite phase was 4 nm. Only traces of nano-hematite could be identified in two vermiculite-supported samples with the highest loading of deposited phase —HS2 (3.36 Fe wt.%) and HS3 (6.72 Fe wt.%). Therefore, it was not possible to determine precise crystal parameters for oxide phase.

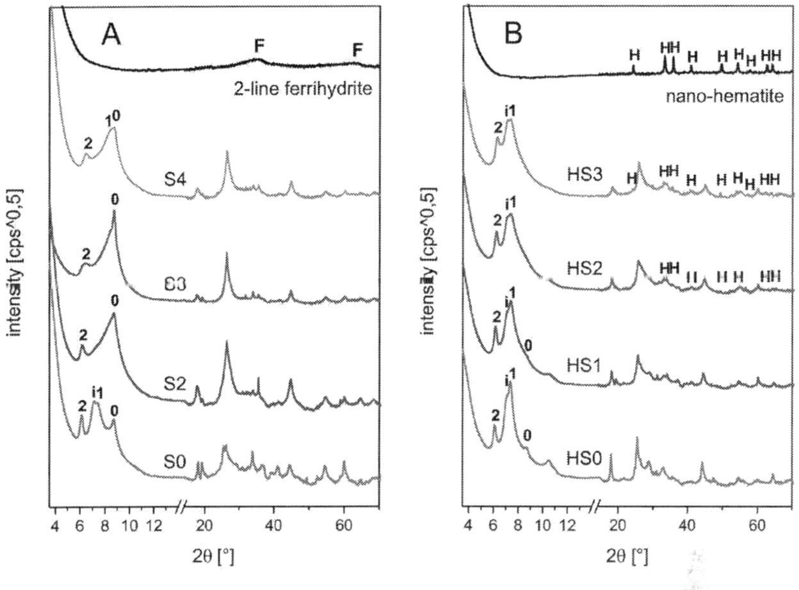

Figure 3. Structure of vermiculite-supported ferrihydrite- (A) and nano-hematite-containing (B) catalysts; F—ferrihydrite; H—nano-hematite; 0, 1, 2—basal reflections of vermiculite corresponding to 0, 1, and 2 layers of interlayer water; i—interstratified vermiculite phases.

Changes in vermiculite structure reflected chemical modifications performed in each synthesis. Starting material (S0) was characterized with complex pattern typical for vermiculites both containing in the interlayer divalent cations and collapsed structure (0 layers of water). Moreover, the interlayer cations are accompanied with 1 or 2 layers of water. Additional peaks below 8° 2θ were assigned to interstratified contracting and non-contracting phases [30].

Upon hydrothermal treatment in the sample HS0 intensity of peak corresponding to one water layer increased, while disappeared peak assigned to the collapsed structure. In the synthesis of 2-line ferrihydrite composite (**Figure 3A**), a first step consisted on an ion exchange of interlayer anions for iron. As a result, peak at 1.40–1.43 nm may be observed; however, it was shifted to lower values for higher loadings of iron oxide: 1.39 and 1.37 nm for S3 and S4, respectively. Described phenomenon is a result of partial dehydration of the interlayer gallery and formation of so-called HIV—hydroxy-interlayered vermiculites [31, 32]. Similar shift was observed also for the sample with the highest nano-hematite content: HS3—1.39 nm.

Application of potassium hydroxide, however, resulted in a deeper rearrangement of interlayer space. Both peaks assigned to interstratification and one water layer almost disappeared. On the other hand, intercalation of K^+ resulted in a large increase in peak intensity at 1 nm [33]. On the contrary, in the samples doped with nano-hematite in hydrothermal conditions (**Figure 3B**), peak assigned to 0 layers of interlayer water decreased with increasing amount of oxide.

In the structure of spent catalysts, traces of hematite were still possible to identify; however, other changes concerning catalyst properties were noticed. Vermiculite support upon reaction in concentrated solution was transformed into Mg^{2+}/Fe^{3+} intercalated structure containing 2 layers in water molecules (**Figure 4A**). In the samples doped with ferrihydrite, only traces of interlayer potassium were preserved, and interlayer spaces were occupied with di- and trivalent cations released from silicate matrix. Hydration state and the number of water molecules strongly depended on the initial amount of iron oxide—the lower doping level the easier rehydration proceeded. Similar dependence was observed also for nano-hematite deposited samples. Such phenomenon should be explained as a result of blocking of interlayer spaces with iron hydroxides. Moreover, iron oxide particles, which were grown near the edges of vermiculite layers, may act as cementing agent, preventing structure swelling. It was also observed that rehydration of the structure depends on the reaction conditions (**Figure 4B**): the higher concentration of phenol and hydrogen peroxide, the easier intercalation of water molecules. As it was shown in **Figure 4C/D**, swelling intensity, which may be expressed as peaks 1.42 (2 layers of water) and 1.20 nm (1 layer of water) intensity ratio, increased at higher concentration of substrates. It cannot be excluded that acidic reaction products also enhanced structural changes of vermiculite support.

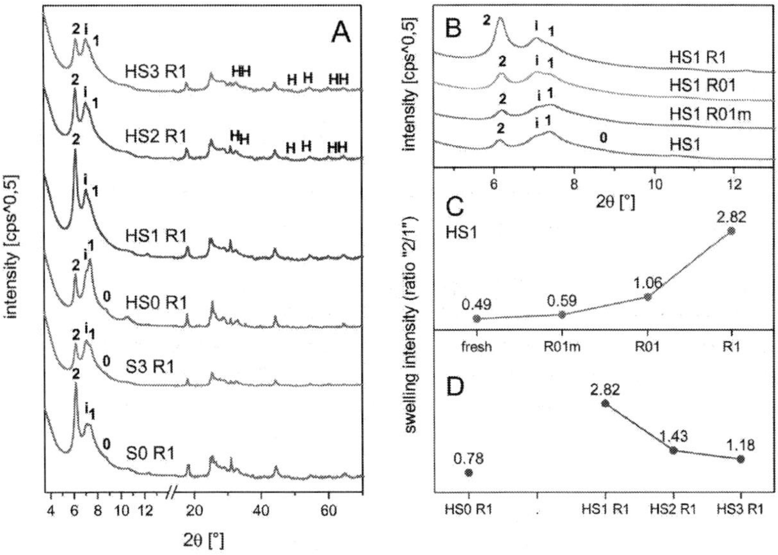

Figure 4. Structure of spent catalysts: A—vermiculite-supported iron oxide catalysts after reaction with phenol concentration 1 g/L; B and C—evolution of basal spacings of HS1 sample at different reaction conditions; D—evolution of basal spacings of nano-hematite-containing catalysts after reaction with phenol concentration 1 g/L; H—nano-hematite; 0, 1, 2—basal reflections of vermiculite corresponding to 0, 1, and 2 layers of interlayer water; i—interstratified vermiculite phases.

http://dx.doi.org/10.5772/65097

Sample name	*d* (nm)	*d* (nm)	*d* (nm)	*d* (nm)	*d* (nm)
Fresh catalysts (precipitated)					
S0	2.40	1.40	1.21	1.17	1.00
S2		1.43			1.01
S3		1.39			1.01
S4		1.37			1.01
Spent catalysts (precipitated)					
S0 H_2O_2	*	1.44	1.25	1.19	1.01
S0 R1	*	1.43	1.25	1.20	1.01
S3 R1	*	1.43	1.25	1.20	1.01
Fresh catalysts (hydrothermal)					
HS0	2.57	1.41	1.22	1.17	1.00
HS1	2.60	1.43	1.24	1.19	1.01
HS2	2.55	1.42	1.24	1.20	
HS3	2.53	1.39	1.22	1.18	
Spent catalysts (hydrothermal)					
HS0 R1	*	1.42	1.24	1.20	
HS1 R01m	*	1.42	1.24	1.19	
HS1 R01	*	1.43	1.25	1.20	
HS1 R1	*	1.43	1.25	1.20	
HS2 R01m	*	1.40	1.23	1.19	
HS2 R1	*	1.43	1.25	1.21	
HS3 R01	*	1.39	1.23	1.19	
HS3 R1	*	1.42	1.25	1.20	
	Interstratification	2 layers of water Mg^{2+}/Fe^{3+} in interlayers	Inter-stratification	1 layer of water Mg^{2+}/Fe^{3+} in interlayers	0 layers of water K^+ in interlayers

*2.4-2.6 nm (low-intensity peak).

Table 3. Interlayer distances of iron oxide-modified vermiculite-based catalysts before and after reaction.

The basal spacings calculated for modified vermiculites (**Table 3**) show that synthesis consisting on 2-line ferrihydrite precipitation resulted in the formation of hydroxy-interlayered phase and disappearance of peaks related to interstratified phases. Moreover, vermiculite was also partially intercalated with potassium. After the reaction, almost all peak positions returned to the initial values similar to the starting material. In nano-hematite modified samples, characteristic peaks for interstratification remained in their positions. However, in the course of phenol oxidation, first peak (~2.5 nm) became less noticeable.

It may be concluded that deposited iron oxide phases changed properties of the support; however, alteration was reversible in reaction conditions. Although XRD patterns do not allow to follow degradation of active phase directly, some indications of that process may be observed through properties of vermiculite.

More data considering properties of the deposited iron oxides were provided by DRS-UV-vis spectroscopy. As it was mentioned before, vermiculite itself contains iron [25] and UV-vis spectrum recorded for solid-state samples consisted of several characteristic bands. Isolated

Fe^{3+} cations in the tetrahedral coordination give rise to peaks at 224 nm in both silicate materials (S0 and HS0), and cations in the octahedral coordination may be identified by the presence of band at 260 nm [34, 35]. The bands at 319 and 358 nm are characteristic for small oligonuclear Fe_xO_y clusters. Formation of bulk Fe_2O_3 particles gave characteristic bands above 400 nm [34].

Ferrihydrite and hematite were characterized by multiple bands, revealed by the second derivative spectra (results not shown), and assigned to the electronic transitions [27, 36]. The spectra of both oxides consisted of peaks at 260–264 nm, which should be assigned to charge transfer. The bands at 401 and 424 nm for ferrihydrite and hematite, respectively, resulted from $^6A_1 \rightarrow {}^4E; {}^4A_1$ transition, 519 and 550 nm—$2(^6A_1) \rightarrow 2(^4T_1)$ (electron pair transition, EPT), 690–718 and 665 nm—$^6A_1 \rightarrow {}^4T_2$. Additionally, for nano-hematite, the following bands were assigned to $^6A_1 \rightarrow {}^4T_1$ transitions at 310 and 840 nm and $^6A_1 \rightarrow {}^4E$ transitions at 384 nm.

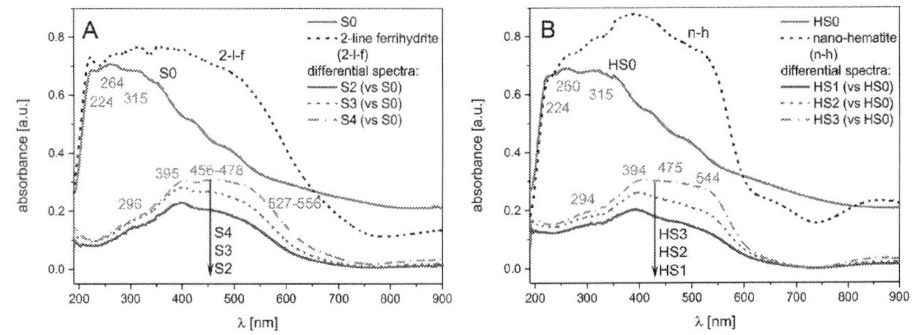

Figure 5. Agglomeration state of iron species in vermiculite-supported ferrihydrite- (A) and nano-hematite-containing (B) catalysts (DRS-UV-vis spectra).

Due to possible release of the cations from vermiculite and the contamination of deposited iron oxides during synthesis, the octahedra may be distorted, and consequently, ligand field and band positions may be changed. For 2-line ferrihydrite-containing catalysts, DRS-UV-vis peaks were shifted to 296, 456–478, 527–556, and 675 nm (**Figure 5A**). Similar result, with peak positions at 294, 442, 476, 544, 679, and 840 nm, was obtained for nano-hematite deposited on silicate (**Figure 5B**).

Further changes in the catalyst structure took place in the course of phenol oxidation. In the spectrum of starting silicate, S0, new band in the range of 360–480 nm was formed (**Figure 6A**). It is possible that adsorbed on the surface and interlayer Fe^{3+} cations present in original material were released and redeposited in the form of larger clusters. Catalysts modified with ferrihydrite after reaction with diluted phenol solution (R01) were depleted with active phase, and DRS-UV-vis spectra have shown minimum at 300 and 480 nm. Much larger minimum was registered in the differential spectrum of sample S3 after reaction with concentrated phenol solution. The shape and positions of minima (390, 453, and 524–550 nm) reflected distribution of absorption peaks in fresh catalyst. It may be expected that degradation of the catalyst is significant, although it is mechanical rather than chemical in nature.

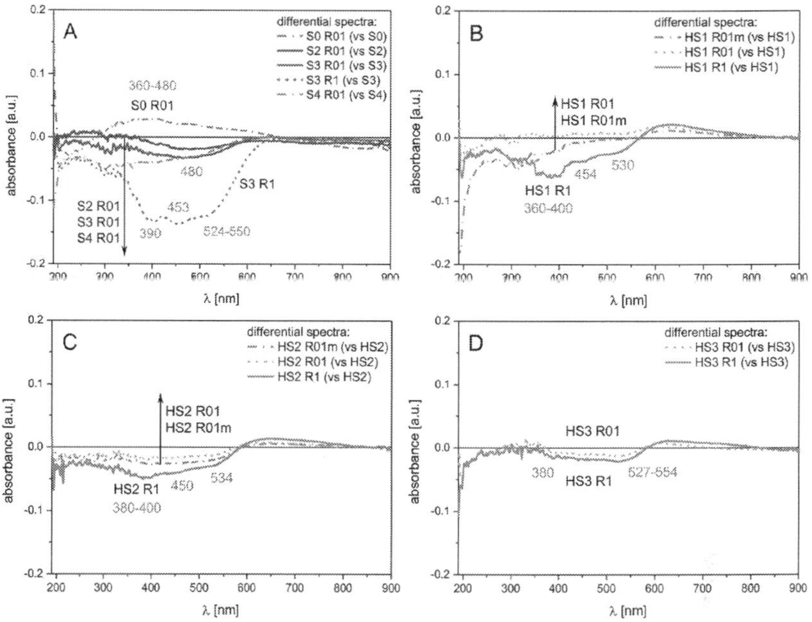

Figure 6. Leaching of iron species from vermiculite-supported iron oxide-containing catalysts upon phenol oxidation reaction (DRS-UV-vis spectra).

In nano-hematite-containing catalysts, degradation proceeded differently for each sample. At the lowest loading of active phase (**Figure 6B**, HS1), leaching was the most noticeable compared to the other samples, which were used in the reaction with concentrated substrates (R1). Moreover, the largest minimum was observed at 360–400 nm, while at 454 and 530 nm, two smaller features were observed. When the amount of hematite was increasing, minima recorded in DRS-UV-vis spectra were smaller and shifted to higher wavelengths (**Figure 6C/D**). Therefore, it may be concluded that optimization of the active phase loading is more important for hematite-containing composites, both in terms of catalyst stability and its activity. Surprisingly, although degradation of the catalysts is less noticeable in the reaction using lower concentration of phenol, the addition of lower excess of oxidant may also increase leaching of active components (e.g., **Figure 6B**). This feature may result in olation-oxolation processes, proceeding differently in the presence of H_2O_2.

On the basis of catalyst characterization, the following model was proposed for more active silicate-based nano-hematite-modified materials (**Figure 7**). In optimum conditions of about 3.36 wt.% of iron, which corresponds to 4.8 wt.% of deposited iron oxide, interlayer spaces of vermiculite are not blocked by hydroxides and are free to accommodate Fe^{3+} cations. On the surface of the layered support, patches of nanocrystalline phase are formed. Below the optimum hematite loading, besides well-defined nanocrystals, also oligomeric clusters of iron oxide are deposited, which may be easily dissolved by the reaction substrates and products in the course of the reaction. The interlayer space of vermiculite is still available for an ion-exchange process. Above the optimum loading of the active phase, interlayer spaces of vermiculite are blocked by hydroxy-compounds, which may be removed during the reaction. Deposited nano-hematite phase remains almost intact during the reaction.

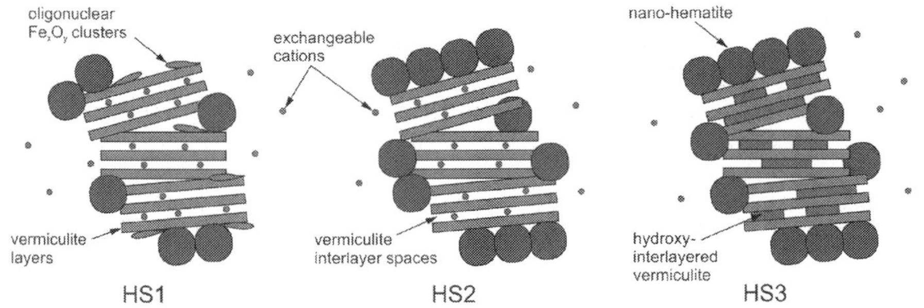

Figure 7. Simplified structure of nano-hematite-containing vermiculite-supported catalysts.

3.3. Catalytic activity vs. catalyst degradation

Changes in the catalyst chemical composition were followed during the reaction and correlated with catalytical results. In **Table 4**, it was presented that ferrihydrite-containing catalysts were more susceptible to Fe leaching. Surprisingly, the lower was oxide doping, the higher percentage of active phase was dissolved. No such straight relationship was observed for nano-hematite-containing catalysts. Apparently, small oligoclusters and interlayered hydroxy-species described in model in Section 3.2, indeed, contributed significantly to dissolved species. It was also observed that catalytic activity should not be attributed completely to homogeneous reaction. Reaction mixtures over ferrihydrite-doped catalysts were characterized with higher concentration of Fe available for homogeneous reaction. Times, required to obtain phenol conversion equal 40 and 50%, were longer for ferrihydrite-containing catalysts in comparison to hematite-doped materials. Moreover, in the latter case Fe concentrations in the reaction mixtures were relatively low. As it was described in Section 3.1, when diluted phenol solution was used for the reaction activity stagnated due to product accumulation. Another explanation could be recombination of radicals formed over the catalysts. Therefore, time for 50% phenol conversion is more or less 10 min delayed compared to 40% conversion. On the other hand, time difference for the reactions performed in concentrated phenol solution is closer to 1–3 min. Another conclusion may be formed on the basis of the analysis of residual phenol concentrations. Within 75 min of the reaction, phenol concentration is reduced to 3–31 and 2–6 mg/L for ferrihydrite- and hematite-containing catalysts, respectively, in reactions using starting solution equal to 100 mg/L. When 1 g/L phenol solution was used, final concentrations were equal to 8–11 and 15–30 mg/L for both iron containing series of catalysts. In this way, it

http://dx.doi.org/10.5772/65097

was confirmed that dispersed pollutants are more difficult to remove efficiently then concentrated.

Sample name	Fe content in catalyst (mg/g)	Fe available (mg/L)*	Fe leached from catalyst (%)**	PhOH residual (mg/L)*	$t_{40\%}$ (min)	$t_{50\%}$ (min)
S0	56.5					
S0 H_2O_2		1.0	1.0			
S0 R01		1.3	1.3	10	10	16
S0 R1		6.3	6.4	577	72	>75
S2	89.1					
S2 R01		5.7	3.6	26	32	44
S2 R1		n.d.	n.d.	11	52	54
S3	109.2					
S3 R01		4.6	2.4	31	34	46
S3 R1		27.1	14.0	21	29	32
S4	154.2					
S4 R01		2.8	1.0	3	40	50
S4 R1		n.d.	n.d.	8	28	29
HS0	59.0					
HS0 R01m		0.9	0.9	92	–	–
HS0 R01		0.5	0.4	35	32	49
HS0 R1		1.3	1.2	863	–	–
HS1	76.7					
HS1 R01m		3.2	2.3	89	–	62
HS1 R01		2.5	1.8	4	29	33
HS1 R1		26.0	19.2	19	41	43
HS2	86.8					
HS2 R01m		3.2	2.1	4	–	–
HS2 R01		1.0	0.7	2	24	26
HS2 R1		13.4	8.7	15	22	25
HS3	119.1					
HS3 R01		5.1	2.4	6	33	41
HS3 R1		23.9	11.4	30	30	33

n.d., not determined.
*In solution after 75 min of reaction.
**Percentage of initial content.

Table 4. Comparison of Fe content in catalysts and reaction solutions, residual concentration of phenol and time of 40 and 50% phenol conversion.

4. Conclusions

Depending on the experimental conditions, a nanocrystalline phase of hematite was formed in the hydrothermal synthesis. On the other hand, precipitation resulted in the formation of ferrihydrite phase. It was demonstrated that the latter phase is less active than nano-hematite; moreover, it was shown that optimum loading of the active phase is required to obtain the highest reaction efficiency: fast and high phenol conversion with minimum amount of side products as well as limited catalyst degradation. Among the transition products, formation of quinones was confirmed using derivative UV-vis spectroscopy. Physicochemical techniques also confirmed that nano-hematite-containing catalysts were more stable in studied reaction —only limited changes were observed in agglomeration state of Fe-containing materials, and leaching of iron was reduced. It was also shown that each group of catalysts is in different extents susceptible to degradation. However, the observed catalytic effect cannot be attributed only to homogeneous reaction. It was confirmed that dispersed pollutants are more resistant to degradation.

Author details

Agnieszka Węgrzyn

Address all correspondence to: a.m.wegrzyn@uj.edu.pl; a.m.wegrzyn@gmail.com

Faculty of Chemistry, Jagiellonian University in Krakow, Kraków, Poland

References

[1] Luck F. Wet air oxidation: past, present and future. Catalysis Today. 1999;53:81–91. DOI: 10.1016/S0920-5861(99)00112-1

[2] Busca G, Berardinelli S, Resini C, Arrighi L. Technologies for the removal of phenol from fluid streams: a short review of recent developments. Journal of Hazardous Materials. 2008;160:265–288. DOI:10.1016/j.jhazmat.2008.03.045

[3] Liotta LF, Gruttadauria M, Di Carlo G, Perrini G, Librando V. Heterogeneous catalytic degradation of phenolic substrates: catalysts activity. Journal of Hazardous Materials. 2009;162:588–606. DOI:10.1016/j.jhazmat.2008.05.115

[4] Rokhina EV, Virkutyte J. Environmental application of catalytic processes: heterogeneous liquid phase oxidation of phenol with hydrogen peroxide. Critical Reviews in Environmental Science and Technology. 2010;41:125–167. DOI:10.1080/10643380802669018

[5] Kim K-H, Ihm S-K. Heterogeneous catalytic wet air oxidation of refractory organic pollutants in industrial wastewaters: a review. Journal of Hazardous Materials. 2011;186:16–34. DOI:10.1016/j.jhazmat.2010.11.011

[6] He J, Yang X, Men B, Wang D. Interfacial mechanisms of heterogeneous Fenton reactions catalyzed by iron-based materials: a review. Journal of Environmental Sciences. 2011;39:97–109. DOI:10.1016/j.jes.2015.12.003

[7] Babuponnusami A, Muthukumar K. A review on Fenton and improvements to the Fenton process for wastewater treatment. Journal of Environmental Chemical Engineering. 2014;2:557–572. DOI:10.1016/j.jece.2013.10.011

[8] Wang N, Zheng T, Zhang G, Wang P. A review on Fenton-like processes for organic wastewater treatment. Journal of Environmental Chemical Engineering. 2016;4:762–787. DOI:10.1016/j.jece.2015.12.016

[9] Xu H-Y, Prasad M, Liu Y. Schorl: a novel catalyst in mineral-catalyzed Fenton-like system for dyeing wastewater discoloration. Journal of Hazardous Materials. 2009;165:1186–1192. DOI:10.1016/j.jhazmat.2008.10.108

[10] Matta R, Hanna K, Chiron S. Fenton-like oxidation of 2,4,6-trinitrotoluene using different iron minerals. Science of the Total Environment. 2007;385:242–251. DOI: 10.1016/j.scitotenv.2007.06.030

[11] Gomes Flores R, Layara Floriani Andersen S, Kenji Komay Maia L, Jorge José H, de Fatima Peralta Muniz Moreira R. Recovery of iron oxides from acid mine drainage and their application as adsorbent or catalyst. Journal of Environmental Management. 2012;111:53–60. DOI:10.1016/j.jenvman.2012.06.017

[12] Rahim Pouran S, Aziz Abdul Raman A, Mohd Ashri Wan Daud W. Review on the application of modified iron oxides as heterogeneous catalysts in Fenton reactions. Journal of Cleaner Production. 2014;64:24–35. DOI:10.1016/j.jclepro.2013.09.013

[13] Zelmanov G, Semiat R. Iron(3) oxide-based nanoparticles as catalysts in advanced organic aqueous oxidation. Water Research. 2008;42:492–498. DOI:10.1016/j.watres.2007.07.045

[14] ElShafei GMS, Yehia FZ, Dimitry OIH, Badawi AM, Eshaq Gh. Ultrasonic assisted-Fenton-like degradation of nitrobenzene at neutral pH using nanosized oxides of Fe and Cu. Ultrasonics Sonochemistry. 2014;21:1358–1365. DOI:10.1016/j.ultsonch.2013.12.019

[15] Pastrana-Martínez LM, Pereira N, Lima R, Faria JL, Gomes HT, Silva AMT. Degradation of diphenhydramine by photo-Fenton using magnetically recoverable iron oxide nanoparticles as catalyst. Chemical Engineering Journal. 2015;261:45–52. DOI:10.1016/j.cej.2014.04.117

[16] Rusevova K, Kopinke F-D, Georgi A. Nano-sized magnetic iron oxides as catalysts for heterogeneous Fenton-like reactions—influence of Fe(II)/Fe(III) ratio on catalytic

performance. Journal of Hazardous Materials. 2012;241–242:433–440. DOI:10.1016/j.jhazmat.2012.09.068

[17] Chun J, Lee H, Lee S-H, Hong S-W, Lee J, Lee C, Lee J. Magnetite/mesocellular carbon foam as a magnetically recoverable fenton catalyst for removal of phenol and arsenic. Chemosphere. 2012;89:1230–1237. DOI:10.1016/j.chemosphere.2012.07.046

[18] Melero JA, Calleja G, Martínez F, Molina R, Pariente MI. Nanocomposite Fe_2O_3/SBA-15: an efficient and stable catalyst for the catalytic wet peroxidation of phenolic aqueous solutions. Chemical Engineering Journal. 2007;131:245–256. DOI:10.1016/j.cej.2006.12.007

[19] Shukla P, Wang S, Sun H, Ang H-M, Tadé M. Adsorption and heterogeneous advanced oxidation of phenolic contaminants using Fe loaded mesoporous SBA-15 and H_2O_2. Chemical Engineering Journal. 2010;164:255–260. DOI:10.1016/j.cej.2010.08.061

[20] Xiang L, Royer S, Zhang H, Tatibouët J-M, Barrault J, Valange S. Properties of iron-based mesoporous silica for the CWPO of phenol: a comparison between impregnation and co-condensation routes. Journal of Hazardous Materials. 2009;172:1175–1184. DOI:10.1016/j.jhazmat.2009.07.121

[21] di Luca C, Massa P, Fenoglio R, Medina Cabello F. Improved Fe_2O_3/Al_2O_3 as heterogeneous Fenton catalysts for the oxidation of phenol solutions in a continuous reactor. Journal of Chemical Technology and Biotechnology. 2014;89:1121–1128. DOI:10.1002/jctb.4412

[22] Aida Zubir N, Yacou C, Zhang X, Diniz da Costa JC. Optimisation of graphene oxide-iron oxide nanocomposite in heterogeneous Fenton-like oxidation of Acid Orange 7. Journal of Environmental Chemical Engineering. 2014;2:1881–1888. DOI:10.1016/j.jece.2014.08.001

[23] Shin S, Yoon H, Jang J. Polymer-encapsulated iron oxide nanoparticles as highly efficient Fenton catalysts. Catalysis Communications. 2008;10:178–182. DOI:10.1016/j.catcom.2008.08.027

[24] Yuan S-J, Dai X-H. Facile synthesis of sewage sludge-derived mesoporous material as an efficient and stable heterogeneous catalyst for photo-Fenton reaction. Applied Catalysis B: Environmental. 2014;154–155:252–258. DOI:10.1016/j.apcatb.2014.02.031

[25] Stawiński W, Freitas O, Chmielarz L, Węgrzyn A, Komędera K, Błachowski A, Figueiredo S. The influence of acid treatments over vermiculite based material as adsorbent for cationic textile dyestuffs. Chemosphere. 2016;153:115–129. DOI:10.1016/j.chemosphere.2016.03.004

[26] Węgrzyn A, Chmielarz L, Zjeżdżałka P, Jabłońska M, Kowalczyk A, Żelazny A, Vázquez Sulleiro M, Michalik M. Vermiculite-based catalysts for oxidation of organic pollutants in water and wastewater. Acta Geodynamica et Geomaterialia. 2013;10(171):341–352. DOI:10.13168/AGG.2013.0033

[27] Cornell RM, Schwertmann U. The Iron Oxides: Structures, Properties, Reactions, Occurrences and Uses. Weinheim: WILEY-VCH Verlag GmbH & Co. KGaA; 2003. DOI: 10.1002/3527602097

[28] Schwertmann U, Cornell RM. Iron Oxides in the Laboratory: Preparation and Characterization. Weinheim: WILEY-VCH Verlag GmbH & Co. KGaA; 2000. DOI: 10.1002/9783527613229

[29] Nakagawa H, Yamaguchi E. Influence of oxalic acid formed on the degradation of phenol by Fenton reagent. Chemosphere. 2012;88:183–187. DOI:10.1016/j.chemosphere.2012.02.082

[30] Harraz HZ, Hamdy MM. Interstratified vermiculite-mica in the gneiss-metapelite-serpentinite rocks at Hafafit area, Southern Eastern Desert, Egypt: from metasomatism to weathering. Journal of African Earth Sciences. 2010;58(2):305–320 DOI:10.1016/j.jafrearsci.2010.03.009

[31] Kalinowski BE, Schweda P. Rates and nonstoichiometry of vermiculite dissolution at 22°C. Geoderma. 2007;142(1–2):197–209. DOI:10.1016/j.geoderma.2007.08.011

[32] Mareschal L, Ranger J, Turpault MP. Stoichiometry of a dissolution reaction of a trioctahedral vermiculite at pH 2.7. Geochimica et Cosmochimica Acta. 2009;73(2):307–319. DOI:10.1016/j.gca.2008.09.036

[33] Abate G, Masini JC. Influence of thermal treatment applied to Fe(III) polyhydroxy cation intercalated vermiculite on the adsorption of atrazine. Journal of Agricultural and Food Chemistry. 2007;55:3555–3560. DOI:10.1021/jf063536y

[34] Kumar, MS, Schwidder, M, Grünert, W, Brückner, A. On the nature of different iron sites and their catalytic role in Fe-ZSM-5 DeNOx catalysts: new insights by a combined EPR and UV/VIS spectroscopic approach. Journal of Catalysis. 2004;227:384–397. DOI: 10.1016/j.jcat.2004.08.003

[35] Pérez-Ramirez, J, Santhosh Kumar, M, Brückner, A. Reduction of N_2O with CO over FeMFI zeolites: influence of the preparation method on the iron species and catalytic behavior. Journal of Catalysis. 2004;223:13–27, DOI:10.1007/s11144-014-0795-y

[36] Torrent J, Barrón V. Diffuse reflectance spectroscopy of iron oxides. In: Encyclopedia of Surface and Colloid Science, vol. 4, ed. Arthur T. Hubbard. Abingdon: Taylor & Francis; 2002. DOI:10.1081/E-ESCS

Section 2

Micropollutants and their Removal Process

Chapter 5

Micropollutants in Wastewater: Fate and Removal Processes

Sreejon Das, Nillohit Mitra Ray, Jing Wan,
Adnan Khan, Tulip Chakraborty and
Madhumita B. Ray

Additional information is available at the end of the chapter

http://dx.doi.org/10.5772/65644

Abstract

The occurrence of micropollutants (MPs) in various streams of municipal wastewater treatment plants (WWTPs), and their fate and removal processes are discussed. The fate of MPs in WWTPs largely depends on adsorption on suspended particulates, primary and secondary sludge and dissolved organic carbon, and removal occurs due to coagulation-flocculation, and biodegradation. The log Kow (>2.5) and pKa are the dominant properties of the MPs, and the concentration, organic fraction, and surface charge of suspended particulates dictate the extent of adsorption of MPs. Most of the conventional WWTPs do not remove complex MPs by biodegradation or biotransformation effectively (kbio ≤0.0042 L/gss/h), and the removal varies widely for different compounds, as well as for the same substance, due to operational conditions such as aerobic, anaerobic, anoxic, sludge retention time (SRT), pH, redox potential, and temperature. Membrane bioreactor performs better for moderately biodegradable compounds due to the diverse nature of microorganisms as well as greater adaptability due to longer SRT. Ozone and UV-based advanced oxidation processes, membrane filtration can be used for tertiary treatment due to their high rate as well as easy implementation. Various partition coefficients and rate constants values for different MPs are also provided for design and application.

Keywords: micropollutants, wastewater, fate and removal, adsorption, coagulation, biodegradation, membrane filtration, advanced oxidation processes

1. Introduction

The widespread presence of micropollutants (MPs) in aquatic systems is a major concern all across the globe. For example, about 143,000 compounds were registered in European market in 2012; many of which would end up in water systems at some point of their lifecycle. Most of them are not eliminated or biotransformed in traditional wastewater treatment plants, can be persistent in aquatic system or form new chemical species reacting with background humic substances in sunlight, can be bioactive, and can bioaccumulate [1–5]. Although they are present in almost undetectable (low to subparts per billion (ppb)) concentrations, their existence in aquatic systems has been connected to various detrimental effects in organisms such as estrogenicity, mutagenicity, and genotoxicity [6].

While no compound-specific regulation exists anywhere for the removal of MPs in wastewater plants, some regulations are there for the presence in water for compounds such as pesticides, lindane, nonylphenol, and synthetic hormones [7]. The MPs fall into several categories as pharmaceuticals, personal care products (PPCPs), household chemicals, and industrial agents. A comprehensive list of 242 chemicals is provided in EU FP7 Project [8] of which about 70% are pharmaceuticals and personal care products and 30% are industrial agents including perfluoro compounds, pesticides, herbicides, and food additives. Since a significant majority of the MPs in municipal wastewater belong to the class of pharmaceuticals and personal care products (PPCP), fate and removal processes of these compounds are discussed in detail in this chapter.

2. Commonly found PPCP in wastewater effluent and surface water

About 70% of the pharmaceuticals in the wastewater originates from household, 20% comes from livestock farming, 5% is from hospital effluent, and rest 5% comes in runoff from nonparticular sources [9]; however, seasonal and geographical variations typically occur. The fate of MPs in wastewater plant depends on the physical properties such as solubility, octanol-water partition coefficient, and Henry's constant. A list of commonly found pharmaceuticals, personal care products, and biocides and their concentration in wastewater effluent and surface water and physical properties are presented in **Table 1**. The solubility of MPs varies in a wide range of 0.15 mg/L (maprotiline, $C_{10} H_{23}$ N, an antidepressant drug) to 588,000 mg/L (acesulfame, $C_4H_4KNO_4S$, and artificial sweetener), which is also in accordance with their concentration in the effluent.

Type	MP	Application	Average concentration (ng/L) [10, 11] Surface water	WWTP effluent	Solubility* (mg/mL)	log K_{ow}*	pKa*	Henry's constant (atm-m^3/mole)*
Disinfectants, pharmaceuticals (prescriptions, over-	Atenolol	β-blocker	205	843	0.3	0.16	9.6	1.37 × E-18
	Azithromycin	Antibiotic	12	175	<1 at 25°C	4.02	8.74	5.30 × E-29

Type	MP	Application	Average concentration (ng/L) [10, 11]		Solubility* (mg/mL)	log K_{ow}*	pKa*	Henry's constant (atm-m³/mole)*
			Surface water	WWTP effluent				
the-counter drugs, veterinary drugs) [10]	Bezafibrate	Lipid-lowering drug	24	139	0.00155	3.97	3.83	
	Carbamazepine	Anticonvulsant	13	482	0.152	2.1	15.96	1.08 × 10^{-10}
	Carbamazepin-10, 11–dihydro-10, 11-dihydroxy	Transformation product	490	1551	–	–	–	–
	Clarithromycin	Antibiotic	30	276	0.00033	3.16	8.99 at 25°C	1.73 × E-29
	Diatrizoate (amidotrizoic acid)	Contrast medium	206	598	0.107	2.89	2.17	–
	Diclofenac	Analgesic	65	647	0.00447	4.98	4	4.73 × E-12
	Erythromycin	Antibiotic	25	42	0.459	2.37	12.44	1.46 × E-29
	Ethinylestradiol	Synthetic estrogen	5	2	0.00677	3.63	10.33	7.94 × E-12
	Ibuprofen	Analgesic	35	394	0.0684	3.5	4.85	1.50 × E-07
	Iomeprol	Contrast medium	275	380	–	–	–	–
	Iopamidol	Contrast medium	92	377	0.117	1.62	4.15	1.14 × E-25
	Iopromide	Contrast medium	96	876	0.0238	−2.05	–	1.00 × E-28
	Mefenamic acids	Analgesic	7	870	0.0137	4.58	3.89	2.57 × E-11
	Metformin	Antidiabetic	713	10347	1.38	−1.8	12.4	–
	Metoprolol	β-blocker	20	166	0.402	1.88	14.09	1.40 × E-13
	Naproxen	Analgesic	37	462	0.0511	3.29	4.19	3.39 × E-10
	Sotalol	β-blocker	63	435	0.782	0.85	10.07	2.49 × E-14
	Sulfamethoxazole	Antibiotic	26	238	0.459	0.79	6.16	6.42 × E-13
	N4-Acetylsulfamethoxazole	Transformation product	3	67	–	–	–	–
	Trimethoprim	Antibiotic	13	100	0.615	1.26	17.33	2.39 × E-14
	Penicillin V	Personal care product	–	28.7	0.454	1.78	3.39	4.42 × E-15
Disinfectants, pharmaceuticals (prescriptions, over-the-counter drugs, veterinary drugs) [11]	Irbesartan	Antihypertensives	–	479.5	0.00884	4.51	7.4	–
	Tramadol	Analgesics	–	255.8	0.75	2.71	13.8	1.54 × E-11
	Risperidone	Neuroleptics	–	6.9	0.171	3.27	8.76	–
	Trihexyphenidyl	Antidementia agents	–	0.2	0.00314	4.93	13.84	4.73 × E-10
	Venlafaxine	Antidepressant	–	118.9	0.23	2.69	14.42	–
	Codeine	Morphine derivates	–	70.6	0.577	1.2	13.78	7.58 × E-14

Type	MP	Application	Average concentration (ng/L) [10, 11]		Solubility* (mg/mL)	log K_{ow}*	pKa*	Henry's constant (atm-m^3/mole)*
			Surface water	WWTP effluent				
	Fluconazole	Antifungal medication	–	108.2	1.39	0.58	12.71	–
	Diphenhydramine	Antihistamine	–	11.7	0.0752	3.44	8.98	3.70 × E-09
	Repaglinide	Antidiabetic medications	–	3.1	0.00294	5.05	3.68	–
	Flecainide	Antiarrhythmic agents	–	45.5	0.0324	2.98	13.68	5.75 × E-13
	Bisoprolol	β-blockers	–	41.6	0.0707	2.3	14.09	2.89 × E-15
	Alfuzosin	Alpha-blockers	–	2.8	0.282	2.02	14.64	–
	Bupropion	Antidepressant	–	1.0	312	3.6	18.29	–
	Ciprofloxacin	Antibiotics	–	96.3	1.35	0.28	6.09	5.09 × E-19
	Oxazepam	Anxiolytics	–	161.7	0.0881	2.24	10.61	5.53 × E-10
	Carbamazepine	Antiepileptic drugs	–	832.3	0.152	2.45	15.96	1.08 × E-10
	Diclofenac	Analgesics	65	647	0.00447	4.98	4	4.73 × E-12
	Orphenadrine	Antihistamine	–	3.9	0.03	3.77	8.91	4.08 × E-09
	Sulfamethoxazole (VITO)	Antibiotics	–	280.2	0.459	0.89	6.16	–
	Haloperidol	Psychiatric medication	–	32.2	0.00446	4.30	8.66	2.26 × E-14
	Citalopram	Antidepressant	–	33.8	–	–	–	–
	Sulfamethoxazole (JRC)	Antibiotics	–	142.3	0.459	0.89	6.16	–
	Fexofenadine	Antihistamine	–	165.0	0.00266	5.6	4.04	–
	Diltiazem	Antiarrhythmic agents	–	10.7	0.0168	3.09	12.86	8.61 × E-17
	Fluoxetine	Antidepressant	–	2.1	0.0017	4.05	9.8	8.90 × E-08
	Terbutaline	Antiasthmatics	–	1.1	5.84	0.90	8.86	1.65 × E-18
	Clindamycin	Antibiotics	–	70.4	3.1	2.16	12.16	2.89 × E-22
	Telmisartan	Antihypertensives	–	367.5	0.0035	7.7	3.65	–
	Eprosartan	Antihypertensives	–	226.8	0.00866	3.9	3.63	–
	Gemfibrozil	Lipid-lowering drugs	–	137.7	0.0278	3.4	4.42	–
	Zolpidem	Hypnotics	–	1.5	0.0313	3.15	6.2	–
	Hydroxyzine	Antihistamine	–	1.1	0.0914	3.43	15.12	–
	Ketoprofen	Analgesics	–	86.0	0.0213	3.12	4.45	2.12 × E-11
	Ranitidine	Antihistamine	–	68	0.0795	0.27	8.08	3.42 × E-15

http://dx.doi.org/10.5772/65644

Type	MP	Application	Average concentration (ng/L) [10, 11]		Solubility* (mg/mL)	log K_{ow}*	pKa*	Henry's constant (atm-m^3/mole)*
			Surface water	WWTP effluent				
	Triclosan	Disinfectants	–	74.8	0.00605	5.53	7.9	4.99 × E-09
	Levamisole	Antihelminthics	–	40.6	1.44	1.84	6.98	4.03 × E-10
	Lincomycin	Antibiotics	–	31.2	29.3	0.56	12.37	3.00 × E-23
	Rosuvastatin	Statins	–	31.0	0.0886	1.47	4	–
	Mianserin	Antidepressant	–	1.5	0.292	3.52	6.92	–
	Clofibric acid	Lipid-lowering drugs	–	5.3	0.583	2.57	−4.9	2.19 × E-08
	Iohexol	Radiocontrast agents	–	158	0.796	−3.05	11.73	2.66 × E-29
	Memantine	Antidementia agents	–	22.8	0.0455	3.28	10.7	1.47 × E-05
	Sertraline	Antidepressant	–	2.1	0.000145	5.06	9.85	–
	Tiamulin	Antibiotics	–	3.3	–	–	–	–
	Clonazepam	Anticonvulsant	–	1.6	0.0106	2.41	11.89	7.02 × E-13
	Alprazolam	Antidepressant	–	1.3	0.0324	2.12	18.3	9.77 × E-12
	Fenofibrate	Lipid-lowering drugs	–	1.1	0.000707	4.86	−4.9	–
	Sulfadiazine	Antibiotics	–	3.5	0.601	−0.09	6.36	1.58 × E-10
	Tilmicosin	Antibiotics	–	3.1	–	–	–	–
	Cyproheptadine	Chemotherapeutic agents	–	3.9	0.0136	4.69	8.05	9.20 × E-09
Detergents, dishwashing liquids, personal care products (fragrances, cosmetics, sunscreens), and food products [11]	Methylbenzotriazole	Personal care product	–	2900	0.366	2.720	8.55	4.13 × E-07
	Gadolinium	Personal care product	–	115.0	–	–	–	–
	Loperamide	Personal care product	–	29.3	0.00086	4.44	13.96	–
	Buprenorphine	Personal care product	–	3.9	0.0168	4.98	8.31 at 25°C	1.76 × E-17
	Maprotiline	Personal care product	–	0.4	0.00015	4.89	10.54	–
	Duloxetine	Personal care product	–	0.1	0.00296	4.72	9.7	–
	Miconazole	Personal care product	–	0.2	0.000763	5.86	6.77	–
	Chlorpromazine	Personal care product	–	0.1	0.00417	5.18	9.3 at 25°C	3.95 × E-11
	Flutamide	Personal care product	–	0.1	0.00566	3.35	13.17	3.73 × E-10
	DEET, N, N'-diethyltoluamide	Personal care product	–	678.1	0.912	2.80		2.08 × E-08
	Caffeine	Food additives	–	191.1	11.0	−0.07	10.4 at 40°C	1.90 × E-19

Type	MP	Application	Average concentration (ng/L) [10, 11]		Solubility* (mg/mL)	log K_{ow}*	pKa*	Henry's constant (atm-m^3/mole)*
			Surface water	WWTP effluent				
	Acesulfame	Food additive	4010	22500	588	−1.33	5.67	–
	Sucralose	Food additive	540	4600	22.7	−1.00	4.2	–
Pesticides [10]	Diazinon	Insecticide	15	173	0.04	3.81	2.6	1.13 × E-07
	Diethyltoluamide (DEET)	Insecticide	135	593	0.912	2.80		2.08 × E-08
	Dimethoate	Insecticide	22	–	25	0.78		1.05 × E-10
	MCPA	Insecticides	–	149.9	0.63	3.25	3.13	1.33 × E-09
	Carbaryl	Insecticide	–	1.6	0.11	2.36	10.4	–
Biocides [10]	2, 4-D	Herbicide	67	13	0.012	2.81	2.73	1.59 × E-07
	Carbendazim	Fungicide	16	81	0.029	1.52	4.2	2.12 × E-11
	Diuron	Herbicide	54	201	0.042	2.68		5.04 × E-10
	Glyphosate	Herbicide	373	–	12	−3.40	0.8	4.08 × E-19
	Irgarol (cybutryne)	Herbicide	3	30	–	–	–	–
	Isoproturon	Herbicide	315	12	0.065	2.87		1.12 × E-10
	MCPA	Herbicide	40	25	0.63	3.25	3.13	1.33 × E-09
	Mecoprop-p	Herbicide	45	424	0.62	3.13	3.1	1.82 × E-08
	Triclosan	Microbiocide	20	116	0.010	4.76	7.9	4.99 × E-09
	Terbutylazine	Herbicide	–	90.6	0.0085	3.21	2	3.72 × E-08
	Atrazine	Herbicide	–	4.2	0.0347	2.61	1.7	2.36 × E-09
	Terbutylazine-desethyl	Herbicide	–	68.8	–	–	–	–
	Isoproturon	Herbicide	–	10.1	0.065	2.87	–	1.12 × E-10
	Bentazone	Herbicide	–	9.6	0.5	2.34	2.92	2.18 × E-09
	Metolachlor	Herbicide	–	12.4	0.53	3.13	–	9 × E-09
	Dichlorprop	Herbicide	–	9.6	0.35	3.43	3.1	8.68 × E-11
	Simazine	Herbicide	–	26.3	0.0062	2.18	1.62	9.42 × E-10
	Atrazine-desethyl	Herbicide	–	13.8	3.2	1.51	–	1.53 × E-09
	Chlortoluron	Herbicide	–	3.2	0.07	2.41	–	–
	Hexazinone	Herbicide	–	0.8	33	1.85	–	2.26 × E-12
	Linuron	Herbicide	–	40.1	0.075	3.20	–	–

Type	MP	Application	Average concentration (ng/L) [10, 11] Surface water	WWTP effluent	Solubility* (mg/mL)	log K_{ow}*	pKa*	Henry's constant (atm-m³/mole)*
	2, 4, 5-T	Herbicide	–	0.3	0.248	3.26	2.88	6.83 × E-09
Hormone active substances (effect on the hormone balance) [10]	Bisphenol A (BPA)	Additive	840	331	0.12	3.32	9.6	1 × E-11
	Estradiol	Natural estrogens	2	3	0.0213	4.01	10.33	3.64 × E-11
	Estrone	Natural estrogens	2	15	0.00394	3.13	10.33	3.8 × E-10
	Nonylphenol	Additive	441	267	0.00635	5.99	10.25	1.1 × E-06
	Perfluorooctane sulfonate (PFOS)	Tenside	–	–	3.1	6.28	0.14	–

"–": Data are not available in the literature. *Solubility, log K_{ow}, pKa, and Henry's law constant for selected micropollutants are found in http://www.drugbank.ca/, http://chem.sis.nlm.nih.gov/chemidplus/ and https://pubchem.ncbi.nlm.nih.gov/.

Table 1. Commonly found MPs in municipal wastewater effluent and surface water.

3. Fate and removal processes of MPs in wastewater

The municipal wastewater treatment plants (WWTP) are designed to remove most of the suspended solids, dissolved organics, and nutrients from the wastewater. WWTPs employ primary, secondary, and occasional tertiary treatment processes to optimally treat the incoming wastewater. In primary treatment, coagulants such as alum, ferric chloride, and polymers and polymeric coagulant aids are used to remove colloidal and suspended particulates. In the process, organics attached with dissolved humic substances and particles can also be removed. In secondary treatment, dissolved organics are removed aerobically by a consortium of microorganisms in suspension. The thickened sludge from both primary and secondary clarifiers is digested anaerobically (biosolids) prior to disposal. In some places, tertiary treatment processes such as activated carbon adsorption, ozonation, or filtration are adopted for final treatment of effluent to remove trace concentration of the organics.

The fate processes for MPs in a typical WWTP include adsorption on suspended particulates, dissolved humic substances, primary and secondary sludge, while the removal processes include coagulation and sedimentation, biodegradation, adsorption, advanced oxidation, and membrane filtration as shown in **Figure 1**. Volatilization of the MPs during any of the treatment steps is negligible due to their very low Henry's constant ($<10^{-5}$ atm-m³/mol) as shown in **Table 1**.

3.1. Fate: adsorption of micropollutants

Adsorption on suspended solids in both primary and secondary treatment units is an important fate process for MPs in wastewater. Adsorption may occur due to the hydrophobic interactions between the aliphatic and aromatic groups of the compounds with the fat and lipid fractions in primary sludge and the lipophilic cell membrane of the microorganisms in secondary sludge, respectively. Electrostatic interactions also occur between the positively charged groups in the MPs and the negatively charged microorganisms in secondary sludge. Many acidic pharmaceuticals are negatively charged at neutral pH, and their sorption on sludge is negligible.

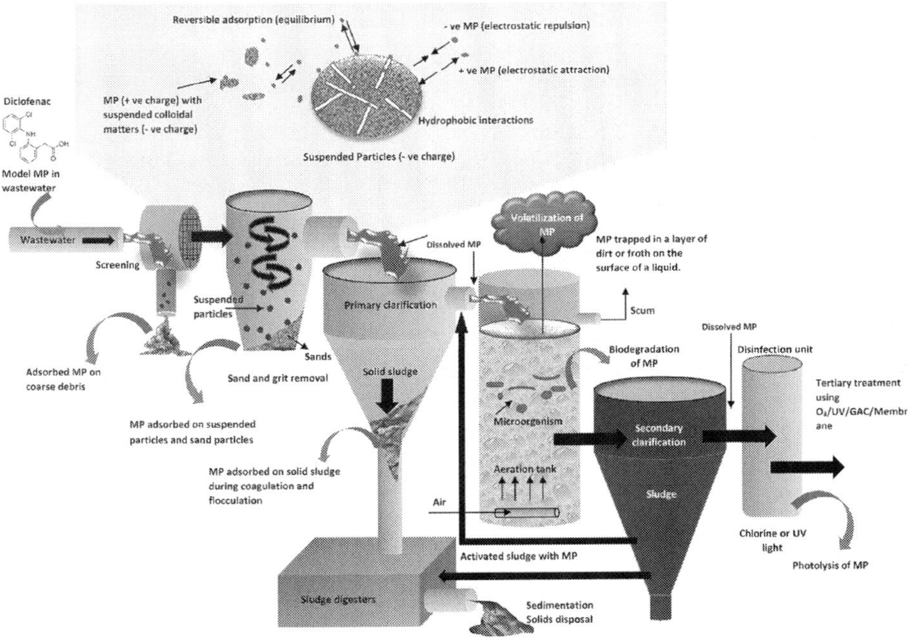

Figure 1. Conceptual model of fate and removal processes of a micropollutant in a typical WWTP.

With a nonpolar core and polar moiety, the properties of pharmaceuticals and antibiotics vary widely, making it difficult to estimate their sorption on sludge. Kinney et al. [12] analyzed nine different biosolids produced by municipal wastewater treatment plants in seven different states in U.S. for 87 different MPs, and the measured concentrations of the contaminants in various sludge were in the range of 64–1811 mg/kg dry weight. Nineteen different pharmaceuticals were detected in these biosolids, representing a wide range of physico-chemical properties, including compounds with low log K_{ow} and high water solubility values. Adsorption of MPs on biosolids did not exhibit any particular trend, and no correlation was found between organic carbon-normalized MPs concentrations in biosolids with log K_{ow}, suggesting that organic carbon content of the biosolids may not be the only factor controlling MPs adsorption. It is generally expected that compounds with low water solubilities and large log K_{ow} values will more likely to be present in organic-rich biosolids compared to highly soluble organics; however, this study indicated significant presence of water soluble pharmaceuticals in all nine biosolids. The 25 MPs detected in all nine biosolids had water solubility ranging from 1.3×10^{-5} to 8.28×10^{4} mg/L, and log K_{ow} from 1.50 to 9.65 indicating complex nature of the process. Other factors, such as the quantity of organics entering the influent stream (which typically varies temporally and spatially), volume of influent, biosolids/water ratio, and sludge retention time (SRT), all affect the distribution of the MPs in different phases. Increasing sludge age had detrimental effect on the adsorption of lindane [13] on activated sludge, adsorption of pentachlorophenol reduced from 40 to 60% at sludge ages below 4 days to less than 10% at sludge ages above 25 days [14].

The concentrations of some of the commonly found MPs in sludge are summarized in **Table 2**.

MP	Type/application	Concentration (mg/kg)	Source	Reference
Triclosan	Personal care product	0.41–46	Sludge (primary, excess activated, anaerobically digested)	Heidler & Halden [15], McAvoy et al. [16]
Triclocarban		4.7–63	Sludge (excess activated, anaerobically digested)	Heidler & Halden [15], Clara et al. [17]
Tonalide		0.4–2.9		
Galaxolide		4.2–21		
Cashmerane		0.022–0.26		
Celestolide		0.023–0.061		
Phantolide		0.010–0.014		
Traesolide		0.29–1.75		
Octocrylene		1.01–1.32		Kupper et al. [18]
Octyl-triazone		2.6–3.04		
Octyl-methoxycinnamate		0.15–1.5		
Pipemidic acid	Antibiotic	0.04 –0.27	Sludge (primary, excess activated, dewatered)	Jia et al. [19]
Fleroxacin		0.02–0.09		
Ofloxacin		0.33–7.79		
Enrofloxacin		0.02–0.07		
Lomefloxacin		0.06–1		
Sarafloxacin		0.39–0.13		
Gatifloxacin		0.09–0.42		
Sparfloxacin		0.01–0.04		
Moxifloxacin		0.17–0.56		

MP	Type/application	Concentration (mg/kg)	Source	Reference
Norfloxacin		1.06–7.23	Sludge (primary, excess activated, anaerobically digested, dewatered)	Jia et al. [19], Golet et al. [20]
Ciprofloxacin		0.22–3.1		
Azithromycin		2.5–64	Sludge (excess activated, anaerobically digested)	Gobel et al. [21]
Clarithromycin		0.7–67		
Erythromycin		0.030–0.041	Sludge, Class A & B biosolids	Kinney et al. [12], Ding et al. [22]
Roxythromycin		0.337–1.446	Anaerobically digested dewatered sludge	Nieto et al. [23]
Sulfamethoxazole		0.019–68	Sludge (excess activated, anaerobically digested), biosolids	Gobel et al. [21], Nieto et al. [23], Ding et al. [22]
Sulfapyridine		0.1–28	Sludge (excess activated, anaerobically digested)	Gobel et al. [21]
Sulfamethazine		0.026–0.128	Anaerobically digested dewatered sludge, biosolids	Nieto et al. [23], Ding et al. [22]
Sulfamerazine		0.112–0.669	Biosolids from sewage sludge	Ding et al. [22]
Chlortetracycline		0.069 –0.346		
Oxytetracycline		0.052–0.743		
Demeclocycline		0.036–0.131		
Tetracycline		0.282–1.914		McCellan & Halden [24], Ding et al. [22]
Doxycycline		0.225–0.966		
Trimethoprim		0.017–41	Sludge (excess activated, anaerobically digested)	Gobel et al. [21], Nieto et al. [23]
Clindamycin		nd–0.006	Municipal sludge	Subedi et al. [25]
Lincomycin		0.006–0.174	Municipal sludge, biosolids	Ding et al. [22], Subedi et al. [25]
Tiamulin		nd–0.7	Agricultural Field soil	Schlusener et al. [26]
Tylosin		1.074–1.958	Anaerobically digested dewatered sludge	Nieto et al. [23]
Acetaminophen	Analgesic	0.013–0.419		
Carbamezipine		0.011–0.042		

MP	Type/application	Concentration (mg/kg)	Source	Reference
Diclofenac		nd–0.087		
Ibuprofen		0.024–0.144		
Naproxen		nd–0.057		
Ketoprofen		0.030 –0.336	Activated sludge	Radjenovic et al. [27, 28]
Codeine		nd–0.022	Sludge, class A biosolids	Kinney et al. [12]
Metoprolol	β-blocker	nd–0.021	Anaerobically digested dewatered sludge	Nieto et al.[23]
Propranolol		0.026–0.044		Radjenovic et al. [28]
Atenolol		0.007–0.084	Sewage sludge	Radjenovic et al. [28]
Caffeine	Psychoactive drug	0.050–0.074	Anaerobically digested dewatered sludge, biosolids	Nieto et al.[23], Ding et al. [22]
Diltiazem	Antihypertension drug	nd–0.059	Sewage sludge, class A biosolids	Kinney et al. [12]
Fluoxetine	Antidepressant	0.072–1.5		Radjenovic et al. [28]
Paroxetine		0.04 –0.62	Sewage sludge	Radjenovic et al. [28]
Gemfibrozil	Lipid lowering drug	0.118–0.420	Sewage sludge, class A biosolids	Kinney et al. [12], Radjenovic et al. [28]
Bezafibrate		nd–0.013	Anaerobically digested dewatered sludge	Nieto et al. [23]
Clofibric acid		0.007 –0.01		
Thiobendazole	Antiparasitic drug	nd–5	Sewage sludge, class A biosolids	Kinney et al. [12]
Warfarin	Anticoagulant	nd – 0.092		
Cimetidine	Antacid	nd–0.071		
Diphenhydramine	Antihistamine	0.015–7		
Miconazole	Antifungal drug	nd–0.46		
Famotidine	Antacid	0.03–0.050	Sewage sludge	Radjenovic et al. [28]
Loratadine	Antiallergic drug	0.052–0.153		
Hydrochlorothiazide	Diuretic drug	0.011–0.060		
Glibenclamide	Antidiabetic drug	0.013–0.127		

nd- not detected

Table 2. Concentrations of commonly found MPs in sludge.

Although a complex process as described above, the extent of MPs adsorption on sludge is traditionally modeled using linear equilibrium model as

$$C_{\text{ads}} = K_d C_{\text{ss}} C_{\text{dis}} \quad (1)$$

where C_{ads} is the adsorbed concentration of the MP (g/L), C_{ss} is the suspended particulate concentration (g/L), C_{dis} (g/L) is the dissolved concentration, and K_d is the adsorption constant (L/gss), which is also known as the partition coefficient of the compound between the solids and water. K_d has been proposed as a relatively accurate indicator of adsorption [29, 30]; for compounds with a K_d value below 300 L/kg (log K_d = 2.48), the sorption onto secondary sludge is insignificant. Polar compounds typically have higher K_d values in secondary sludge compared to primary sludge. Typical K_d values are presented in **Table 3**. K_d of a compound can be correlated to more fundamental properties such as K_{ow}.

Micropollutants	log K_{ow}*	log K_d	log K_{oc}	Ref[#].	Micropollutants	log K_{ow}	log K_d	log K_{oc}	Ref[#].
Diclofenac	4.98	1.2041	–	b	Estradiol	4.01	2.2304	–	c
Ibuprofen	3.5	0.8513	–	b	Estriol	2.45	1.7324	–	c
DEET	2.18	1.91	2.27	a	Diphenhydramine	3.27	2.5	2.86	a
Clofibric acid	2.57	0.6812	–	b	Estrone	3.13	2.2304	–	c
Ifosfamide	0.86	0.1461	–	b	Ethinylestradiol	3.67	2.4997	–	c
Carbamazepine	2.45	1.95	2.31	a	Fenoprofen	3.1	1.415	–	c
Hydrocodone	2.16	2.03	2.38	a	Fluoxetine	4.05	0.699	–	c
Cyclophosphamide	0.63	0.3802	–	b	Amitriptyline	4.92	2.87	3.21	a
Gemfibrozil	4.77	2.11	2.47	a	Gemfibrozil	3.4	1.2856	–	c
Diazepam	2.82	1.3222	–	b	Hydrocodone	1.2	2.0294	–	c
Diazepam	2.82	2.14	2.53	a	Fluoxetine	4.05	3.08	3.43	a
Ethinylestradiol	3.9	2.5428	–	b	Indomethacine	4.27	1.4472	–	c
Naproxen	3.2	2.16	2.56	a	Ketoprofen	3.12	1.2041	–	C
Perfluorooctanoic acid	6.3	2.3424	–	c	Mefenamic acid	5.12	2.6375	–	C
Diclofenac	4.51	2.18	2.54	a	Methadone	3.93	1.8808	–	C
Perfluorononanoic acid	5.48	3.0934	–	c	Metoprolol	1.88	1.8129	–	C

Micropollutants	log K_{ow}*	log K_d	log K_{oc}	Ref#.	Micropollutants	log K^*_{ow}	log K_d	log K_{oc}	Ref#.
Perfluoroundecanoic acid	6.9	3.3581	–	c	Morphine	0.89	1.0792	–	C
Ketoprofen	3.12	2.25	2.64	a	Naproxen	3.18	1	–	C
Bisphenol A	3.32	2.28	2.64	a	Primidone	0.91	1.699	–	C
Amoxycillin	0.87	0.0253	–	c	Propranolol	3.48	2.5353	–	C
Amitriptyline	4.92	2.8698	–	c	Risperidone	2.5	2.73	–	C
Trimethoprim	4.9	2.3	2.65	a	Roxithromycin	1.7	1.7076	–	C
Androstenedione	2.75	2.1271	–	c	Sotalol	0.24	1.2553	–	C
Aspirin	1.19	0.3464	–	c	Sulfadimethoxine	1.63	0.4771	–	C
Ibuprofen	3.97	2.32	2.64	a	Sulfamethazine	0.89	1.301	–	C
Atorvastatin	5.7	1.9685	–	c	Sulfamethoxazole	0.89	1.0414	–	C
Azithromycin	4.02	2.4472	–	c	Sulfapyridine	0.35	0	–	C
Bezafibrate	3.97	1.9395	–	c	Testosterone	3.32	2.1335	–	C
Benzophenone	3.18	2.1335	–	c	Tramadol	2.4	1.6721	–	C
Bisoprolol	1.87	1.6021	–	c	Trimethoprim	0.91	1.4048	–	C
Dilantin	2.47	2.49	2.84	a	Triclosan	4.76	3.59	3.95	A
Celiprolol	2.29	1.9294	–	c	Triclocarban	4.9	4.41	4.76	A
Clarithromycin	3.16	2.415	–	c	Diazepam	2.82	1.301	–	C
Clofibric acid	2.84	0.699	–	c	Diphenhydramine	3.27	2.4997	–	C
Codeine	1.19	1.1461	–	c	Erythromycin	2.37	1.4456	–	C

"–": Data are not available in the literature. *log K_{ow} for selected MPs are found in http://www.drugbank.ca/. # log K_d and log K_{oc} values are collected from references (Ref.) as follows: (a) [31], (b) [30], (c) [32].

Table 3. log K_d and log K_{oc} values of some commonly found MPs.

As mentioned before, the sorption to sludge is not significant for compounds with log K_{ow} < 2.5, moderate sorption for log K_{ow} between 2.5 and 4, and high sorption for log K_{ow} > 4.0 is expected. In absence of experimental data, to relate K_d with $K_{ow,}$ Eqs. (2) and (3) are given by Matter-Muller et al. [33] and Dobbs et al. [34], respectively:

$$\log K_d = 0.67 \times \log K_{ow} + 0.39 \quad (2)$$

$$\log K_d = 0.58 \times \log K_{ow} + 1.14 \quad (3)$$

K_d can also be estimated using Eq. (4) (Fetter [35]) and Eq. (5) (Jones et al. [36]) if the fraction of organic carbon of the solids is known as

$$K_d = f_{oc} \times \frac{10^{0.72 \times \log K_{ow} + 0.49}}{1000} \quad (4)$$

$$K_d = f_{oc} \times 0.41 \times K_{ow} \quad (5)$$

values of K_d and K_{oc} versus K_{ow} for MPs from the literature are plotted in **Figure 2** showing slightly lower linear dependence of K_d and K_{oc} on K_{ow} as compared to Eqs. (2) and (3). In addition, the goodness of fit as indicated by R^2 is in the range of 0.45–0.48, indicating possible influence of other parameters than only K_{oc} or K_{ow}.

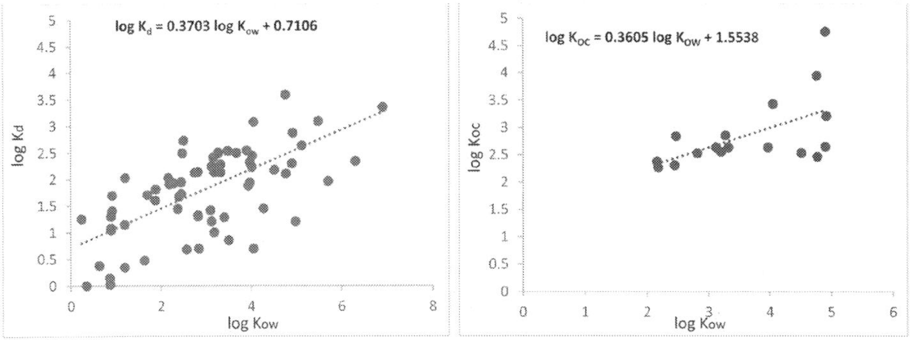

Figure 2. Correlation between log K_d versus log K_{ow} and log K_{oc} versus log K_{ow} for MPs listed in **Table 3**.

MP adsorption on sludge mostly follow linear isotherm such as Fruendlich:

$$q_e = K_f . C_e^{1/n} \quad (6)$$

where q_e = mass adsorbed per unit mass of adsorbent at equilibrium (mg/g)

C_e = concentration of MP in water at equilibrium (mg/L)

$\frac{1}{n}$ = strength of adsorption (dimensionless)

K_f = adsorption capacity at unit concentration $(mg/g)(L/mg)^{1/n}$

The values of K_f and $1/n$ for MPs on sludge varied from 0.0052 to 4.40 (mg/g) $(L/mg)^{1/n}$ and 0.51 to 1.0076, respectively [37–40]. Larger K_f values indicate higher affinity of adsorption for a particular sludge and closer the value of $1/n$ around 1.0, greater is the indication of comparatively strong adsorption bond. Typically, adsorption equilibrium is achieved within 24 hours with almost 90% removal from dissolved phase occurs in an hour; for example, at 3.6 g/L mixed liquor suspended solids (MLSSs) concentration, 95% of oxytetracycline was removed from water within only 1 hour and the concentration at equilibrium remained unchanged over 24 hours [40].

Colloidal particles are a relatively small fraction of the total waterborne particle mass (<10%) in typical wastewater but possess large surface areas which can enable covalent, electrostatic, and hydrophobic binding of MPs depending on their polarity. The magnitude of sorption depends on the molecular weight distribution and aromatic content of the colloids fraction, which also depends on the sewage composition, strength, and sludge age [41]. Similar to adsorption on suspended particulates, adsorption on colloidal particles can be quantified using a distribution coefficient K_{coc}. Holbrook et al. [41] determined K_{coc} using pyrene as a model MP and colloidal fractions from two biological wastewater plants; sorption coefficients (K_{coc}) for pyrene ranged from 1×10^3 L/kg colloids to 80×10^3 L/kg colloids and were comparable to values obtained in the literature for natural organic matter. Good correlation was obtained between K_{coc} and the aromaticity of the colloidal particles.

3.2. Removal processes

3.2.1. Coagulation and sedimentation of micropollutants

Coagulation-flocculation processes are typically used for improving efficiency of wastewater treatment plants promoting the removal of suspended solids, colloids, and some dissolved organics, which do not settle spontaneously. The coagulation process works by destabilizing the colloids/emulsions using coagulants such as metal salts and/or synthetic organic polymers following any of the mechanisms such as double-layer compression, adsorption and charge neutralization, entrapment of particles in precipitate, adsorption and interparticle bridging. The parameters that affect the performance of coagulation are coagulant dosage, pH, and ionic strength of the solution. Based on the type of coagulant such as aluminum sulfate, ferrous sulfate, and ferric chloride, optimum pH range for coagulation varies between 4.0 and 8.5. In case of polymeric coagulants, the active group (carboxyl, amino group, etc.) present on the polymer influences the change of charge with pH [42].

In general, removal of MPs by coagulation-flocculation processes is not very effective for most of the compounds studied with a few exceptions. Earlier studies on removal of MPs by coagulation were reported for simulated drinking water treatment processes [43–47], and percent removal of various MPs varied from 15 to 75% using alum and iron salts, and excess lime/soda ash softening. Vieno et al. [46] evaluated the role of dissolved organic matter, mainly the humic substances in the coagulation process. In the presence of dissolved humic matter, diclofenac, ibuprofen, and bezafibrate could be removed by ferric sulfate coagulation. The removal of diclofenac reached a maximum of 77%, while 50% of ibuprofen, and 36% of bezafibrate were removed. Hence, a high amount of high-molecular-weight dissolved organic matter enhanced the removal of ionizable pharmaceuticals. However, contradictory results were reported by Choi et al. [43] where removal of seven tetracycline classes of antibiotic (TAs) from synthetic and river water using coagulation was achieved. TAs were assumed to be removed through the charge neutralization of zwitterionic or negative TAs by cationic Al (III) and sweep coagulation using poly-aluminum chloride (PACl). Aluminum hydroxide precipitates were formed in the presence of sufficient alkalinity, and TAs were removed by being enmeshed into or adsorbed onto the precipitates. It was suggested that the presence of dissolved organic matter, especially the low-molecular-weight fractions, resulted in possible inhibition of MP removal. This was due to preferential removal of the organic matter by the coagulant.

Huerta-Fontela et al. [48] performed coagulation with alum-coagulants, flocculation with a diallyldimethyl ammonium chloride homopolymer (poly-DADMAC), followed by clarification through sand filters. Of the 55 pharmaceutical compounds present, only five compounds (chlordiazepoxide, zolpidem, bromazepam, clopidogrel, and doxazosin) were completely removed, while warfarin, betaxolol, and hydrochlorothiazide accounted for removals higher than 50%. For some pharmaceuticals such as irbesartan, losartan, or carbamazepine epoxide, negligible removals were obtained.

Suarez et al. [49] evaluated the performance of coagulation-flocculation process for the pretreatment of hospital effluent, both in a batch mode and continuous pilot scale. Highest removal efficiency (>90%) was reported for PPCPs such as galaxolide, tonalide, and synthetic musk (ADBI); these are lipophilic compounds, carrying high negative charge, which facilitates their coagulation in the presence of higher fat content in wastewater. Asakura and Matsuto [50] studied the effect of coagulation for treating landfill leachate. Out of the various EDCs, only nonylphenol showed a removal of >90%, whereas diethylhexylphthalate (DEHP) removal was about 70%. Other EDCs such as diethylphthalate (DEP), dibutylphthalate (DBP), butylbenzylphthalate (BBP), 4-t-octylphenol (4tOP), and 4-n-octylphenol (4nOP) showed poor removal (<50%) by coagulation, with the lowest removal of 20% for bisphenol A.

Few studies have reported the removal of MPs due to coagulation and flocculation in wastewater (**Table 4**). Matamoros and Salvadó [51] evaluated several MPs removal in a coagulation/flocculation-lamellar clarifier for treating secondary effluent. The hydrophobicity of the compounds (log K_{ow}) was found to be a major factor in determining the removal efficiency with coagulation-flocculation. The highest removal of 20–50% was observed for the compounds with log $K_{ow} \geq 4$ at pH 7–8. Since adsorption of MPs on the suspended solids and colloids is the precursor step for their removal during coagulation, the removal efficiency can be tied with the removal efficiency of suspended solids as

http://dx.doi.org/10.5772/65644

$$\% \text{removal} = \frac{K_d C_{ss}}{1 + K_d C_{ss}} E_{Tss} \quad (7)$$

where E_{TSS} is the efficiency of TSS removal (%) during coagulation.

Carballa et al. [52] observed that during coagulation-flocculation of primary wastewater, lipophilic compounds such as musks were adsorbed in the lipid fractions of the sludge with two different fat concentrations of 60 and 150 mg/L, while acidic compounds such as diclofenac were adsorbed due to electrostatic interaction. Compounds with high sorption properties (galaxolide and tonalide) and diclofenac were significantly removed during coagulation-flocculation with efficiencies around 70%. Compounds with lower K_d values, such as diazepam, carbamazepine, ibuprofen, and naproxen, were reduced to a lesser extent (up to 25%).

Coagulant	Dosage(ppm) with pH	compound	Source	Removal (%)	Reference
Ferric chloride/ aluminum sulfate	25, 50–pH 7	Ibuprofen	Hospital wastewater	12.0 ± 4.8	Suarez et al. [49]
		Diclofenac		21.6 ± 19.4	
		Naproxen		31.8 ± 10.2	
		Carbamazepine		6.3 ± 15.9	
		Sulfamethoxazole		6.0 ± 9.5	
		Tonalide		83.4 ± 14.3	
		Galaxolide		79.2 ± 9.9	
Ferric chloride	100, 200–pH(4, 7, 9)	Bisphenol A	Landfill leachate	20	Asakura and Matsuto [50]
		DEHP		70	
		Nonylphenol		90	
	Not mentioned	Sulfamethoxazole	Drinking water treatment plant	33	Stackelberg et al. [47]
		Acetaminophen		60	
		Cholesterol		45	
		Diazenon		34	
		Metachlor		28	
Aluminum sulfate	200–pH 7	Aldrin	Surface water	46	Thuy et al. [53]
	100–pH 7	Bentazon		15	
	78–pH 6.8	Estradiol	Drinking water treatment plant	2	Westerhoff et al. [45]
		Estrone		5	
		Progesterone		6	
		Fluoxetine		15	
		Hydrocodone		24	
		Chlordane		25	
		Benzanthracene		26	
		Chrysene		33	
		Erythromycin		33	
		DDT		36	

Coagulant	Dosage(ppm) with pH	compound	Source	Removal (%)	Reference
		Heptachlor		36	
		Aldrin		49	
		Benzofluoranthine		70	
		Benzopyrene		72	
Ferric sulfate	78.5–pH 4.5	Dichlofenac	Lake water with dissolved humic acid	77	Vieno et al. [46]
		Ibuprofen		50	
		Bezafibrate		36	
		Carbamazepine		<10	
		Sulfamethoxazole		<10	
–	–	Celestolide	Secondary effluent from WWTP	50	Matamoros and Salvadó [51]
		Tricholsan		24	
		Octylphenol		50	
		Tonalide		24	
		DMP		19	
		Galaxolide		16	
		Ibuprofen		4	
		Carbamazepine		2	

"–": Data are not available in the literature. *log K_{ow} for selected MPs are found in http://www.drugbank.ca/. # log K_d and log K_{oc} values are collected from references (Ref.) as follows: (a) [31], (b) [30], (c) [32].

Table 4. Removal of MPs by coagulation/flocculation process from various effluents.

3.2.2. Biodegradation of micropollutants in secondary treatment

Most of the conventional municipal WWTPs do not remove complex MPs by biodegradation and/or biotransformation effectively. Observed removal efficiencies vary in a wide range for different compounds, as well as for the same substance, due to operational conditions such as aerobic, anaerobic, anoxic, sludge retention time (SRT), pH, redox potential, and water temperature. Membrane bioreactors (MBRs) seem to be more effective than conventional-activated sludge (CAS) process as MBR process combines biological treatment with membrane filtration (micro and ultrafiltration). In addition, due to higher SRT at MBRs compared to CAS, biodiversity of the microorganisms in MBR is greater than CAS, and opportunity for adaptation of specific microorganisms to the persistent compounds is greater in MBR than in CAS. Removal of 29 antibiotics in a CAS process was reviewed by Verlicchi et al. [54], where removal of compounds such as sulfamethoxazole, ciprofloxacin, roxithromycin, norfloxacin, erythromycin, etc., varied in a wide range of 0 (spiramycin) and 98% (cefaclor) in CAS and between 15 (azithromycin) and 94% (ofloxacin) in MBRs. Only 1 (azithromycin) out of 10 compounds investigated in both systems exhibited higher average removal efficiency in CAS than in MBR. Trinh et al. [55] traced 48 MPs including steroidal hormones, xenoestrogens, pesticides, caffeine, pharmaceuticals, and personal care products (PPCPs) in a MBR with >90% removal for many of the compounds. However, amitriptyline, carbamazepine, diazepam, diclofenac, fluoxetine, gemfibrozil, omeprazole, sulphamethoxazole, and trimethoprim were only partially removed in MBR with the removal efficiencies of 24–68% [55]. Similar results were obtained in a pilot-scale MBR operated for a Swiss hospital effluent for 1 year [56, 57]. Among the 56 pharmaceuticals, an overall load elimination of all pharmaceuticals and metabolites in the MBR was only 22% due to the presence of persistent iodinated contrast media (almost 80% of the total organic load). Weiss and Reemtsma [58] reported that major advantage of MBR lies for the compounds with moderate removal in CAS; MBR showed no advantages for both well-degradable and recalcitrant compounds. For polar compounds, MBR does not provide significant benefits, because effluent quality is improved only gradually and the most critical components of high aerobic stability remain almost unaltered [58].

Longer SRT as required for nitrogen removal also played an important role in reducing the concentrations of certain MPs [59, 60], and a SRT > 10 days was recommended. Longer SRTs resulted in diverse growth of the microbial community including the growth of nitrifying bacteria. Nitrifying bacteria had shown potential for cometabolic degradation of MPs [61, 62]. However, much longer SRT (49 days) was required for 61% removal of iopromide compared to zero removal in CAS [61]. Mixed bacterial cultures also have proved to be quite effective in removing MPs such as triclosan, BPA, and ibuprofen in river [63, 64] and WWTP [65, 66]. While MPs such as quaternary ammonium compounds are biodegraded as single compound, their biodegradation is inhibited in a mixture using *Pseudomonas* sp. isolated from returned activated sludge [67].

Although an important process variable, hydraulic retention time (HRT) shows varied results for the removal of MPs in WWTP indicating that further research is required on this. A study conducted by Wever et al. [57] reported that decreasing the HRT in a CAS resulted in increasing the concentrations of MPs such as 2, 6 and 1, 6 NDSA; however, it did not affect the percent removal of these compounds in a MBR. In case of pharmaceutical and fragrance compounds, Joss et al. [29] reported that HRT played a very minor role when considering a time period of 0.7 hours for fixed bed reactor, 13 hours for a MBR, and 17 hours for a CAS process.

Solution pH plays a significant role in the removal of MPs as the highly acidic or highly basic solutions affect the solubility of the MPs and also hinder growth of the microbial community [68]. As listed in **Table 1**, MPs exhibit a wide range of pKa values. At pH range of 6–8, as found in most wastewater, many antibiotics and other MPs with pKa values in this range will be ionized. For example, about 40% of pharmaceuticals contain at least one functional group with pKa values in the range of 5–10 [69]. The degree of speciation of such ionizable compounds and their subsequent adsorption and biotransformation will be affected by pH.

The microbial growth and activity, as well as solubility and other physicochemical properties of MPs, are significantly affected by temperature. Temperature variability has been related to deterioration in bulk water quality and system instability; it has also been linked to sludge deflocculation and decreased sludge metabolic activity [70]. Vieno et al. [71] reported that the removal of ibuprofen, diclofenac, benzafibrate, ketoprofen, and naproxen increased during the summer (average temperature 17°C) and decreased in the winter (average temperature 7°C). However, Lesjean et al. [72] reported that in a conventional WWTP, temperature variation between 12 and 25°C brought about little or no change to the degradation process of MPs whereas for a MBR the removal rates were greatly affected by the seasonal changes. Hai et al. [70] reported that the removal of most hydrophobic compounds (log $K_{ow} > 3.2$) in a MBR was stable in the temperature range of 10–35°C, while for less hydrophobic compounds, significant variation occurred in the lower temperature regimes (10–35°C). Lower and more variable removal efficiency at 10°C was observed for certain hydrophilic compounds, which have been reported to be moderately recalcitrant in MBR treatment.

No quantitative relationship between structure and activity can be found for the biological transformation. Overall, it can be concluded that for compounds with a sorption coefficient (K_d) below 300 L/kg, sorption onto secondary sludge is not relevant, and their transformation can consequently be assessed simply by comparing influent and effluent concentrations.

At low dissolved concentration, the kinetics of biodegradation/biotransformation of MPs follow first order as

$$\text{rate} = K_{\text{bio}}\, C_{\text{ss}} C_{\text{dis}} \tag{8}$$

where K_{bio} is the biodegradation rate constant, C_{ss} is the suspended solids concentration, and C_{dis} is the dissolved concentration of MPs_{ss}. Typically, complex aromatic structure with more than one benzene ring and/or with chlorine and nitro groups are not efficiently biodegraded [32, 73]. The aerobic biodegradation constants of 20 aromatic species using activated sludge were reported, and the kinetic constants were correlated to the structure of the molecules [73]. The normalized first-order rate constants K_{bio} (L/gss/h) using C_{ss} (g/L) were 0.003, 0.02, and 3.80 for 3, 5 dinitrobenzoic acid, 2, 6 dichlorophenol, and benzoic acid, respectively. Pomiesa et al. [32] summarized a list of both aerobic and anaerobic rate constants for 20 pharmaceuticals including antibiotics, and other compounds such as bisphenol A and nonylphenol, and the aerobic K_{bio} (L/gss/h) varied from 0.0025 to 7.08 with carbamazepine being the lowest, and galaxolide (a synthetic fragrance) being the highest biodegradable compound. The difference in rate constants for aerobic and anaerobic conditions is less than 15% for some substances (e.g., celestolide and galaxolide) or can be much higher in some other cases (e.g., >50% for estradiol and roxithromycin). Compounds with $k_{\text{bio}} < 0.0042$ L/gss/h are not removed significantly (<20%), whereas compounds with $k_{\text{bio}} > 0.4$ L/gss/h can be transformed by >90%. Therefore, with the existing biological treatment schemes in municipal wastewater, 90% of the MPs are not removed or biotransformed. Many of the plant data do not distinguish between adsorption and biotransformation due to challenging chemical analyses. In most cases, overall removal is estimated based on the influent and effluent concentrations, and information about the intermediate steps is either missing or not reliable [74]. Other challenges are the fate of metabolites, transformation products of pharmaceuticals, and complex chemistry involving these compounds with background water quality, which are all unknown at this point.

Tertiary treatment of wastewater using various combinations of membrane processes, activated carbon adsorption, and advanced oxidation are being performed or characterized in various jurisdictions with stringent water quality requirements. Above technologies all work well for the removal of trace concentration of organics in lab studies and will be described below.

3.2.3. *Activated carbon adsorption*

Adsorption as a unit operation using either granular- or powder-activated carbon (GAC and PAC) to remove organics from water metrics is well established. The mechanism of adsorption, relevant parameters, and adsorption models discussed in the section of adsorption on sludge are applicable for GAC and PAC adsorption. In absence of experimental data on adsorption isotherm, a correlation developed by Crittenden et al. [75] combining Polanyi potential theory and linear solvation energy relationships (LSERs) can be used.

Activated carbon adsorption for the removal of MPs has been applied in both secondary and tertiary treatment units. Simultaneous adsorption of sulfamethoxazole and carbamazepine to powdered-activated carbon (PAC) in a membrane bioreactor (MBR) was reported at PAC dosage of 0.1–1 g/L [76–78]. Altmann et al. [77] compared the performance of PAC and ozonation for seven MPs from four different wastewater plants. Typical dosages were about 20 mg/L of PAC and 5–7 mg/L of ozone, respectively, and the performances of both technologies were very much dependent on the type of pollutants. Hydrophobic compounds with log K_{ow} > 5 have much better removal potential by adsorption than polar compounds, with the exceptions of some protonated bases and deprotonated acids. Empty bed contact time (EBCT) for a biological-activated carbon filter for the removal of numerous MPs for three full-scale reclamation plants varied from 9 to 45 min.

3.2.4. *Membrane processes*

Membrane-based process systems can be classified as direct membrane-based, integrated membrane-based, and combined direct and integrated membrane system. Pressure-driven membrane filtration processes, such as nanofiltration (NF), ultrafiltration (UF), microfiltration (MF), and reverse osmosis (RO), are routinely used for various effluent treatments. While MF and UF are low-pressure processes, NF and RO are high-pressure processes. In tertiary treatment of wastewater for MPs, UF and NF can be effectively used. The removal of MPs by membrane depends on many different factors including characteristics of membrane, MP, aqueous media/solute characteristics, operating conditions, and membrane fouling. The fundamental mechanism of membrane filtration is size exclusion, although adsorption due to hydrophobic interactions, electrostatic repulsion, and adsorption on fouling layer all can play a part [79–82]. Size exclusion mechanism is mostly applicable to noncharged MPs, however, shape of the molecule should also be taken into consideration. Hydrophobic interaction and hydrogen bonding contribute to the adsorption of MPs on the membrane surface. Membrane fouling and the presence of dissolved organic carbon could also increase adsorption by changing the membrane surface characteristics and pore size. For charged MP, electrostatic interaction between the compound and membrane surface gives rise to electrostatic exclusion for membrane surfaces with like charges. **Figure 3** shows the four mechanisms of MP removal by membrane processes. Membrane-based processes have several advantages such as good adaptability, high removal rate, robustness, and no harmful intermediates are formed. An overview of research at laboratory, pilot and full-scale applications of MPs removal is presented in **Table 5**.

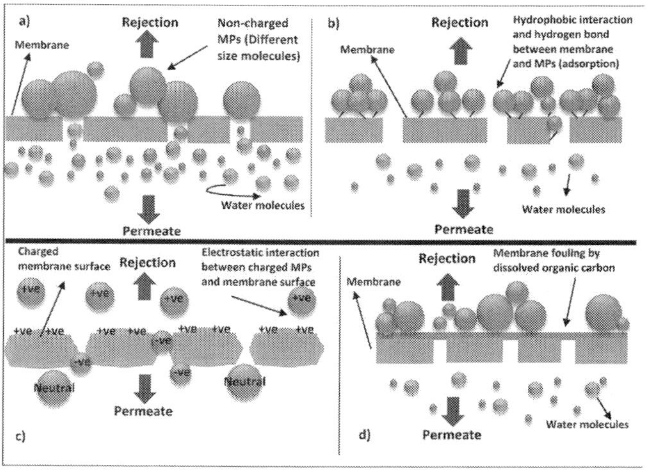

Figure 3. Micropollutants removal mechanism in polymeric membranes. (a) size exclusion, (b) adsorption (hydrophobic interaction), (c) electrostatic repulsion, and (d) adsorption (fouling layer interaction) (concept adopted from Ojajuni et al. [83]).

MPs	% Removal	Remarks	Reference
11 MPs 500 μg/L, (pharmaceuticals and pesticides)	>70%	UF and NF; laboratory scale; secondary effluent	Acero et al. [84]
80 MPs; Metals 18–265 μg/L, VOC 0.65–7.10 μg/L, PAH 0.23–0.67 μg/L, and HVOC 1.45–12.17 μg/L	~40–50% removal for metals	UF; full scale; secondary clarified effluent	Battistoni et al. [85]
Macrolides, roxithromycin (ROX), clarythromycin (CLA), erythromycin (ERY), sulfonamides, and trimethoprim:sulfamethazine (SMZ), sulfamethoxazole (SMX), and trimethoprim (TMP)	45– 94%	Full scale UF; raw sewage of WWTP	Sahar et al. [86]
Pharmaceutically active contaminants (PhACs): sulfamethoxazole, carbamazepine, and Ibuprofen (500 μg/L)	50–85%	NF; laboratory scale; spiked synthetic solution	Nghiem et al. [87]
EDCs–estrone, estradiol, and salicin at initial concentration of 1 mg/L	85±/4% for estradiol, 65±/3% for estrone, 91±/1% for salicine	NF; laboratory scale; spiked synthetic solution	Braeken and Van der Bruggen [88]

http://dx.doi.org/10.5772/65644

MPs	% Removal	Remarks	Reference
Pesticide endosulfan (10–100 μg/L)	84–96%	NF; laboratory scale; spiked synthetic solution	De Munari et al. [89]
11 neutral EDCs and PhACs at initial concentration of 100 μg/L	0–91%	RO; laboratory scale; synthetic solution	Kimura et al. [90]
22 EDCs and pharmaceutically active compounds (PhAC)- ~ 1 μg/L	variable removal in NF; >90% removal in RO	Loose and tight NF; RO; bench scale; surface water; effluent of MBR of WWTP	Comerton et al. [91]
PhACs: carbamazepine, diclofenac, and ibuprofen (IBU) l concentration 0.025–0.1 μg/L	31–39% removal for carbamazepine; 55–61% removal of ionic diclofenac and ibuprofen	NF; laboratory; drinking water	Vergili [92]
22 compounds representing pharmaceutically active compounds, pesticides, hormones and industrial chemicals; 5 μg/L	80–99%	MBR; laboratory; spiked synthetic municipal wastewater	Hai et al. [70]
bisphenol A (750 μg/L), sulfamethoxazole (750 μg/L)	90% removal for Bisphenol A; 50% for sulphamethoxazole	MBR (submerged); laboratory; secondary effluent spiked	Nghiem et al. [93]
40 organic compounds	above 85% for hydrophobic compounds; less than 20% for the rest	MBR; laboratory;secondary effluent spiked	Tadkaew et al. [80]
Ionisable trace organics :sulfamethaxozale, ibuprofen, ketoprofen, and diclofenac at 2 μg/L	Removal dependent on mixed liquor pH.	MBR (submerged); laboratory; synthetic wastewater	Tadkaew et al. [94]
56 pharmaceuticals, 10 metabolites, and two corrosion inhibitors at concentration from 0.1 μg/L to 2.6 mg/L	Removal varies	MBR; pilot scale; wastewater directly from the hospital sewer collection system	Kovalova et al. [56]
11 emerging contaminants: acetaminophen, metoprolol, caffeine, antipyrine, sulfamethoxazole, flumequine, ketorolac, atrazine, isoproturon, 2-hydroxybiphenyl, and diclofenac(all at 0.5 mg/L)	UF with GAC posttreatment performed better than UF with PAC pretreatment.	UF combined with PAC (pretreatment) and GAC (posttreatment), secondary effluent spiked	Acero et al. [95]
6 antibiotics, 3 pharmaceuticals (ibuprofen, salicyclic acid, and diclofenac) and Bisphenol A	>90%	MBR-RO, pilot plant, real wastewater	Sahar et al. [96]
PPCPs; acetaminophen, atenolol, carbamazepine, clopidogrel, diclofenac, dilantin, ibuprofen, iopromide, glimepiride, naproxen, and sulfamethoxazole	Up to 95%	MBR-NF; laboratory; real wastewater	Chon et al. [81]
10 micropollutants detected in wastewater including carbamazepine, ibuprofen, and caffeine	>76.9%	MBR-NF and MBR-RO; pilot plant; real wastewater	Cartagna et al. [97]
9 pharmaceuticals, bezafibrate, carbamazepine, clofibric acid, diclofenac, gemfibrocil, ibuprofen, ketoprofen, naproxen, and fenofibric acid	60–80%	MBR-PAC (submerged); pilot plant; WWTP primary pollutant	Lipp et al. [98]

Table 5. Membrane systems for micropollutants removal in different scales.

3.2.5. Advanced oxidation processes

Advanced oxidation processes (AOPs) using hydroxyl radicals (OH•) are increasingly used for tertiary treatment of municipal wastewater and for water recycling. These processes are fast, nonselective, and effective for recalcitrant compounds. Among numerous combinations of AOPs, UV-, hydrogen peroxide-, and ozone-based processes are easy to implement for tertiary treatment of WWTP effluent. In a comprehensive research, removal efficiency of 220 MPs with postozonation was studied at full scale for a WWTP [1]. Compounds with activated aromatic moieties, amine functions, or double bonds such as sulfamethoxazole, diclofenac, or carbamazepine had second-order rate constants for ozonation >10^4/M/s at pH 7 (fast reacting) were eliminated to concentrations below the detection limit for an ozone dose of 0.47 g O_3/g DOC. Higher ozone dosage of 0.6 g O_3/g DOC was needed for more recalcitrant compounds such as atenolol and benzotriazole for >85%. Rahman et al. [99] summarized the second-order ozone and OH• oxidation constants for commonly found EDCs and pharmaceuticals in pure water, which varied from 0.8 to 7×10^9 and 1.2×10^9 to 9.8×10^9 /MS, respectively. In wastewater, rates will be somewhat lower due to the competition of background organics, suspended particulates, and radical scavengers. However, the effect of background organics competition was found to be minimal for estrone degradation in wastewater by Sarkar et al. [100]. The overall cost of ozonation was found to be lower than that of UV/H_2O_2 process for estrone degradation, although electrical energy per order was lower for UV/H_2O_2. AOPs are effective in a wide range of pH (i.e., 4–11) depending on the type of target compounds; although ozonation is more effective in alkaline pH. In some cases, transformation products that form due to AOPs may be even more toxic compared to parent compounds. For example, intermediates of UV/H_2O_2 oxidation of bisphenol A exhibited different estrogenic activity depending on the treatment conditions [101]. Whole effluent analysis methods are better for assessing the toxicity of resulting water instead of time- and labor-intensive chemical analyses.

4. Conclusion

Fate and removal processes of micropollutants (MPs) in wastewater treatment are complex, and difficult to assess due to tedious and cost-intensive analyses. However, these processes can be somewhat estimated based on their physical properties such as log K_{ow}, pKa, and solubility. Adsorption on colloidal and suspended particles and subsequent removal in sludge may occur for compounds with log $K_{ow} > 4.0$. Majority of the MPs are not removed in conventional-activated sludge process, although better removal for some cases occurs in membrane bioreactors due to greater diversity and adaptability of microorganisms. Compounds with biological degradation constant <0.0042 L/gss/h are not removed significantly (<20%), whereas compounds with rate constants >0.4 L/gss/h can be transformed by >90%. Tertiary treatment of wastewater effluent using activated carbon adsorption, membrane filtration, and advanced oxidation processes are capable of removing MPs with varying degrees of success, although both lab and pilot-scale studies are required to establish their rates of removal. In the case of intermediates or transformation, products are produced during a treatment, whole effluent analysis using a bioassay is a better method to evaluate the quality of effluent instead of conducting compounds specific chemical analyses.

Author details

Sreejon Das, Nillohit Mitra Ray, Jing Wan, Adnan Khan, Tulip Chakraborty and Madhumita B. Ray*

*Address all correspondence to: mray@eng.uwo.ca

Department of Chemical and Biochemical Engineering, University of Western Ontario, London, Ontario, Canada

References

[1] Hollender J, Zimmermann SG, Koepke S, Krauss M, McArdell CS, Ort C, Singer H, Gunten UV, Siegrist H. Elimination of organic micropollutants in a municipal wastewater treatment plant upgraded with a full-scale post-ozonation followed by sand filtration. Environmental Science & Technology. 2009;43(20):7862–9.

[2] Ternes TA. Occurrence of drugs in German sewage treatment plants and rivers. Water Research. 1998;32(11):3245–60.

[3] Fromme H, Küchler T, Otto T, Pilz K, Müller J, Wenzel A. Occurrence of phthalates and bisphenol A and F in the environment. Water Research. 2002;36(6):1429–38.

[4] Heberer T. Occurrence, fate, and removal of pharmaceutical residues in the aquatic environment: a review of recent research data. Toxicology Letters. 2002;131(1):5–17.

[5] Kreuzinger N, editor. Occurrence of highly discussed pollutants in the stretch of the Austrian Danube related to the Catchment Area. Oral Presentation at SETAC Europe 12th Annual Meeting, Vienna, Austria, 2002.

[6] Baronti C, Curini R, D'Ascenzo G, Di Corcia A, Gentili A, Samperi R. Monitoring natural and synthetic estrogens at activated sludge sewage treatment plants and in a receiving river water. Environmental Science & Technology. 2000;34(24):5059–66.

[7] Eggen RI, Hollender J, Joss A, Schã±¥r M, Stamm C. Reducing the discharge of micropollutants in the aquatic environment: the benefits of upgrading wastewater treatment plants. Environmental Science & Technology. 2014;48(14):7683–9.

[8] Schüth PDC. Demonstrating Managed Aquifer Recharge as a Solution to Water Scarcity and Drought: An EU FP7 Project. http://www.marsol.eu/files/marsol_d14-1_list-of-micropollutants.pdf (accessed 31 May 2016).

[9] Gerly Hey, Relevant studies related to the presence of micropollutants in the environment, Published date: May 18, 2016. http://micropollutants.com/About-micropollutants (accessed 31 May 2016).

[10] Kase R, Eggen R, Junghans M, Götz C, Hollender J. Assessment of micropollutants from municipal wastewater-combination of exposure and ecotoxicological effect data for Switzerland. InTech-Open Access Publisher, Lausanne, Switzerland, 2011.

[11] Loos R, Carvalho R, António DC, Comero S, Locoro G, Tavazzi S, Paracchini B, Ghiani M, Lettieri T, Blaha L and Jarosova B. EU-wide monitoring survey on emerging polar organic contaminants in wastewater treatment plant effluents, Water research. 2013; 47(17): 6475-6487.

[12] Kinney CA, Furlong ET, Zaugg SD, Burkhardt MR, Werner SL, Cahill JD, Jorgensen GR. Survey of organic wastewater contaminants in biosolids destined for land application. Environmental Science & Technology. 2006;40(23):7207–15.

[13] Kipopoulou A, Zouboulis A, Samara C, Kouimtzis T. The fate of lindane in the conventional activated sludge treatment process. Chemosphere. 2004;55(1):81–91.

[14] Jacobsen BN, Nyholm N, Pedersen BM, Poulsen O, Østfeldt P. Removal of organic micropollutants in laboratory activated sludge reactors under various operating conditions: sorption. Water Research. 1993;27(10):1505–10.

[15] Heidler J, Halden RU. Fate of organohalogens in US wastewater treatment plants and estimated chemical releases to soils nationwide from biosolids recycling. Journal of Environmental Monitoring. 2009;11(12):2207–15.

[16] McAvoy DC, Schatowitz B, Jacob M, Hauk A, Eckhoff WS. Measurement of triclosan in wastewater treatment systems. Environmental Toxicology and Chemistry. 2002;21(7): 1323–9.

[17] Clara M, Gans O, Windhofer G, Krenn U, Hartl W, Braun K, Scharf S, Scheffknecht C. Occurrence of polycyclic musks in wastewater and receiving water bodies and fate during wastewater treatment. Chemosphere. 2011;82(8):1116–23.

[18] Kupper T, Plagellat C, Brändli R, De Alencastro L, Grandjean D, Tarradellas J. Fate and removal of polycyclic musks, UV filters and biocides during wastewater treatment. Water Research. 2006;40(14):2603–12.

[19] Jia A, Wan Y, Xiao Y, Hu J. Occurrence and fate of quinolone and fluoroquinolone antibiotics in a municipal sewage treatment plant. Water Research. 2012;46(2):387–94.

[20] Golet EM, Xifra I, Siegrist H, Alder AC, Giger W. Environmental exposure assessment of fluoroquinolone antibacterial agents from sewage to soil. Environmental Science & Technology. 2003;37(15):3243–9.

[21] Göbel A, Thomsen A, McArdell CS, Joss A, Giger W. Occurrence and sorption behavior of sulfonamides, macrolides, and trimethoprim in activated sludge treatment. Environmental Science & Technology. 2005;39(11):3981–9.

[22] Ding Y, Zhang W, Gu C, Xagoraraki I, Li H. Determination of pharmaceuticals in biosolids using accelerated solvent extraction and liquid chromatography/tandem mass spectrometry. Journal of Chromatography A. 2011;1218(1):10–6.

[23] Nieto A, Borrull F, Pocurull E, Marcé RM. Occurrence of pharmaceuticals and hormones in sewage sludge. Environmental Toxicology and Chemistry. 2010;29(7):1484–9.

[24] McClellan K, Halden RU. Pharmaceuticals and personal care products in archived US biosolids from the 2001 EPA national sewage sludge survey. Water Research. 2010;44(2): 658–68.

[25] Subedi B, Lee S, Moon H-B, Kannan K. Emission of artificial sweeteners, select pharmaceuticals, and personal care products through sewage sludge from wastewater treatment plants in Korea. Environment International. 2014;68:33–40.

[26] Schlüsener MP, Spiteller M, Bester K. Determination of antibiotics from soil by pressurized liquid extraction and liquid chromatography–tandem mass spectrometry. Journal of Chromatography A. 2003;1003(1):21–8.

[27] Radjenović J, Petrović M, Barceló D. Fate and distribution of pharmaceuticals in wastewater and sewage sludge of the conventional activated sludge (CAS) and advanced membrane bioreactor (MBR) treatment. Water Research. 2009;43(3): 831–41.

[28] Radjenović J, Jelić A, Petrović M, Barceló D. Determination of pharmaceuticals in sewage sludge by pressurized liquid extraction (PLE) coupled to liquid chromatography-tandem mass spectrometry (LC-MS/MS). Analytical and Bioanalytical Chemistry. 2009;393(6–7):1685–95.

[29] Joss A, Keller E, Alder AC, Göbel A, McArdell CS, Ternes T, Siegrist H. Removal of pharmaceuticals and fragrances in biological wastewater treatment. Water Research. 2005;39(14):3139–52.

[30] Ternes TA, Herrmann N, Bonerz M, Knacker T, Siegrist H, Joss A. A rapid method to measure the solid–water distribution coefficient (Kd) for pharmaceuticals and musk fragrances in sewage sludge. Water Research. 2004;38(19):4075–84.

[31] Hyland KC, Dickenson ER, Drewes JE, Higgins CP. Sorption of ionized and neutral emerging trace organic compounds onto activated sludge from different wastewater treatment configurations. Water Research. 2012;46(6):1958–68.

[32] Pomiès M, Choubert J-M, Wisniewski C, Coquery M. Modelling of micropollutant removal in biological wastewater treatments: a review. Science of the Total Environment. 2013;443:733–48.

[33] Matter-Muller C, Gujer W, Giger W, Stumm W. Non-biological elimination mechanisms in a biological sewage treatment plant. Progress in Water Technology. 1980;12:299–314.

[34] Dobbs RA, Wang L, Govind R. Sorption of toxic organic compounds on wastewater solids: correlation with fundamental properties. Environmental Science & Technology. 1989;23(9):1092–7.

[35] Fetter C. Contaminant Hydrogeology. Macmillan Publishing Co.: New York, NY, 1993.

[36] Jones O, Voulvoulis N, Lester J. Aquatic environmental assessment of the top 25 English prescription pharmaceuticals. Water Research. 2002;36(20):5013–22.

[37] Liu J, Wang X, Fan B. Characteristics of PAHs adsorption on inorganic particles and activated sludge in domestic wastewater treatment. Bioresource Technology. 2011;102(9):5305–11.

[38] Lenz K, Koellensperger G, Hann S, Weissenbacher N, Mahnik SN, Fuerhacker M. Fate of cancerostatic platinum compounds in biological wastewater treatment of hospital effluents. Chemosphere. 2007;69(11):1765–74.

[39] Yu J, Hu J. Adsorption of perfluorinated compounds onto activated carbon and activated sludge. Journal of Environmental Engineering. 2011;137(10):945–51.

[40] Kim S, Eichhorn P, Jensen JN, Weber AS, Aga DS. Removal of antibiotics in wastewater: effect of hydraulic and solid retention times on the fate of tetracycline in the activated sludge process. Environmental Science & Technology. 2005;39(15):5816–23.

[41] Holbrook RD, Love NG, Novak JT. Investigation of sorption behavior between pyrene and colloidal organic carbon from activated sludge processes. Environmental Science & Technology. 2004;38(19):4987–94.

[42] Luo Y, Guo W, Ngo HH, Nghiem LD, Hai FI, Zhang J, Liangc S, Wang XC. A review on the occurrence of micropollutants in the aquatic environment and their fate and removal during wastewater treatment. Science of the Total Environment. 2014;473:619–41.

[43] Choi K-J, Kim S-G, Kim S-H. Removal of antibiotics by coagulation and granular activated carbon filtration. Journal of Hazardous Materials. 2008;151(1):38–43.

[44] Adams C, Wang Y, Loftin K, Meyer M. Removal of antibiotics from surface and distilled water in conventional water treatment processes. Journal of Environmental Engineering. 2002;128(3):253–60.

[45] Westerhoff P, Yoon Y, Snyder S, Wert E. Fate of endocrine-disruptor, pharmaceutical, and personal care product chemicals during simulated drinking water treatment processes. Environmental Science & Technology. 2005;39(17):6649–63.

[46] Vieno N, Tuhkanen T, Kronberg L. Removal of pharmaceuticals in drinking water treatment: effect of chemical coagulation. Environmental Technology. 2006;27(2):183–92.

[47] Stackelberg PE, Gibs J, Furlong ET, Meyer MT, Zaugg SD, Lippincott RL. Efficiency of conventional drinking-water-treatment processes in removal of pharmaceuticals and other organic compounds. Science of the Total Environment. 2007;377(2):255–72.

[48] Huerta-Fontela M, Galceran MT, Ventura F. Occurrence and removal of pharmaceuticals and hormones through drinking water treatment. Water Research. 2011;45(3):1432–42.

[49] Suarez S, Lema JM, Omil F. Pre-treatment of hospital wastewater by coagulation–flocculation and flotation. Bioresource Technology. 2009;100(7):2138–46.

[50] Asakura H, Matsuto T. Experimental study of behavior of endocrine-disrupting chemicals in leachate treatment process and evaluation of removal efficiency. Waste Management. 2009;29(6):1852–9.

[51] Matamoros V, Salvadó V. Evaluation of a coagulation/flocculation-lamellar clarifier and filtration-UV-chlorination reactor for removing emerging contaminants at full-scale wastewater treatment plants in Spain. Journal of Environmental Management. 2013;117:96–102.

[52] Carballa M, Omil F, Lema JM. Removal of cosmetic ingredients and pharmaceuticals in sewage primary treatment. Water Research. 2005;39(19):4790–6.

[53] Thuy PT, Moons K, Van Dijk J, Viet Anh N, Van der Bruggen B. To what extent are pesticides removed from surface water during coagulation–flocculation? Water and Environment Journal. 2008;22(3):217–23.

[54] Verlicchi P, Al Aukidy M, Zambello E. Occurrence of pharmaceutical compounds in urban wastewater: removal, mass load and environmental risk after a secondary treatment—a review. Science of the Total Environment. 2012;429:123–55.

[55] Trinh T, Van Den Akker B, Stuetz R, Coleman H, Le-Clech P, Khan S. Removal of trace organic chemical contaminants by a membrane bioreactor. Water Science and Technology. 2012;66(9):1856–63.

[56] Kovalova L, Siegrist H, Singer H, Wittmer A, McArdell CS. Hospital wastewater treatment by membrane bioreactor: performance and efficiency for organic micropollutant elimination. Environmental Science & Technology. 2012;46(3):1536–45.

[57] De Wever H, Weiss S, Reemtsma T, Vereecken J, Müller J, Knepper T, Rördend O, Gonzaleze S, Barceloe D, Hernando MD. Comparison of sulfonated and other micropollutants removal in membrane bioreactor and conventional wastewater treatment. Water Research. 2007;41(4):935–45.

[58] Weiss S, Reemtsma T. Membrane bioreactors for municipal wastewater treatment–a viable option to reduce the amount of polar pollutants discharged into surface waters? Water Research. 2008;42(14):3837–47.

[59] Clara M, Kreuzinger N, Strenn B, Gans O, Kroiss H. The solids retention time—a suitable design parameter to evaluate the capacity of wastewater treatment plants to remove micropollutants. Water Research. 2005;39(1):97–106.

[60] Göbel A, McArdell CS, Joss A, Siegrist H, Giger W. Fate of sulfonamides, macrolides, and trimethoprim in different wastewater treatment technologies. Science of the Total Environment. 2007;372(2):361–71.

[61] Batt AL, Kim S, Aga DS. Enhanced biodegradation of iopromide and trimethoprim in nitrifying activated sludge. Environmental Science & Technology. 2006;40(23):7367–73.

[62] Wahman DG, Henry AE, Katz LE, Speitel GE. Cometabolism of trihalomethanes by mixed culture nitrifiers. Water Research. 2006;40(18):3349–58.

[63] Kang J-H, Kondo F, Katayama Y. Human exposure to bisphenol A. Toxicology. 2006;226(2):79–89.

[64] Kang J-H, Katayama Y, Kondo F. Biodegradation or metabolism of bisphenol A: from microorganisms to mammals. Toxicology. 2006;217(2):81–90.

[65] Kanda R, Griffin P, James HA, Fothergill J. Pharmaceutical and personal care products in sewage treatment works. Journal of Environmental Monitoring. 2003;5(5):823–30.

[66] Thompson A, Griffin P, Stuetz R, Cartmell E. The fate and removal of triclosan during wastewater treatment. Water Environment Research. 2005;77(1):63–7.

[67] Khan AH, Topp E, Scott A, Sumarah M, Macfie SM, Ray MB. Biodegradation of benzalkonium chlorides singly and in mixtures by a Pseudomonas sp. isolated from returned activated sludge. Journal of Hazardous Materials. 2015;299:595–602.

[68] Cirja M, Ivashechkin P, Schäffer A, Corvini PF. Factors affecting the removal of organic micropollutants from wastewater in conventional treatment plants (CTP) and membrane bioreactors (MBR). Reviews in Environmental Science and Bio/Technology. 2008;7(1):61–78.

[69] Gulde R, Helbling DE, Scheidegger A, Fenner K. pH-dependent biotransformation of ionizable organic micropollutants in activated sludge. Environmental Science & Technology. 2014;48(23):13760–8.

[70] Hai FI, Tessmer K, Nguyen LN, Kang J, Price WE, Nghiem LD. Removal of micropollutants by membrane bioreactor under temperature variation. Journal of Membrane Science. 2011;383(1):144–51.

[71] Vieno NM, Tuhkanen T, Kronberg L. Seasonal variation in the occurrence of pharmaceuticals in effluents from a sewage treatment plant and in the recipient water. Environmental Science & Technology. 2005;39(21):8220–6.

[72] Lesjean B, Gnirss R, Buisson H, Keller S, Tazi-Pain A, Luck F. Outcomes of a 2-year investigation on enhanced biological nutrients removal and trace organics elimination in membrane bioreactor (MBR). Water Science and Technology. 2005;52(10–11):453–60.

[73] Andreozzi R, Cesaro R, Marotta R, Pirozzi F. Evaluation of biodegradation kinetic constants for aromatic compounds by means of aerobic batch experiments. Chemosphere. 2006;62(9):1431–6.

[74] Jelic A, Gros M, Ginebreda A, Cespedes-Sánchez R, Ventura F, Petrovic M, et al. Occurrence, partition and removal of pharmaceuticals in sewage water and sludge during wastewater treatment. Water Research. 2011;45(3):1165–76.

[75] Crittenden JC, Sanongraj S, Bulloch JL, Hand DW, Rogers TN, Speth TF, Ulmer M. Correlation of aqueous-phase adsorption isotherms. Environmental Science & Technology. 1999;33(17):2926–33.

[76] Li X, Hai FI, Nghiem LD. Simultaneous activated carbon adsorption within a membrane bioreactor for an enhanced micropollutant removal. Bioresource Technology. 2011;102(9):5319–24.

[77] Altmann J, Ruhl AS, Zietzschmann F, Jekel M. Direct comparison of ozonation and adsorption onto powdered activated carbon for micropollutant removal in advanced wastewater treatment. Water Research. 2014;55:185–93.

[78] Bolong N, Ismail A, Salim MR, Matsuura T. A review of the effects of emerging contaminants in wastewater and options for their removal. Desalination. 2009;239(1): 229–46.

[79] Nghiem LD, Hawkes S. Effects of membrane fouling on the nanofiltration of pharmaceutically active compounds (PhACs): mechanisms and role of membrane pore size. Separation and Purification Technology. 2007;57(1):176–84.

[80] Tadkaew N, Hai FI, McDonald JA, Khan SJ, Nghiem LD. Removal of trace organics by MBR treatment: the role of molecular properties. Water Research. 2011;45(8):2439–51.

[81] Chon K, KyongShon H, Cho J. Membrane bioreactor and nanofiltration hybrid system for reclamation of municipal wastewater: removal of nutrients, organic matter and micropollutants. Bioresource Technology. 2012;122:181–8.

[82] Bellona C, Drewes JE, Xu P, Amy G. Factors affecting the rejection of organic solutes during NF/RO treatment—a literature review. Water Research. 2004;38(12):2795–809.

[83] Ojajuni O, Saroj D, Cavalli G. Removal of organic micropollutants using membrane-assisted processes: a review of recent progress. Environmental Technology Reviews. 2015;4(1):17–37.

[84] Acero JL, Benitez FJ, Teva F, Leal AI. Retention of emerging micropollutants from UP water and a municipal secondary effluent by ultrafiltration and nanofiltration. Chemical Engineering Journal. 2010;163(3):264–72.

[85] Battistoni P, Cola E, Fatone F, Bolzonella D, Eusebi AL. Micropollutants removal and operating strategies in ultrafiltration membrane systems for municipal wastewater

treatment: preliminary results. Industrial & Engineering Chemistry Research. 2007;46(21):6716–23.

[86] Sahar E, Messalem R, Cikurel H, Aharoni A, Brenner A, Godehardt M, et al. Fate of antibiotics in activated sludge followed by ultrafiltration (CAS-UF) and in a membrane bioreactor (MBR). Water Research. 2011;45(16):4827–36.

[87] Nghiem LD, Schäfer AI, Elimelech M. Role of electrostatic interactions in the retention of pharmaceutically active contaminants by a loose nanofiltration membrane. Journal of Membrane Science. 2006;286(1):52–9.

[88] Braeken L, Van der Bruggen B. Feasibility of nanofiltration for the removal of endocrine disrupting compounds. Desalination. 2009;240(1):127–31.

[89] De Munari A, Semiao AJC, Antizar-Ladislao B. Retention of pesticide endosulfan by nanofiltration: influence of organic matter–pesticide complexation and solute–membrane interactions. Water Research. 2013;47(10):3484–96.

[90] Kimura K, Toshima S, Amy G, Watanabe Y. Rejection of neutral endocrine disrupting compounds (EDCs) and pharmaceutical active compounds (PhACs) by RO membranes. Journal of Membrane Science. 2004;245(1):71–8.

[91] Comerton AM, Andrews RC, Bagley DM, Hao C. The rejection of endocrine disrupting and pharmaceutically active compounds by NF and RO membranes as a function of compound and water matrix properties. Journal of Membrane Science. 2008;313(1):323–35.

[92] Vergili I. Application of nanofiltration for the removal of carbamazepine, diclofenac and ibuprofen from drinking water sources. Journal of Environmental Management. 2013;127:177–87.

[93] Nghiem LD, Tadkaew N, Sivakumar M. Removal of trace organic contaminants by submerged membrane bioreactors. Desalination. 2009;236(1):127–34.

[94] Tadkaew N, Sivakumar M, Khan SJ, McDonald JA, Nghiem LD. Effect of mixed liquor pH on the removal of trace organic contaminants in a membrane bioreactor. Bioresource Technology. 2010;101(5):1494–500.

[95] Acero JL, Benitez FJ, Real FJ, Teva F. Coupling of adsorption, coagulation, and ultrafiltration processes for the removal of emerging contaminants in a secondary effluent. Chemical Engineering Journal. 2012;210:1–8.

[96] Sahar E, David I, Gelman Y, Chikurel H, Aharoni A, Messalem R, Brenner A. The use of RO to remove emerging micropollutants following CAS/UF or MBR treatment of municipal wastewater. Desalination. 2011;273(1):142–7.

[97] Cartagena P, El Kaddouri M, Cases V, Trapote A, Prats D. Reduction of emerging micropollutants, organic matter, nutrients and salinity from real wastewater by

combined MBR–NF/RO treatment. Separation and Purification Technology. 2013;110:132–43.

[98] Lipp P, Groß H-J, Tiehm A. Improved elimination of organic micropollutants by a process combination of membrane bioreactor (MBR) and powdered activated carbon (PAC). Desalination and Water Treatment. 2012;42(1–3):65–72.

[99] Rahman M, Yanful E, Jasim S. Endocrine disrupting compounds (EDCs) and pharmaceuticals and personal care products (PPCPs) in the aquatic environment: implications for the drinking water industry and global environmental health. Journal of Water and Health. 2009;7(2):224–43.

[100] Sarkar S, Ali S, Rehmann L, Nakhla G, Ray MB. Degradation of estrone in water and wastewater by various advanced oxidation processes. Journal of Hazardous Materials. 2014;278:16–24.

[101] Chen P-J, Linden KG, Hinton DE, Kashiwada S, Rosenfeldt EJ, Kullman SW. Biological assessment of bisphenol A degradation in water following direct photolysis and UV advanced oxidation. Chemosphere. 2006;65(7):1094–102.

Chapter 6

PCDDs/PCDFs and PCBs in Wastewater and Sewage Sludge

Magdalena Urbaniak and Anna Wyrwicka

Additional information is available at the end of the chapter

http://dx.doi.org/10.5772/66204

Abstract

The chapter includes the information concerning the wastewater treatment plants (WWTPs) functioning in respect to polychlorinated dibenzo-*p*-dioxins (PCDDs)/polychlorinated dibenzofurans (PCDFs) and polychlorinated biphenyls (PCBs). In particular, the chapter describes the occurrence and fate of PCDDs/PCDFs and PCBs in WWTPs, at different treatment stages, including the tertiary wastewater treatment (e.g. constructed, wetlands biofilters) and factors affecting the removal of these micropollutants during treatment process. Considering the production of growing amounts of sewage sludge as an end product of the wastewater treatment process, the chapter describes also the occurrence and fate of above-mentioned compounds in sewage sludge and the ways of their utilization with the special emphasis on agricultural uses, bioremediation and phytoremediation processes. With regard to the agricultural use of sewage sludge, the impact of sludge-born PCDDs/PCDFs and PCBs on plant growth and plant metabolism is described, together with the current state of knowledge on the accumulation and translocation of the studied compounds in plant tissues.

Keywords: PCDDs/PCDFs, PCBs, wastewater, sewage sludge, phytoremediation, plant growth, plant metabolism

1. Introduction

Rapid growth in global population has been observed from approximately 5.3 billion in 1992 [1] to about 6.97 billion in 2011 [2]. United Nation predicted that in 2030, the global population reach over 8 billion, whereas in 2050 exceed 9 billion. The growing population affect the consumption of water and consequential production of wastewater. The projections concerning influent wastewater flow in USA estimate its rise from 100,000,000 m^3/day in 1996 to 170,000,000 m^3/day in 2025 [3–5].

An increased usage of water around the world led to an increased concern about the outgoing wastewater quality from municipal wastewater treatment plants (WWTPs) [6]. Usually, quantification of wastewater quality is based on monitoring of traditional parameters which can be analysed in easy and inexpensive way and are regulated by the European Urban Wastewater Directive (91/271/EEC). These parameters include biochemical oxygen demand (BOD), chemical oxygen demand (COD), nitrates, phosphates and total suspended solids [7]. Nevertheless, these routine chemical analyses cannot give a complete overview of the threat to the water environment posed by other substances released through the WWTPs effluents such as polychlorinated dibenzo-*p*-dioxins (PCDDs), polychlorinated dibenzofurans (PCDFs) and polychlorinated biphenyls (PCBs), which are toxic, carcinogenic and known endocrine disrupters posing a serious risk for living organisms [6]. According the Directive of the European Parliament and the Council 2013/39/EC of 12 August 2013 amending Directive 2000/60/EC and 2008/105/EC in respect of priority substances in the field of water policy, PCDDs/PCDFs and PCBs have been identified as priority hazardous substances, which need to be eliminated from the water environment.

Considering the above, the present chapter reviews the available data concerning the occurrence and fate of PCDDs/PCDFs and PCBs in wastewater (point 2) and sewage sludge (point 3) with the special emphasis of the ways of sewage sludge utilization and impact of sludge-born PCDDs/PCDFs and PCBs on the plant growth and plant metabolism.

2. The occurrence and fate of PCDDs/PCDFs and PCBs in WWTPs

WWTPs represent an obligatory and final step prior to the release of wastewater into the environment. Hence, an emerging task for WWTPs would be to act as a barrier for micropollutants, preventing the emission of potentially harmful substances into the aqueous environment. WWTPs use different kinds of methods including biological, physical and chemical processes, to fulfil the regulatory standards regarding the quality of the effluent discharges. Regardless of the methods used at any particular WWTP, all the treatment processes can be generally divided into three categories: (1) primary, (2) secondary and (3) advanced tertiary treatment [5]. The primary treatment removes large objects from incoming wastewater through floatation, settling and screening mechanisms and the smaller objects such as sand are removed in grit chambers and sedimentations tanks. The secondary treatment is designed to substantially degrade organic matter and dissolved nutrients using trickling filters and activated sludge. The purpose of the tertiary treatment is to further improve the effluent quality before it is discharged to the receiving environment and include filtration, chlorination and UV radiation.

The occurrence of PCDDs/PCDFs and PCBs in the untreated wastewater and sewage sludge has been studied very intensively during recent decades and revealed their very high concentrations with a predominance of highly chlorinated congeners [8–15].

http://dx.doi.org/10.5772/66204

The available literature data indicate that conventional wastewater treatment systems are not able to sufficiently remove hydrophobic contaminants, which have adverse effects on the receiving water ecosystem [5, 7, 13]. Thus, organic compounds are detected in the river water worldwide [13, 16–22]. This is due to the fact that for many years, quantification of wastewater effluents and receiving river water pollution were restricted to monitor biochemical oxygen demand (BOD), chemical oxygen demand (COD), nitrogen and phosphorus concentrations and total suspended solids [7]. However, as shown in the work of Urbaniak et al. [13] and Urbaniak and Kiedrzyńska [14], significant concentrations of PCDDs/PCDFs and PCBs may be present in treated wastewater, with the highest values in the smallest WWTPs. All WWTPs studied by Urbaniak et al. [13] were found to discharge toxic PCDD/PCDF and PCB compounds into their receiving rivers. This is the effect of insufficient regulation of the discharge of toxic congeners of PCDDs/PCDFs by municipal WWTPs: the existing regulations only apply to municipal WWTPs with a population equivalent (p.e.) of 100,000. In consequence, the release of PCDDs/PCDFs in treated wastewater from the studied WWTPs is not regulated, as the plants are below this p.e. This, together with the increasing number of municipal WWTPs, and the results presented by Sztamberek-Gola et al. [23] and Oleszek-Kudlak et al. [24], which demonstrate increases in the concentrations of the lower chlorinated, and hence, more toxic, PCDDs/PCDFs in WWTP outlet water, may result in lower quality of the receiving waters. Data presented by Sztamberek-Gola et al. [23] and Oleszek-Kudlak et al. [24] obtained on the basis of three WWTP analyses, revealed total and toxic equivalency (TEQ) concentrations within the range of 107.26–219.19 pg/m^3 for total PCDDs, from 201.75 to 736.50 pg/m^3 for total PCDFs and from 14.70 to 116.40 pg I-TEQ/m^3 for TEQ. Moreover, the authors observed increased PCDD and PCDF concentrations to be related to increased daily wastewater flow: the lowest values were noted in effluents from the smallest WTP, with a daily flow of 20,000 m^3, whereas samples coming from WWTPs with twice the flow (40,000 and 45,000 m^3) were found to have concentrations about two times higher. Considering the above results, the authors note that wastewater treatment affects the fate of PCDDs/PCDFs, with increased amounts of lower chlorinated, and thus more toxic, congeners in the outlet effluents. As a consequence, the International-TEQ (I-TEQ) concentrations are more than five times higher in the outgoing treated effluent than the incoming wastewater. Moreover, the authors observe a predominance of PCDFs over PCDDs in the outgoing effluents. Also other studies confirm the presence of PCDDs/PCDFs in wastewater effluents. The study of Rappe et al. [10] showed that the TEQ and PCDD/PCDF concentrations in wastewater effluents from publicly owned treatment works ranged between 0.264 and 3.84 pg TEQ/L. Urbaniak et al. [13] examined 17 outflows of treated wastewater from municipal wastewater treatment plants. Sewage treatment plants were divided into three classes based on their p.e. size, that is: class I (0–1999 p.e.), class II (2000–9999 p.e.), class III (10,000–14,999 p.e.) and class IV (15,000–99,999 p.e.). The analysis of the treated wastewater collected at the sewage outlets revealed that toxic PCDDs/PCDFs and dl-PCBs were present at a range of concentrations from 32.30 to 732.79 pg/L. The mean values at high water flow and during stable hydrological conditions were respectively 81.96 and 216.92 pg/L for class I wastewater treatment plants, 80.47 and 74.30 pg/L for class II, and 69.82 and 137.06 pg/L for class IV. These results indicate that small wastewater treatment plants had higher concentrations of the studied compounds than the larger ones. In the case of the concentrations measured as TEQs, the obtained values were less diverse, amounting to 4.38 and 3.81 pg TEQ/L for high and stable flow in class I wastewater treatment plants, 4.72 and 3.97 pg TEQ/L for class II, and 3.94 and 3.15 pg TEQ/L for class IV.

With reference to the PCBs, only a few publications refer to their concentrations in the WWTPs effluents. Katsoyiannis and Samara [25, 26] demonstrated the occurrence of the sum of indicator PCBs (∑7PCBs) in the raw urban wastewater and wastewater after primary and secondary treatment steps. The authors showed decreasing mean concentrations of ∑7PCBs from 1,000,000 through 631,000 to 250,000 pg/L in raw wastewater and effluents from primary and secondary treatment stage, respectively. Another research conducted by Blanchard et al. [27] in the outflow from the Montreal WTP (Canada) showed much lower concentrations (measured as sum of 13 PCB congeners) ranged from 20 to 860 pg/L with the mean value of 310 pg/L, whereas Pham and Prolux [28] found a concentration of ∑13PCBs in the treated wastewater from the same WTP of 1400 pg/L. The study of Bergqvist et al. [6] conducted in two WTPs in Umea (Sweden) and in Siauliai (Lithuania) showed higher ∑7PCBs ranged from 1000 to 6000 pg/L. The authors also demonstrated a rapid increase of the ∑7PCBs during the treatment process (ranged from 300 to 1000 pg/L and from 1000 to 6000 pg/L in the case of Umea and Siauliai WTP, respectively). However, other authors suggest that treatment processes such as sorption to the sludge remove up to 70% of PCBs, whereas volatilization led to eliminate of about 50% of the Aroclor 1254 [29, 30]. According to Pham and Prolux, 1997, the removal rates ranged from 33% (for PCB: 101) up to 100% (for PCB: 194) with the average value for the∑13PCBs of 67%. Despite the above, Urbaniak and Kiedrzyńska [14] note that the treated wastewater effluent of the smallest wastewater treatment plants, class I, is characterized by dl-PCB values more than double those of medium and large wastewater plants. This phenomenon was not noted for TEQ values, which were found in the narrow range of 0.31–0.37 pg TEQ/L. The study of Urbaniak and Kiedrzyńska [14] demonstrates a significant problem with the maintenance of the proper purification efficiency in all the studied WWTPs and in this way effluents quality which have the potential to affect the quality of river water.

In order to enhance the removal of PCDDs/PCDFs and PCBs from wastewater effluents and the receiving river waters, the land-water ecotones constructed in a river valley with different kinds of plants and micro-organisms may be applied. Such structures may partially purify the inflowing surface water and groundwater contaminated by PCDDs/PCDFs and PCBs through their capturing, immobilization and/or degradation [31–33]. Wetlands are another promising solution towards wastewater purification due to their intrinsic function to transform and store organic matter and nutrients [34, 35] and associated micropollutants such as PCDDs/PCDFs and PCBs. Due to these properties, wetlands have been used for water quality improvement worldwide [36]. Constructed wetlands were first used for treatment of wastewater in the 1950s, while in last years, they are also used for treatment of runoff water from city areas and agriculture. Constructed wetlands exploit natural processes to remove pollutants in a sustainable cost and in an energy effective way with minimal operation and maintenance cost.

Moreover, their usage as tools in the treatment of polluted waters has been gaining popularity as an ecological engineering alternative over conventional, chemical-based methods.

The promising solution provides ecohydrology [37], through the use of cascade system of biofilters for the purification of wastewater, runoff water, leachate, etc. The biofilters which consist with zone of intensive sedimentation, which facilitate the deposition of matter, nutrients and micropollutants and their further biodegradation by existing microbiota and macrophyte zone where an intensive phytodegradation processes occur, are considered to be one of the most effective solutions for pollutant removal. Our earlier results obtained on the basis of such systems functioning in the urban area and receiving the untreated sewage and storm water (Sokołówka River, Poland) showed the removal efficiency reaching 95% for mineral matter, 86% for organic matter, 81% for total nitrogen and 86% for total phosphorus [38]. At the same time, removal efficiency of biofilter located in rural area (Asella lake, Ethiopia) was 67%, 36%, 76% and 93% for mineral matter, organic matter, total nitrogen and total phosphorus, respectively [38]. Moreover, results from the biofilter located in Asella (Central Ethiopia) demonstrated a 70% reduction of the lake sediment TEQ after one year of biofilter implementation (data not published). The implementation of such biofiltration system enabled a reduction in the input of micropollutants into the river recipients through sedimentation and acceleration of biodegradation and phytodegradation processes and in this way indicates the positive role of such systems in the quality of water ecosystems and in consequence of human health.

3. The occurrence and fate of PCDDs/PCDFs and PCBs in sewage sludge and sewage sludge amended soil

The occurrence of PCDDs/PCDFs and PCBs in inflowing wastewater causes considerable problems for the WWTPs because conventional biological and chemical processes are insufficient for removing them. What is more is scarce data exist to explain how wastewater treatment affects the behaviour and fate of PCDDs/PCDFs and PCBs. Since they have a very high sorption potential [39], they are expected to partition into the sewage sludge part of the wastewater during treatment processes. In addition, the majority of treatment processes are very conducive to volatilization; hence, low volatilization potentials of PCDDs/PCDFs reduce their loss [24].

Various studies confirm that sewage sludge contains a very high level of PCDDs/PCDFs and PCBs ranging between 2.26 and 1270 ng I-TEQ/kg in the United States [10], from 19 to 225 ng I-TEQ/kg in UK [40], from 7 to 160 ng I-TEQ/kg in Spain [41], and between 16.85 and 74.56 ng I-TEQ/kg in Poland [24, 42]. Our study from the Lodz Wastewater Treatment plant showed the concentration of 17 toxic congeners PCDDs/PCDF in sewage sludge equal to 3270.07 ng/kg and the TEQ concentration equal to 29.71 ng TEQ/kg.

These findings confirm that the majority of PCDDs, PCDFs and PCBs are deposited in sludge. This in turn implicates problems with the further use of such contaminated sludge as a fertilizer especially because PCDDs/PCDFs and PCBs toxicity is further enhanced by their accumulation in soil, and bioaccumulation and biomagnification within food chains.

Concerning the above, the further part of the chapter is focused on the methods dedicated to safe disposal and utilization of sewage sludge and the fate of sludge-born PCDDs/PCDFs and PCBs in the environment.

4. Sewage sludge utilization

The main methods of sewage sludge utilization include storage, natural resources and agricultural land use, and burning. At present, most often the sewage sludge is stored on sludge lagoons. This practice became insufficient because (1) the storage has a limited capacity and (2) sludge could be used as a potential recyclable material; however, this method requires drying of sewage sludge to the content of 58–95% dry weight (d.w.), which is high energy-consuming.

The use of thermal processes removes organic compounds associated with the sewage sludge but leave contaminated fly ash. Moreover, this kind of sewage sludge utilization led to air pollution and airborne diseases among human population due to smoke production which may contain toxic compounds like heavy metals. The use of efficient equipment led to the reduction of emissions of harmful elements to the atmosphere but at the same time move the problem of pollutant emissions to the captured ashes. Additionally, during the incineration process as an end by-product, the hydrogen (H_2), methane (CH_4), carbon monoxide (CO) and carbon dioxide (CO_2) are produced. This led to increased production of greenhouse gases, which are the main concern of the Kyoto Protocol regarding climate change.

The alternative method of sewage sludge utilization is their use as a soil and plant fertilizer. This way of their utilization is possible thanks to high organic matter content and high levels of nitrogen and phosphorus which are required for plant growth [43, 44]. Moreover, the organic matter increases the water capacity influences by this way positive on the structure, texture and microbial activity of the soil. The use of sewage sludge as a fertilizer is widespread [9]. The amount of sludge used for agricultural purpose is 25% in Germany and up to 90% in Sweden [9].

The use of sewage sludge as plant fertilizer is not only the method of sewage sludge management but also the method for implementation of the Renewable Energy Sources Directive—when using energetic crops (2001/77/EC) and the Kyoto Protocol (OJ L 203 of 2005, p.1684). Moreover, crops may be used for the reduction of enhanced pollutants levels in soil after sludge application.

Following application, the sludge is present in relatively thin film on the soil surface. Nevertheless, it should be stated that due to high persistence of the PCDDs/PCDFs and PCBs, the addition of these compounds to the soil through the application of sludge must lead to an increase in soil contamination. This is important because Regulation of the Ministry of Environment of Poland from 16 April 2002 recommended to not exceed the 20 ng PCB/g d.w. in agricultural soil (PCB: 28, 52, 101, 118, 138, 153 and 180) and 2000 ng PCB/g d.w. in industrial soil (OJ 2002, 63 item 634). Nevertheless, there is no law regulation concerning the concentrations of 17 toxic congeners of PCDD, PCDF and 12 toxic dl-PCB in soil. In contrast, in Germany, the limit for PCB is 0.4, 2.0, 0.8 and 40.0 mg/kg d.w. for playgrounds, parks, residential and industrial areas, respectively [45]. The limit for PCDD/PCDF is the following: 100, 1000, 1000 and 10,000 ng TEQ/kg d.w. for playgrounds, parks, residential and industrial areas, respectively [46].

In case of sewage sludge, the Directive 86/278/EEC on the protection of the environment, and in particular of the soil when sewage sludge is used in agriculture, does not provide any limit values or requirements for organic compounds in sewage sludge. Thus, several national regulations on the use of sludge have added specifications on organic compounds. This is the case in particular of Austria, France and Germany which have all included limit values for some organic compounds in the relevant regulation for the use of sludge, for example, in Austria, the limit values for PCDDs/PCDFs in sewage sludge are 100 ng TEQ/kg d.w. in Lower and Upper Austria and Burgenland and 50 ng TEQ/kg d.w. in Carinthia [47]. The limit of 100 ng TEQ/kg d.w. is also valid in Germany [48]. The limits for PCBs in sewage sludge are the following: 0.2 mg/kg d.w. in Lower and Upper Austria and Burgenland and 1.0 mg/kg d.w. in Burgenland. In France, the limit for sum of seven principal PCBs (PCB 28, 52, 101, 118, 138, 153, 180) is 0.8 mg/kg d.w.; in Germany is 0.2 mg/kg d.w. for each of the six PCB congeners; and in Sweden is 0.4 mg/kg d.w. In Poland, according to Ministerial Decree (OJ 2009, 27 item 169), PCBs should be completely removed from the sewage during their treatment; nevertheless, there is no limits of the aforementioned compounds in sewage sludge.

Despite the above national regulation, European Union proposed some limit values for concentrations of organic compounds and PCDDs/PCDFs in sludge for use on land. The mentioned proposed limit values are following: 0.8 mg/kg of dry matter for PCBs (sum of PCBs 28, 52, 101, 118, 138, 153 and 180) and 100 ng TEQ/kg of dry matter for PCDDs/PCDFs [49]. Also the US EPA proposed the limit of 300 ng TEQ/kg of dry matter for 17 toxic PCDDs/PCDFs and 12 coplanar PCBs [50].

5. Bioremediation and phytoremediation of sludge originated PCDDs/PCDFs and PCBs in soil

Reports on the biodegradation of chlorinated dioxins in the soil are contradicting. On the one hand, there are studies that indicate that chlorinated PCDDs/PCDFs are persistent. One of such studies considered chlorinated PCDDs/PCDFs that were introduced into soil through land application of sewage sludge [51]. According to this study, the PCDDs/PCDFs concentrations did not change significantly after 260 days of monitoring. On the other hand, the evidence obtained in other experiments suggests that PCDDs/PCDFs are degraded in soil, for example, the concentration of 2,3,7,8-TCDD was monitored over 10 years in the soil and was shown to be significantly decrease. Biodegradation was also observed in the soil spiked with one to 100 ppm of 2,3,7,8-TCDD. Between 37 and 44% of added 2,3,7,8-TCDD was eliminated during 1 year [52].

The above results are connected with the soil microbial transformation of micropollutants. Many literature data suggest that microbial biodegradation is the critical event determining the fate and persistence of PCDDs/PCDFs in the soil [53].

Biodegradation is dependent on micro-organism enzymes which modify toxic compounds into less toxic forms. Biodegradation can be carried as two processes: mineralization, when organic compound uses a sole source of carbon and energy by micro-organisms, and co-metabolism where transformation of given pollutant depends on the presence of other substrate. Products of this process can be further mineralized; otherwise, incomplete degradation occurs, leading to a formation and accumulation of metabolites more toxic than parent substrates.

In this place, there is a need to underline the role of humic acids—major components of soil organic matter which consist of complex polymers of hydroxyphenols, hydroxybenzoic and methoxybenzoic acids and other aromatic structures with linked peptides, amino sugar compounds, fatty acids and possibly other constituents. Hydroquinone/quinone-type couples are perceived to affect the redox properties of humic acid and to act either as electron transfer mediators or as direct donors of electrons. Thus, the amount of humic acids may determine the microbial dechlorination of PCDDs/PCDFs [54].

The use of anaerobic and aerobic micro-organisms is the only known process of PCDDs/PCDFs degradation in soil and aquatic systems, leading to a removal of chlorine atoms from the biphenyl molecule and theoretically releasing CO_2, chlorine and water. Highly chlorinated congeners have been found to be reductively dechlorinated under anaerobic conditions through a preferential meta- and para-chlorine removal and production of less chlorinated congeners, which can then be used in aerobic transformations. Thus, complete degradation of PCDDs/PCDFs can be achieved by a sequential exposure to anaerobic and aerobic micro-organisms [55].

The fungi, similarly to bacteria, are also capable to degrade $PCDD_s/PCDF_s$ in the presence of oxygen using both processes: mineralization and co-metabolism. The fungi use specific enzymes named lignin peroxidase or manganese peroxidase which enable to oxidize the pollutant molecule. The fungal aerobic biodegradation was first reported by Bumpus et al. [56]. The authors documented the mineralization of [^{14}C] 2,3,7,8-TCDD to $^{14}CO_2$ by *Phanerochaete chrysosporium* within 30 days. *P. chrysosporium* also successfully been used to degrade 2,7-DCDD. It should also be mentioned that the biodegradation activity of fungi is not limited to less chlorinated congeners, for example, *P. chrysosporium* is able to remove 34 and 48% of a mixture of PCDD/PCDF congeners containing from 5 to 8 chlorine atoms in the molecule during 7 to 14 days [57].

It was estimated that the highest rate of microbial degradation of pollutants occurs in the plant rhizosphere [58, 59]. Rhizodegradation of organic micropollutants is one of the most effective remediation processes due to existing interactions in the rhizosphere between plant roots, plant exudates, soil and micro-organisms. Moreover, plants are able to store in their rhizosphere up to 40% of aminoacids, carbohydrates and other photosynthesis products. This influences on the availability of carbon used by micro-organisms in the co-metabolism process.

Whipps [60] demonstrated that 1 g of planted soil contains 10^{12} higher amount of micro-organisms in comparison with non-planted one. Rhizosphere microbiota plays also an intrinsic role in the protection of plants against pathogens and stress caused by too high concentration of pollutants and eases the uptake of biogenic substances by a given plant [61]. The effectiveness of rhizosphere biodegradation depends on the ability of micro-organisms to adapt to a given pollution concentration and effectiveness of colonization of roots [61]. The study of Kuiper et al. [62] demonstrated that naturally occurred rhizosphere biodegradation may be enhanced by an addition of micro-organisms to the rhizosphere.

The study with application of plants for phytoremediation/rhizoremediation of soil contaminated with organic compounds showed the decline in the concentration of organochlorine compounds of 30% during 2 years of plant cultivation. At the same time, the unplanted soil demonstrated the reduction of about two times lower [63]. On the basis of 21-month study, Nedunuri et al. [64] showed the decrease of aromatic compounds concentrations of about 42 and 50% in soil cultivated with fibre flax (*Lolium annual*) and St. Augustine grass (*Stenotaphrum secundatum*), respectively. Other examples showed phytoremediation of soil contaminated with crude oil using combination of grass and fertilizers [64–66]. Despite grasses, the shrubs and trees can be also used as effective phytoremediation tools. The example can be the study of Vervaeke et al. [67] who reported 57% reduction of aromatic compounds and mineral oils during 1.5 years of willow (*Salix viminalis*) cultivation.

With respect to the removal of sludge-born PCBs, Wyrwicka et al. [68] demonstrated that the use of cucumber (*Cucumis sativus* L. var. Cezar) resulted in a decrease in PCB concentrations by an average of 38.63%. However, the efficiency of PCB removal decreased as the dose of sludge increased in sludge-treated soil (41.28, 38.39 and 36.22% PCB reduction at doses of 3, 9 and 18 tonnes/ha). Urbaniak et al. [43] demonstrated that the use of other plant from the *Cucurbitaceae* family—*Cucurbita pepo* L. cv Atena Polka—reduced total PCDDs/PCDFs and TEQ concentration by 37 and 68%, respectively, in soil amended with sewage sludge. The comparative study of the use of *Cucurbita pepo* L. cv Atena Polka (zucchini) and *Cucumis sativus* L. var. Cezar (cucumber) showed that zucchini was more efficient in sludge-born PCDDs/PCDFs removal, while cucumber demonstrated higher efficiency in soil phytotoxicity alleviation [44]. Presented studies demonstrate that cultivation of the plants from the *Cucurbitaceae* family plays a positive role in reducing the PCDDs/PCDFS in soil amended with sewage sludge.

The above data confirm the positive role of plant-bacteria systems in the removal of PCDDs/PCDFs and PCBs from soil contaminated through agricultural utilization of sewage sludge.

6. Impact of PCDDs/PCDFs on plant growth and plant metabolism

There is limited data on the impact of PCDDs/PCDFs on the plant growth and biomass production. The literature on this issue mainly comes from studies on the effects of sewage sludge on plant growth and metabolism [69–71]. Application of sewage sludge as soil organic amendment and as a source of macronutrients and micronutrients can contribute not only to restore the soil cover and vegetation on devastated land [72] but also can be used in the organization and maintenance of green areas in cities and recreational facilities. An important aspect of the use of sewage sludge is to improve soil fertility, of low quality class, which can be used, for example, for energy crops (biomass extraction). The addition of sewage sludge may have beneficial effects on plants and soil expressing itself by improving the physico-chemical properties of the soil, increased nutrient content for plants, increased production of plant biomass, and increased activity of soil enzymes and soil micro-organisms [73]. However, the presence of pollutants in sewage sludge may have a negative impact on the growth and development of plants. The content of heavy metals, toxic organic compounds including PCDDs/PCDFs and microbiological contaminants may contribute to the occurrence of secondary oxidative stress [74–76]. The occurrence of environmental stresses can lead to an imbalance in cellular redox state and predominance of oxidation reaction over reduction reactions. The reactive oxygen species (ROS) are highly reactive and toxic and can damage important from the biological point of view molecules such as nucleic acids, proteins and lipids [77]. It is well known that oxidative stress is a common plant reaction to numerous biotic [78] and abiotic stresses including drought [79], high salinity [80], temperature extremes [81, 82], anoxia [83], mineral nutrients' deficiencies and metal toxicity [84], increased UV-B radiation [85], gaseous pollutants [86], acid rain [87] and PCDD/PCDFs [88].

The enzymatic and non-enzymatic antioxidant systems present in the plant tissues prevent the accumulation of ROS caused by stress factors. The enzymatic free radical scavengers include, among others, superoxide dismutase (SOD), catalase (CAT) and peroxidases; ascorbate peroxidase (APx), glutathione peroxidase (GSH-Px), phenolic peroxidase (POx). Non-enzymatic, low molecular weight antioxidants mainly include ascorbic acid, glutathione, carotenoids, flavonoids, α-tocopherol and the phenolic compounds [89, 90].

Currently, there is little literature concerning the impact of PCDDs/PCDFs and PCBs on plant antioxidative system, and usually, information are related to multistress associated with the presence of organic pollutants and heavy metals. The plants belonging to the *Cucurbitaceae* family are known to accumulate high levels of PCDDs/PCDFs and PCBs compared with other plant species. However, the studies showed also that plant belonging to cucurbits: zucchini and cucumber, activate the antioxidative system and detoxification mechanisms as an effect of application of sewage sludge containing high levels of POPs including PCDDs/PCDFs and PCB [68, 91]. Obtained results indicate that signs of sewage sludge toxicity were greater in zucchini than in cucumber plants. Visible symptoms of leaf blade damage after sewage sludge application occurred only on the zucchini plants. Activity of peroxidases such as ascorbate peroxidase (APx) and guaiacol peroxidase (POx) increased in zucchini plants significantly with increasing of sewage sludge dose, but they decreased in cucumber plants. Moreover, both in zucchini and cucumber plants, the relationship between peroxidases activity and catalase (CAT) activity was inverse. Activity of detoxifying enzyme—glutathione S-transferase (GST)—increased progressively with the sludge concentration in both the zucchini and cucumber leaves. Moreover, the increase in GST activity was greater in zucchini plants and was visible at the lowest dose used. Concentration of α-tocopherol, a lipophilic antioxidant, increased with sewage sludge dose in both investigated species.

Other research focused on the influence of light soil fertilization using sewage sludge on soil toxicity showed its negative impact on growth and development of three plant species *Lepidium sativum*, *Sorgo saccharatum* and *Sinapis alba* [92].

7. Accumulation and translocation of PCDDs/PCDFs and PCBs in plant tissue

Plants are the organisms, which are the first stage in the food chain. Widely distributed at low concentration in the environment, extraordinary toxic PCDDs/PCDFs and PCBs have the ability to bioaccumulate in the food chain. For these reasons, accumulation of these compounds by plants is an important step for the transfer of PCDDs/PCDFs and PCBs into the higher trophic levels and biomagnification. Understanding the mechanisms of uptake and translocation of PCDDs/PCDFs and PCBs allows to control the risk of unexpected contamination of important vegetative plants. On the other hand, this knowledge can be used as a tool for selecting plants that have high phytoremediative potential [59, 93].

The accumulation of persistent pollutants such as chlorinated pesticides, chlorobenzenes, PCBs and PAH as well as PCDDs/PCDFs in vegetation has been demonstrated in several investigations carried out in the 1980s of the twentieth century [94–96]. In recent years, our understanding of the uptake of PCDDs/PCDFs by plants increased considerably [97, 98] but the pathway by which above-mentioned organic pollutants enter to the plant tissues still remain under discussion. Early evidence suggested that organic compounds were unlikely to be taken up from soil and translocated within the plants due to their hydrophobicity [99]. Nevertheless, according to many publications, the absorption from soil vapour may be the major pathway by which PCDDs/PCDFs from soil enter into the aerial plant tissues [100–102]. Other studies also evidence that dry gaseous deposition is the dominant pathway of PCDDs/PCDFs in plant tissue, such as lettuce, potato, apple, pear, rice, pea and oilseed rape [103–108]. However, more recent studies have shown that some species of plant have the ability to mobilize and accumulate significant concentrations of several organic compounds from soil. Generally, it is estimated that there are several pathways of PCDDs/PCDFs and PCBs accumulation in plants: (1) adsorption to the root surface, (2) root uptake trough absorption from soil vapour or water phases of soil and translocation to upper plant organs, (3) contamination of the foliage and fruits by soil particles which develop in contact with or in close proximity to the ground, (4) absorption of volatilized from soil PCDDs/PCDFs and PCBs by aerial plants parts, (5) atmospheric deposition of airborne PCDDs/PCDFs and PCBs for both gas and particle phase [97, 109].

Uptake and distribution of PCDDs/PCDFs and PCBs are a function of chemical and physical properties of particular pollutant, such as hydrophobicity, water solubility and vapour pressure, as well as environmental conditions, such as temperature, pH, organic carbon content of the soil and plant species [110].

The most important property which determines the possibility of absorption of various compounds from the soil by roots is hydrophobicity. Usually, it is expressed as the 1-octanol/

water partition coefficients (K_{ow}) and extends over a wide range for different organic compounds [111]. K_{ow} values vary over several orders of magnitude and are expressed as log K_{ow}. More hydrophobic substances that having a higher log K_{ow} value are sorbed more strongly to soil organic particles. In consequences, if the log K_{ow} value of the compound is around 2, the compound could be easily absorbed by plants, whereas if the log K_{ow} value is over 5, the compound is hardly absorbed [112, 113]. The value closely related to the hydrophobicity is solubility. Water solubility describes the amount of a chemical which can dissolve in a known quantity of water. The solubility of the chemical compound is dependent on temperature and is pH-dependent. Another feature describing the tendency of the substance to move from the aqueous phase to the gas phase is Henry's constant (H_c). This parameter can be useful in predicting the ability of chemical to volatilize from soil, water or plant surfaces into the atmosphere. Although chemical properties are important predictors of uptake, the physiology and composition of the plant root itself are also a significant influence. One explanation for such difference in uptake potential is the varying types and amounts of lipids in roots cells.

The uptake of organic chemicals by plants is also influenced by soil properties. Transfer of organic pollutants from soil to plant roots might be carried out by the uptake of soil pore water during plant transpiration. Non-ionized organic pollutants, which usually are lipophilic, are principally sorbed or bound to several components in soil including clays, iron oxides and onto the organic fraction of the soil's solid phase. The latter usually exerts the strongest influence on the organic chemicals pore water concentration [114]. Similarly, compounds with a high log K_{ow} associate with particulates in the wastewater and become incorporated into sewage sludge during sedimentation, and thus, substances with a log K_{ow} of <2.0 appear less frequently in sewage sludge. It is considered that with the increase in the organic matter content of a soil, the proportion of the chemical in the pore water decreases, and consequently, its uptake by plant also decreases. Moreover, it should be noted that the increase of the amount of organic carbon fraction reduces the optimum of K_{ow} for uptake into plants.

Deposition of non-ionic organic compounds on leaves and its sorption at the leaf surface or rapid movement into the leaf depends on diffusion through the cuticle or stomata. The concentration of all these compounds on leaves is mainly due to adsorption from the gaseous phase. Accumulation of PCDDs/PCDFs in above-ground plant parts mainly results from atmospheric deposition in the gaseous state alone. The contribution of particle-bound deposition may be, despite areas of extreme particle loading, of secondary importance [94].

The pathways of PCDDs/PCDFs accumulation in rice plants were carefully examined by Uegaki et al., [105] who estimated that dioxins were not absorbed from the soil by growing plants, but its uptake from atmosphere has the greatest importance. They reported that dioxin levels in rice plants were strongly influenced by soil adhesion but only at the early growth stage of brown rice plants grown in three different soils: dioxin-contaminated soil, paddy soil and upland soil. In the later stage of growth, over the time of experiment, predominant influence on dioxin level in rice leaf and steam was attributed to concentrations of these compounds in atmospheric gas phase. This remarks remain in agree with results other investigations which indicate that approximately 70% of 2,3,7,8-TCDD added to the growth solution, but only 3% of 2-chlorobiphenyl was adsorbed by the roots of soybean and corn [108], and the most important mechanism of foliar contamination is connected with volatility of 2,3,7,8-TCDD from the growth solution.

Taking above relations into consideration, the hydrophobic nature of PCDDs/PCDFs and PCBs (log K_{ow} values between 4.8 and 10.5) and their consequent strong adsorption to soil particles renders them to largely immobile and generally unavailable to plants [115, 116]. The majority of available evidence nevertheless suggests that the adsorption or absorption of PCDDs/PCDFs and PCBs into plant roots and their subsequent translocation into other parts of the plant structure is minimal. However, the notable exceptions are several plants of the genus *Cucurbita*, which readily take up PCDDs/PCDFs from soil and translocate them to leaves and fruits [97, 117]. It was also found that *Cucurbita* plants (e.g. zucchini, pumpkin and squash) can phytoextract polychlorinated biphenyls (PCBs) [118, 119], p,p′ DDE [120, 121] and chlordane [122, 123] from soil and translocate some quantities to aerial tissues, as well as it was found that there is remarkable diversity in the uptake and transportation of persistent organic pollutants (POPs) among subspecies [119, 121, 123]. In case of willow, study of Oleszczuk and Baran [124] demonstrated the uptake of 16 Polycyclic aromatic hydrocarbons (PAHs) by willow from the soil amendment with the contaminated sewage sludge. The authors showed that soil total (PAHs) content decreased significantly within the first half year, followed by minimal changes over the subsequent 3 years of treatment. The authors showed that the total content of (PAHs) in control ranged between 3.6–7.3 μg/kg in shoots and 13–27 μg/kg in leaves, whereas treated plant demonstrated higher concentrations ranged from 5.5 to 17.6 and 13.5 to 33.8 μg/kg in shoots and leaves, respectively.

8. Conclusions

PCDDs/PCDFs and PCBs pose one of the most challenging problems in environmental science and technology. Their discharges via insufficiently treated wastewater are responsible for their occurrence in river ecosystems, both water and bottom sediments. The administration of sewage sludge, as end products of purification processes, additionally generates problems with the occurrence of PCDDs/PCDFs and PCBs in the environment and creates risk for ecosystem functioning and human well-being. Despite the above, the available literature data concerning PCDDs/PCDFs and PCBs removal from the environment using range of bio- and phytoremediation technologies demonstrate a promising tool towards safe and effective elimination of the compounds and in this way improvement of ecosystems quality.

Acknowledgements

The research is financed within the project 'Impact of sludge originated PCDDs/PCDFs on soil contamination and Salix sp. Metabolism' (UMO—2013/09/D/ST10/04043) and the project funded under the Ministry of Science and Higher Education programme under the name 'Iuventus Plus' for the years 2015–2017: project no. IP2014 049273.

Author details

Magdalena Urbaniak[1,2*] and Anna Wyrwicka[3]

*Address all correspondence to: m.urbaniak@unesco.lodz.pl

1 European Regional Centre for Ecohydrology of the Polish Academy of Sciences, Łódz, Poland

2 Department of Applied Ecology, Faculty of Biology and Environmental Protection, University of Łódz, Łódz, Poland

3 Department of Plant Physiology and Biochemistry, Faculty of Biology and Environmental Protection, University of Łódz, Łódz, Poland

References

[1] UNEP. The UN-Water Status Report on the Application of Integrated Approaches to Water Resources Management. 2012.

[2] United Nations, Department of Economic and Social Affairs, Population Division. World Population Prospects: The 2010 Revision, Highlights and Advance Tables. Working Paper No. ESA/P/WP.220. 2011.

[3] US EPA. Progress in water quality: an evaluation for the national investment in municipal wastewater treatment. EPA-832-R-00-008. Washington, DC, USA. 2000.

[4] US EPA. Clean watersheds needs survey 2004—report to congress. (http://water.epa.gov/scitech/datait/databases/cwns/upload/2008_01_09_2004rtc_cwns2004rtc.pdf) 2004.

[5] Carey RO and Migliccio KW. Contribution of wastewater treatment plant effluents to nutrient dynamics in aquatic systems; a review. Environ Manag 2009;44:205–217.

[6] Bergqvist PA, Augulyte L, Jurjoniene V. PAH and PCB removal efficiencies in Umea (Sweden) and Sialial (Lithuania) municipal wastewater treatment plants. Water Air Soil Pollut 2006;175:291–303.

[7] Cirja M, Ivashechkin P, Schäffer A, Corvini PFX. Factors affecting the removal of organic micropollutants from wastewater in conventional treatment plants (CTP) and membrane bioreactors (MBR). Rev Environ Sci Biotechnol 2008;7:61–78.

[8] Alcock RE and Jones KC. Pentachlorophenol (PCP) and chloranil as PCDD/Fs sources to sewage sludge and sludge amended soils in the UK. Chemosphere 1997;35:2317–2330.

[9] McLachlan MS, Horstmann M, Hinkel M. Polychlorinated dibenzo-p-dioxins and dibenzofurans in sewage sludge: sources and fate following sludge application to land. Sci Total Environ 1996;185:109–123.

[10] Rappe C, Bergek S, Fiedler H, Cooper K. PCDD and PCDF contamination in catfish feed from Arkansas, USA. Chemosphere 1998;36:2705–2720.

[11] Eljarrat E, Caixach J, Rivera J. Comparison of TEQ contributions from PCDDs, PCDFs and dioxin-like PCBs in sewage sludge from Catalonia (Spain). Chemosphere 2003;51:595–601.

[12] Koch M, Knoth W, Rotard W. Source identification of PCDD/Fs in a sewage treatment plant of s German village. Chemosphere 2001;43:737–741.

[13] Urbaniak M, Kiedrzyńska E, Kiedrzyński M, Mendra M, Grochowalski A. The impact of point sources of pollution on the transport of micropollutants along the River continuum. Hydrol Res 2014;45(3):391–410.

[14] Urbaniak M and Kiedrzynska E. Concentrations and toxic equivalency of polychlorinated biphenyls in polish wastewater treatment plant effluents. Bull Environ Contam Toxicol 2015;95:530–535.

[15] Urbaniak M, Kiedrzyńska E, Kiedrzyński M, Zieliński M, Grochowalski A. The role of hydrology in the polychlorinated dibenzo-p-dioxin and dibenzofuran distributions in a lowland river. J Environ Qual 2015;44(4):1171–1182. doi:10.2134/jeq2014.10.0418

[16] Kakimoto H, Oka H, Miyata Y, Yonezawa Y, Niikawa A, Kyudo H, Tang N, Toriba A, Kizu R, Hayakawa K. Homologue and isomer distribution of dioxins observed In water samples collected from Kahokugata Lagoon and inflowing rivers, Japan. Water Res 2006;40:1929–1940.

[17] Chi KH, Hsu Sh-Ch, Lin Ch-Y, Kao ShJ, Lee TY. Deposition fluxes of PCDD/Fs in a reservoir system in northern Taiwan. Chemosphere 2011;83:745–752.

[18] Minomo K, Ohtsuka N, Hosono Sh, Nojiri K, Kawamura K. Seasonal change of PCDDs/PCDFs/DL-PCBs in water of Ayase River, Japan: pollution sources and their contributions to TEQ. Chemosphere 2011;85:188–194.

[19] Kowalewska G, Konat-Stepowicz J, Wawrzyniak-Wydrowska B, Szymczyk-Żyła M. Transfer of organic contaminants to the Baltic in the Odra Estuary. Mar Pollut Bull 2003;46:703–718.

[20] Wolska I, Galer K, Namieśnik J. Transport and speciation of PAHs and PCBs in a river ecosystem. Pol J Environ Stud 2003;12(10):105–110.

[21] Urbaniak M, Kiedrzyńska E, Zalewski M. The role of a lowland reservoir in the transport of micropollutants, nutrients and the suspended particulate matter along the river continuum. Hydrol Res 2012;43(4):400–411.

[22] Urbaniak M, Skowron A, Zieliński M, Zalewski M. Hydrological and environment al conditions as key drivers for spatial and seasonal changes in PCDD/PCDF concentra-

tions, transport and deposition along urban cascade reservoirs. Chemosphere 2012;88:1358–1367.

[23] Sztamberek-Gola I, Grochowalski A, Chrząszcz R. Monitoring of PCDDs, PCDFs and PAHs In waste-water with use the semipermeable membrane devices (SPMD). Organohalogen Compd 2003;60:45–48.

[24] Oleszek-Kudlak S, Grabda M, Czaplicka M, Rosik-Dulewska Cz, Shibata E, Takashi N. Fate of PCDD/PCDF turing mechanical-biological sludge treatment. Chemosphere 2005;62:389–397.

[25] Katsoyiannias A, Samara C. Persistent organic pollutants (POPs) in the sewage treatment plant in Thessaloniki, Northern Greece: occurrence and removal. Water Res 2004;38:2685–2698.

[26] Katsoyiannis A, Samara C. Persistent organic pollutants (POPs) in the conventional activated sludge treatment process: fate and mass balance. Environ Res 2005;97:245–257.

[27] Blanchard M, Teil MJ, Ollivon D, Garban B, Chesterikoff C, Chevreuil M. Origin and distribution of polyaromatic hydrocarbons and polychlorobiphenyls in urban effluents to wastewater treatment plants of the Paris area (France). Water Air Soil Poll 2001;35:3679–3687.

[28] Pham Th-Th, Proulx S. PCBs and PAHs in the Montreal urban community (Quebec, Canada) wastewater treatment plant and in the effluent plume in the St. Lawrence River. Water Res 1997;31(8):1887–1896.

[29] Morris S, Lester JN. Behavior and fate of polychlorinated biphenyls in a pilot wastewater treatment plant. Water Res 1994;28:1553–1561.

[30] Petrasek AC, Kugelman T J, Austern BM, Pressley AT, Winslow LA, Wise RH. Fate of toxic organic compounds in wastewater treatment plants. J Wat Pollut Control Fed 1983;55:1286–1296.

[31] Naiman RJ, Decamps H. (Eds.) The Ecology and Management of Aquatic–Terrestrial Ecotones. UNESCO, MAB, Parthenon, Paris. 1990.

[32] Schiemer F, Zalewski M, Thorpe JE. (Eds) The Importance of Aquatic–Terrestrial Ecotones for Freshwater Fish. Developments in Hydrobiology, 105. Dordrecht, Boston, London: Kluwer Academic Publisher. 1995.

[33] Susarla S, Medina VF, McCutcheon SC. Phytoremediation: an ecological solution to organic chemical contamination. Ecol Eng 2002;18:647–658.

[34] Mitsch WJ, Gosselink JG. Wetlands, 4th edn. New York: John Wiley & Sons. 2007.

[35] Kiedrzyńska E, Zalewski M. Water quality improvement through an integrated approach to point and non-point sources pollution and management of river floodplain

wetlands. In: Voudouris K, Voutsa D. (Eds.). Ecological Water Quality—Water Treatment and Reuse. INTECH Open Access, Rijeka, Croatia, 325–342. 2012.

[36] Mitsch WJ, Jørgensen SE. Ecological Engineering and Ecosystem Restoration. New York: Wiley. 2004.

[37] Zalewski m, Robarts R. Ecohydrology—a new Paradigm for Integrated Water Resources Management. SIL News 40, Sep. 2003:1–5.

[38] Zerihun Negussie Y, Urbaniak M, Szklarek S, Lont K, Gągała I, Zalewski M. Efficiency analysis of two sequential biofiltration systems in Poland and Ethiopia—the pilot study. Ecohydrol Hydrobiol 2012;12(4):271–285.

[39] Mackay D, Shiu WY, Ma K-Ch, Lee SC. Handbook of Physical-Chemical Properties and Environmental Fate for Organic Chemicals, Second Edition, Taylor & Francis Group, LLC, Boca Raton, USA, 2006.

[40] Sewart A, Harrad SJ, McLachlan MS, McGrath SP, Jones KC. PCDDs/Fs and non-o-PCBs in digested UK sewage sludges. Chemosphere 1995;30:51–67.

[41] Eljarrat E, Caixach J, Rivera J. Decline in PCDD and PCDF levels in sewage sludges from Catalonia (Spain). Environ Sci Technol 1999;33:2493–2498.

[42] Dudzińska MR, Czerwiński J. PCDD/F levels in sewage sludge from MWTP in South-Eastern Poland. Organohalogen Comp 2002;57:305–308.

[43] Urbaniak M, Wyrwicka A, Zieliński M, Mankiewicz-Boczek J. Potential for phytoremediation of PCDD/PCDF-contaminated sludge and sediments using *Cucurbitaceae* plants: a pilot study. Bull Environ Contam Toxicol 2016;97(3):401–406. doi:10.1007/s00128-016-1868-6

[44] Urbaniak M, Zieliński M, Wyrwicka A. The influence of the *Cucurbitaceae* on mitigating the phytotoxicity and PCDD/PCDF content of soil amended with sewage sludge. Int J Phytoremediation 2016. doi:10.1080/15226514.2016.1207606

[45] The Federal Ministry of the Environment: Organic pollutants in sewage sludge. The report of the Federal Ministry of the Environment about the need to define additional specific rules-waste list, 1999; 4/99:10–13.

[46] The publication issued by the Ministry of the Environment of Baden-Württemberg (1996): Fourth Administrative Regulation of the Ministry of Environment for the protection of soil, in particular, the identification and classification of the level of organic pollutants in soil (Organic Pollutants) Common Official Journal of Baden-Württemberg of 14 February 1996, n. 2, 87–94, Stuttgart.

[47] Regulation of Lower Austria on sewage sludge and compost waste, Journal of Laws No. 6160 / 1–0 1994. Lower Austria.

[48] The regulation on sewage sludge on 15.04.1992, the Federal Journal of Laws 1992. Part 1, 912–934.

[49] EU Working Document on Sludge 3rd Draft. Unpublished, 2000; 19 p.

[50] USEPA, 40 CFR Part 503. Standards for the use or disposal of sewage sludge as amended. Proposed Rule 1999.

[51] Wilson SC, Alcock RE, Sewart AP, Jones KC. Persistence of organic contaminants in sewage sludge-amended soil, a field experiment. J Environ Qual 1997;26:1467–1477.

[52] Kearney PC, Woolson EA, Ellington CP. Persistence and metabolism of chlorodioxins in soils. Environ Sci Technol 1972;6:1017–1019.

[53] Field JA, Sierra-Alvarez R. Microbial degradation of chlorinated dioxins. Chemosphere 2008;71:1005–1018.

[54] Barkovskii AL, Adriaens P. Impact of humic constituents on microbial dechlorination of polychlorinated dioxins. Environ Toxicol Chem 1998;17:1013–1020.

[55] Urbaniak M. Biodegradation of PCDD/PCDF and dl- PCB. [In:]: Chamy R, Rosenkran F (Eds) Biodegradation—Engineering and Technology. INTECH Publisher, Rijeka, Croatia, ISBN 978-953-51-1153-5, 73–100, 2013.

[56] Bumpus M, Tien D, Wright SD. Oxidation of persistent environmental-pollutants by a white root fungi. Science 1985;228:1434–1436.

[57] Takada S, Nakamura M, Matsueda T, Kondo R, Sakai K. Degradation of polychlorinated dibenzo-p-dioxins and polychlorinated dibenzofurans by the white root fungus Phanerochaete sordida YK-624. Appl Environ Microbiol 1996;62:4323–4328.

[58] Mackova M, Prouzova P, Stursa P, Ryslava E, Uhlik O, Beranova K, Rezek J, Kurzawova V, Demnerova K, Macek T. Phyto/rhizoremediation studies using long-term PCB contaminated soil. Environ Sci Pollut Res 2009;16:817–829.

[59] Macek T, Macková M, Káš J. Exploitation of plants for the removal of organics in environmental remediation. Biotechnol Adv 2000;18:23–34.

[60] Whipps JM. Carbon economy, [In:] J.M. Lynch (Eds.). The Rhizosphere. New York: Wiley, 1990;59–97.

[61] Lugtenberg BJJ, Dekkers L, Bloemberg GV. Molecular determinants of rhizosphere colonization by *Pseudomonas*. Annu Rev Phytopathol 2001;39:461–490.

[62] Kuiper I, Lagendijk EL, Bloemberg GV, Lugtenberg BJJ. Rhizoremediation: a beneficial plant-microbe interaction. Mol Plant Microbe Interact 2004;17:6–15.

[63] Siciliano SD, Germida JJ, Banks K, Greer CW. Changes in microbial community composition and function during a polyaromatic hydrocarbon phytoremediation field trial. Appl Environ Microbiol 2003;69:483–489.

[64] Nedunuri KV, Govindaraju RS, Banks MK, Schwab AP, Chen Z. Evaluation of phytoremediation for field-scale degradation of total petroleum hydrocarbons. J Environ Eng 2000;126:483–490.

[65] Robinson SL, Novak JT, Widdowsen MA, Crosswell SB, Fetterolf GJ. Field and laboratory evaluation of the impact of tall fescue on polyaromatic hydrocarbon degradation in aged creosote contaminated surface oil. J Environ Eng 2002;129:232–240.

[66] Banks MK, Kulakow P, Schwab AP, Chen Z, Rathbone K. Degradation of crude oil in the rizosphere of sorghum bicolor. Int J Phytoremed 2003;5:225–234.

[67] Vervaeke P, Luyssaert S, Mertens J, Meers E, Tack FM, Lust N. Phytoremediation prospects of willow stands on contaminated sediments: a field trial. Environ Pollut 2003;126:7–282.

[68] Wyrwicka A, Steffani S, Urbaniak M. The effect of PCB-contaminated sewage sludge and sediment on metabolism of cucumber plants (*Cucumis sativus* L.). Ecohyd Hydrobiol 2014;14:75–82.

[69] Antolín MC, Pascual I, García C, Polo A, Sánchez-Díaz M. Growth, yield and solute content of barley in soils treated with sewage sludge under semiarid Mediterranean conditions. Field Crops Res 2005;94:224–237.

[70] Fernández JM, Hockaday WC, Plaza C, Polo A, Hatcher PG, Effects of long-term soil amendment with sewage sludges on soil humic acid thermal and molecular properties. Chemosphere 2008;73:1838–1844.

[71] Lakhdar A, ben Achiba W, Montemurro F, Jedidi N, Abdelly Ch. Effect of municipal solid waste compost and farmyard manure application on heavy-metal uptake in wheat. Commun Soil Sci Plan. 2009;40:3524–3538.

[72] Gawron M. The use of sewage sludge for the restoration of degraded land on the background of planning decisions and legal requirements. Environmental Engineering 2007; 13:103–109.

[73] Singh RP, Agrawal M. Potential benefits and risks of land application of sewage sludge. Waste Manage 2008;28:347–358.

[74] Aki C, Güneysu E, Acar O. Effect of industrial wastewater on total protein and the peroxidase activity in plants. Afr J Biotechnol 2009;8:5445–5448.

[75] Antolín MC, Muro I, Sánchez-Díaz M. Application of sewage sludge improves growth, photosynthesis and antioxidant activities of nodulated alfalfa plants under drought conditions. Environ Exper Bot 2010;68:75–82.

[76] Antolín MC, Muro I, Sánchez-Díaz M. Sewage sludge application can induce changes in antioxidant status of nodulated alfalfa plants. Ecotoxicol Environ Saf 2010;73:436–442.

[77] Mittler R. Oxidative stress, antioxidants and stress tolerance. Trends Plant Sci. 2002;7:405–410.

[78] Kuzniak E, Skłodowska M. The effect of *Botrytis cinerea* infection on the antioxidant profile of mitochondria from tomato leaves. J Exper Bot 2004;55:605–612.

[79] Mittler R, Zilinskas B. Regulation of pea cytosolic ascorbate peroxidase and other antioxidant enzymes during the progression of drought stress and following recovery from drought. Plant J 1994;5:397–405.

[80] Hernández JA, Campillo A, Jiménez A, Alarcón JJ, Sevilla F. Response of antioxidant systems and leaf water relations to NaCl stress in pea plants. New Phytol 1999;141:241–251.

[81] Kang H-M, Saltveit ME. Activity of enzymatic antioxidant defence systems in chilled and heat shocked cucumber seedling radicles. Physiol Plant 2001;113:548–556.

[82] Foyer CH, Vanacker H, Gomez LD, Harbinson J. Regulation of photosynthesis and antioxidant metabolism in maize leaves at optimal and chilling temperatures: review. Plant Physiol Biochem 2002;40:659–668.

[83] Blokhina OB, Chirkova TV, Fagerstedt KV. Anoxic stress leads to hydrogen peroxide formation in plant cells. J Exper Bot 2001;52:1179–1190.

[84] Gajewska E, Skłdowska M. Effect of nickel on ROS content and antioxidative enzyme activities in wheat leaves. BioMetals 2007;20:27–36.

[85] Mackerness SA. Plant responses to ultraviolet-B (UV-B: 280–320 nm) stress: what are the key regulators? Plant Growth Regul 2000;32:27–39.

[86] Chernikova T, Robinson JM, Lee EH, Mulchi CL. Ozone tolerance and antioxidant enzyme activity in soybean cultivars. Photosynth Res 2000;64:15–26.

[87] Wyrwicka A, Skłodowska M. Intercompartmental differences between cytosol and mitochondria in their respective antioxidative response and lipid peroxidation levels in acid rain stress. Acta Physiol Plant 2014;36:837–848.

[88] Zhang B, Zhang H, Jin J, Ni Y, Chen J. PCDD/Fs-induced oxidative damage and antioxidant system responses in tobacco cell suspension cultures. Chemosphere 2012;88:798–805.

[89] Gill SS, Tuteja N. Reactive oxygen species and antioxidant machinery in abiotic stress tolerance in crop plants. Plant Physiol Biochem 2010;48:909–930.

[90] Foyer CH, Noctor G. Redox homeostasis and antioxidant signaling: a metabolic interface between stress perception and physiological responses. Plant Cell 2005;17:1866–1875.

[91] Wyrwicka A, Urbaniak M. The different physiological and antioxidative responses of zucchini and cucumber to sewage sludge application. PLoS One 2016;11:e0157782. doi: 10.1371/journal.pone.0157782

[92] Oleszczuk P. Testing of different plants to determine influence of physico-chemical properties and contaminants content on municipal sewage sludges phytotoxicity. Environ Toxicol 2010;25:38–47.

[93] Materac M, Wyrwicka A, Sobiecka E. Phytoremediation techniques in wastewater treatment. Environ Biotechnol 2015;11:10–13.

[94] Reischl A, Reissinger M, Thoma H, Hutzinger O. Uptake and accumulation of PCDD/F in terrestrial plants: basic considerations. Chemosphere 1989;19:467–474.

[95] Buckley EH. Accumulation of airborne polychlorinated biphenyls in foliage. Science 1982;216:520–522.

[96] Gaggi C, Bacci E. Accumulation of chlorinated hydrocarbon vapours in pine needles. Chemosphere 1985;14:451–456

[97] Hülster A, Müller JF, Marschner H. Soil-plant transfer of polychlorinated dibenzo-p-dioxins and dibenzofurans to vegetables of the cucumber family (Cucurbitaceae). Environ Sci Technol 1994;28:1110–1115.

[98] Inui H, Sawada M, Goto J, Yamazaki K, Kodama N. A major latex-like protein is a key factor in crop contamination by persistent organic pollutants. Plant Physiol 2013;161:2128–2135.

[99] Briggs GG, Bromilow RH, Evans AA. Relationships between lipophilicity and root uptake and translocation of non-ionised chemicals by barley. Pesticide Sci 1982;13:495–504.

[100] Kew GA, Schaum JL, White P, Evans TT. Review of plant uptake of 2,3,7,8-TCDD from soil and potential influences of bioavailability. Chemosphere 1989;18:1313–1318.

[101] Trapp S, Matthies M. Generic one-compartment model for uptake of organic chemicals by foliar vegetation. Environ Sci Technol 1995;29:2333–2338.

[102] Trapp S, Matthies M. Modeling volatilization of PCDD/F from soil and uptake into vegetation. Environ Sci Technol 1997;31:71–74.

[103] Welsch-Pausch K, McLachlan M, Umlauf G. Determination of the principal pathways of polychlorinated dibenzo-p-dioxins and dibenzofurans to *Lolium multiflorum* (Welsh Ray Grass). Environ Sci Technol 1995;29:1090–1098.

[104] Hülster A, Marschner H. Transfer of PCDD/PCDF from contaminated soils to food and fodder crop plants. Chemosphere 1993;27:439–446.

[105] Uegaki R, Seike N, Otani T. Polychlorinated dibenzo-*p*-dioxins, dibenzofurans, and dioxin-like polychlorinated biphenyls in rice plants: possible contaminated pathways. Chemosphere 2006;65:1537–1543.

[106] Müller JF, Hülster A, Päpke O, Ball M, Marschner H. Transfer pathways of PCDD/PCDF to fruits. Chemosphere 1993;27:195–201.

[107] Müller JF, Hülster A, Päpke O, Ball M, Marschner H. Transfer of PCDD/PCDF from contaminated soils into carrots, lettuce and peas. Chemosphere 1994;29:2175–2181.

[108] McCrady JK, McFarlanea C, Ganderb LK. The transport and fate of 2,3,7,8-TCDD in soybean and corn. Chemosphere 1990;21:359–376.

[109] Wild SR, Jones KC. Organic chemicals entering agricultural soils in sewage sludges: screening for their potential to transfer to crop plants and livestock. Sci Total Environ 1992;119:85–119.

[110] Wagrowski DM, Hites RA. Polycyclic aromatic hydrocarbon accumulation in urban, suburban, and rural vegetation. Environ Sci Technol 1997;31:279–282.

[111] Trapp M, McFarlane C. Plant Contamination. Boca Raton: Lewis Publishers; 1995.

[112] Briggs GG. Theoretical and experimental relationships between soil adsorption, octanol-water partition coefficients, water solubilities, bioconcentration factors, and the parachor. J Agric Food Chem 1981;29:1050–1059.

[113] Briggs GG, Bromilow RH, Evans AA, Williams M. Relationships between lipophilicity and the distribution of non- ionised chemicals in barley shoots following uptake by the roots. Pesticide Sci 1983;14:492–500.

[114] Collins C, Fryer M, Grosso A. Plant uptake of non-ionic organic chemicals. Environ Sci Technol 2006;40:45–52.

[115] Govers H, Krop H. Partition constants of chlorinated dibenzofurans and dibenzo-*p*-dioxins. Chemosphere 1998;37:2139–2152.

[116] Lovett AA, Foxall CD, Creaser CS, Chewe D. PCB and PCDD/DF congeners in locally grown fruit and vegetable samples in Wales and England. Chemosphere 1997;34:1421–1436.

[117] Engwall M, Hjelm K. Uptake of dioxin-like compounds from sewage sludge into various plant species—assessment of levels using a sensitive bioassay. Chemosphere 2000;40:1189–1195.

[118] White JC, Parrish ZD, Isleyen M, Gent MPN, Iannucci-Berger W, Eitzer BD, Kelsey JW, Mattina MI. Influence of citric acid amendments on the availability of weathered PCBs to plant and earthworm species. Int J Phytoremediation 2005;8:63–79.

[119] Inui H, Wakai T, Gion K, Kim YS, Eun H. Differential uptake for dioxin-like compounds by zucchini subspecies. Chemosphere. 2008;73:1602–1607.

[120] White JC. Differential bioavailability of field-weathered *p,p'*-DDE to plants of the *Cucurbita* and *Cucumis* genera. Chemosphere 2002;49:143–152.

[121] White JC, Wang X, Gent MPN, Iannucci-Berger W, Eitzer BD, Schultes NP, Arienzo M, Mattina MI. Subspecies-level variation in the phytoextraction of weathered *p,p'*-DDE by *Cucurbita pepo*. Environ Sci Technol 2003;37:4368–4373.

http://dx.doi.org/10.5772/66204

[122] Mattina MI, Eitzer BD, Iannucci-Berger W, Lee WY, White JC. Plant uptake and translocation of highly weathered, soil-bound technical chlordane residues: data from field and rhizotron studies. Environ Toxicol Chem 2004;23:2756–2762.

[123] Mattina MJI, Iannucci-Berger W, Dykas L. Chlordane uptake and its translocation in food crops. J Agric Food Chem 2000;48:1909–1915.

[124] Oleszczuk P, Baran S. Polycyclic aromatic hydrocarbons content in shoots and leaves of willow (*Salix viminalis*) cultivated on the sewage sludge-amended soil. Water, Air Soil Pollut 2005;168:91–111.

Section 3

Resource Recovery and Management

Phosphorus Recovery by Struvite Crystallization from Livestock Wastewater and Reuse as Fertilizer: A Review

Tao Zhang, Rongfeng Jiang and Yaxin Deng

Additional information is available at the end of the chapter

http://dx.doi.org/10.5772/65692

Abstract

In China, the intensive livestock farming produces massive livestock wastewater with high concentration of phosphorus. Discharge of these compounds to surface water not only causes water eutrophication but also wastes phosphorus resources for plant growth. Therefore, it's necessary combining the removal of phosphorus from livestock wastewater with its recovery and reuse as fertilizer. As a valuable slow-release mineral fertilizer, struvite crystallization has become a focus in phosphorus recovery. In this chapter, struvite crystallization mechanism, reaction factors, crystallizers, and the applications of struvite as fertilizer are discussed. Two steps of nucleation and crystal growth for struvite crystallization from generation to growth are introduced. The reaction factors, including molar ratio of magnesium and phosphate, solution pH, coexisting substances, and seeding assist, of struvite crystallization are summarized. Several innovate types of crystallizer, which relate to the shape and size of harvest struvite to realize the phosphorus recycling, are demonstrated. Due to the influence of toxic or harmful impurities in struvite on its reuse as fertilizer, the environmental risk evaluation of struvite application is introduced. In conclusion, struvite crystallization is a promising tool for recovering phosphorus from livestock wastewater.

Keywords: phosphorus, struvite, livestock wastewater, fertilizer, review

1. Introduction

Phosphorus is a key factor causing water eutrophication, on the other hand, it is also a nonrecyclable, nonrenewable, and quite valuable resource. According to the Mineral Commodity Summaries 2015 [1] from the United States Geological Survey (USGS), the reserve of phosphate rock in China is 3.7 billion tons in 2014, which is in second place in the world. However, with a total of 43–48% of the world's phosphate rock production over the last 3 years [2], the phosphate rock might run out in less than 40 years. So phosphate rock has been one of the 20 minerals that could not meet the demand of the national economy development after 2010 as reported by the Ministry of Land and Resources in China.

On the other side, the intensive livestock farming is a pillar industry in agricultural economy and an important way to increase rural incomes in China [3]. However, it usually produces large amount of livestock wastewater containing high concentration of phosphorus [4]. If this wastewater was not treated reasonably, it would not only lead to the pollution of water eutrophication, but also waste nonrenewable resources and would become one of the major contributors to phosphorus loss [5]. According to the first national sources of pollution survey [6] in China in 2008, nonpoint source pollution in agriculture is a major cause of eutrophication. It accounts for 34.24% of the total phosphorus emission amount, including the livestock and poultry industry. Therefore, it is quite valuable to combine nutrient recycling with environmental pollution control to recover losing phosphorus from livestock wastewater [7].

Numerous phosphorus recovery technologies have been developed, such as biological phosphorus removal, chemical precipitation, electrolysis, adsorption, and crystallization. Biological phosphorus removal utilizes polyphosphate-accumulating organisms to capture phosphorus in their cells. However, this method is limited by the lack of carbon sources and the difficulty of culturing pure bacteria [8]. Chemical precipitation process may consume expensive chemicals and produce large amounts of chemical sludge [9]. Electrolysis is restricted by the small capacity of handling wastewater and the frequent renewal of electrodes [10, 11]. Recovering phosphorus from wastewater using chemical adsorbents is expensive, so cheaper and more efficient adsorbents are necessary for research [12].

Recovering phosphorus by crystallization, by contrast, is a more economical and efficient way. As long as the crystallization conditions are suitable, the struvite crystal would be generated just by adding magnesium (Mg^{2+}) in the raw wastewater which has high concentrations of $H_nPO_4^{n-3}$ and NH_4^+-N. This technology can remove nitrogen at the same time and its production can be used as fertilizer. So it had been studied in many kinds of wastewater, such as multiple wastewater [13], industrial wastewater [14, 15], municipal landfill leachate [16], biogas slurry [17], and effluent of sewage sludge [18], and livestock wastewater is no exception.

2. Struvite characteristic

Magnesium ammonium phosphate, also known as struvite, is a white crystal generated in neural or mild alkali condition, for which the chemical formula is $Mg(NH_4)PO_4{\cdot}6H_2O$. Struvite consists of one molecule of magnesium(Mg^{2+}), one molecule of ammonium (NH_4^+),

http://dx.doi.org/10.5772/65692

one molecule of phosphate (PO_4^{3-}), and six molecules of water (H_2O), whose relative molar mass is 245.43 g/mol. It is only slightly soluble in water but soluble in acid solution [19]. Struvite is a light crystal with low relative density of 1.65–1.7. It is not easy to be rush off by rainfall [20]. Pure struvite belongs to orthorhombic crystals which consists of regular PO_4^{3-} octahedron, distorted $Mg(H_2O)_6^{2+}$ octahedron, and groups of NH_4^+ connected by hydrogen bonding [21], but shows rod-like structure [22] or irregular structure [23] sometimes (**Figure 1**). And struvite of rod-like structure is of low purity, because of the coprecipitation with foreign ions.

Actually, struvite had been widely studied as early as 1937 for the congestion in the pipes of the sludge anaerobic digester [24]. The general struvite forming reaction equation is shown below:

$$NH_4^+ + Mg^{2+} + PO_4^{3-} + 6\,H_2O \rightarrow MgNH_4PO_4 \cdot 6\,H_2O \tag{1}$$

$$NH_4^+ + Mg^{2+} + HPO_4^{2-} + 6\,H_2O \rightarrow MgNH_4PO_4 \cdot 6\,H_2O + H^+ \tag{2}$$

$$NH_4^+ + Mg^{2+} + H_2PO_4^- + 6\,H_2O \rightarrow MgNH_4PO_4 \cdot 6\,H_2O + 2\,H^+ \tag{3}$$

When Mg^{2+}, NH_4^+, and $H_nPO_4^{n-3}$ (n = 0, 1, or 2) exist in the solution and the product of their concentrations are bigger than the solubility product constant (K_{sp}) of struvite, the crystal would be generated spontaneously. And the calculation formula of struvite's K_{sp} is shown below:

$$K_{sp} = [Mg^{2+}] \cdot [NH_4^+] \cdot [PO_4^{3-}] \tag{4}$$

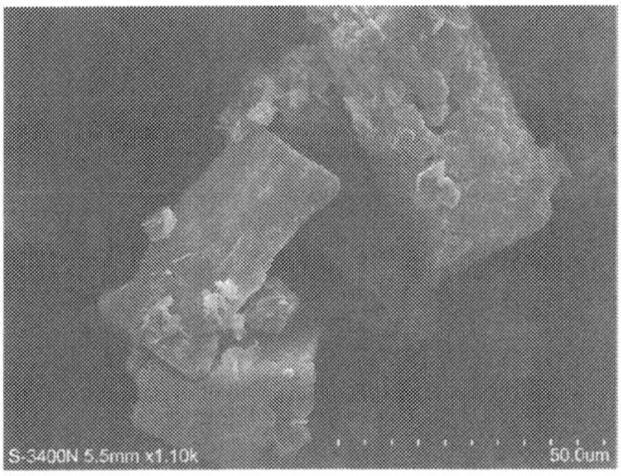

Figure 1. The SEM figure of magnesium ammonium phosphate crystal.

where $[Mg^{2+}]$, $[NH_4^+]$, and $[PO_4^{3-}]$ are concentrations of Mg^{2+}, NH_4^+, and PO_4^{3-} in the solution, respectively. As the molar ratio of Mg^{2+}, NH_4^+, and PO_4^{3-} is 1:1:1 in struvite, so C* is used to present the same concentration of these three ions, which means $C^* = [Mg^{2+}] = [NH_4^+] = [PO_4^{3-}]$. So the calculation formula of struvite's K_{sp} also can be shown as:

$$K_{sp} = (C^*)^3 \tag{5}$$

Snoeyink et al. [25] got the K_{sp} of struvite is $10^{-12.6}$ as early as 1980. Ohlinger et al. [26] corrected it to $10^{-13.26}$ in 1999. And then Bhuiyan et al. [27] corrected it again to $10^{-13.36}$ in 2007, which is widely used now. However, K_{sp} of struvite is hard to get in the real wastewater for the negative impact of the soluble coexisting ions. Therefore, in the estimation of the saturability of the real wastewater, ionic activity coefficient (K_{so}) is more widely useful than K_{sp}. Considering the impact of ionic strength (I) and the ionic activity (Ai) in estimating the K_{so}, the value of K_{so} is bigger than K_{sp}. And the calculation formula of struvite's K_{so} is shown below:

$$K_{so} = \alpha_{Mg^{2+}} \times \alpha_{NH_4^+} \times \alpha_{PO_4^{3-}} \tag{6}$$

$$\alpha_i = \gamma_i [C_i] \tag{7}$$

where α_i presents the ionic activity (Ai), γ_i presents the activity coefficient of the ionic strength (I), and $[C_i]$ presents the concentration of the ion. Only when the value of γ_i is 1, K_{sp} is able to represent K_{so}. Therefore, it is necessary to eliminate the interruptions of the soluble coexisting ions (like Ca^{2+}, CO_3^{2-}, and SO_4^{2-}) and clear of the ionic activities of Mg^{2+}, NH_4^+, and PO_4^{3-} in the specific pH condition. **Table 1** shows the ionization equations and pKa value in magnesium ammonium phosphate solution at 25°C, which is helpful to estimate the distribution of these ions and predict the probability to generate struvite under such environment of solution.

No.	Ionization equation	pK_a
1	$NH_4^+ \rightleftharpoons NH_3\,(aq) + H^+$	9.26
2	$H_3PO_4 \rightleftharpoons H_2PO_4^- + H^+$	2.12
3	$H_2PO_4^- \rightleftharpoons HPO_4^{2-} + H^+$	7.20
4	$HPO_4^{2-} \rightleftharpoons PO_4^{3-} + H^+$	12.36
5	$MgNH_4PO_4{\cdot}6H_2O \rightleftharpoons Mg^{2+} + NH_4^+ + PO_4^{3-} + 6H_2O$	12.70
6	$MgOH^+ \rightleftharpoons Mg^{2+} + OH^-$	2.56
7	$MgH_2PO_4^+ \rightleftharpoons H_2PO_4^- + Mg^{2+}$	0.45
8	$MgHPO_4 \rightleftharpoons HPO_4^{2-} + Mg^{2+}$	2.91
9	$MgPO_4^- \rightleftharpoons PO_4^{3-} + Mg^{2+}$	4.80

Table 1. The ionization equations and pKa value in magnesium ammonium phosphate solution at 25° [30, 31].

3. Mechanism of struvite crystallization

Nucleation and crystal growth are two classical steps in the process of struvite crystallization from generation to development. As shown in **Figure 2**, nucleation is the first step of the struvite crystallization. When Mg^{2+}, NH_4^+, and PO_4^{3-} meet under the proper pH value, nucleation occurs. And the nucleation time is the time required to form a saturated solution to the beginning of the nucleation. It is mainly influenced by the pH of solution, mixing energy, coexisting ions, and saturation index (SI). The ion activity affected by the value of pH significantly leads to differentiation of combine speed of free ions [28]. Weak ion activity means slow combine speed and longer nucleation time indirectly. When the rate of struvite nucleation and growth is greater than or equal to the rate of mixing magnesium to the solution, there needs additional mixing energy. Kim et al. [15] emphasized that mixing energy could influence the quantity and size of struvite strongly. However, the greatest impaction of struvite nucleation is saturation index (SI) of solution which decides the development of crystal to homogeneous or heterogeneous directly [26]. SI is used to describe the saturation state of the reaction system of struvite. And the SI calculation of struvite is shown as follows:

$$SI = \log \frac{IAP}{K_{sp}} \tag{8}$$

where IAP and K_{sp} represent ionic activity product and the thermodynamic solubility product of struvite, respectively [29]. The homogeneous crystallization that we want happens on metastable region in the solution. In this region, nucleation is not spontaneous, which differentiated between the process of crystallization and precipitation, and avoids the occurrence of undesirable spontaneous nucleation to a great extent [28]. However, mestastable state of solution is very difficult to control. Therefore, SI, as the indicator for metastable state, is very important. Bonurophoulos et al. [30] found that the threshold between homogeneous and heterogeneous precipitation is the condition where SI ≈ 2.0 and the nucleation rate of 1 nucleus/($cm^3 \cdot s$). When the SI is less than 1.716, the struvite crystals are in heterogeneous precipitation and vice versa. Bhuiyan et al. [31] and Mehta et al. [32] also got the threshold at SI = 1.83 and SI = 1.7 at the special nucleation rate, respectively. In addition, Durrant et al. [33] emphasized the great influence of SI on the shape of struvite as early as 1999. And it also has a SI threshold between rhombic structure and rod-like structure of struvite.

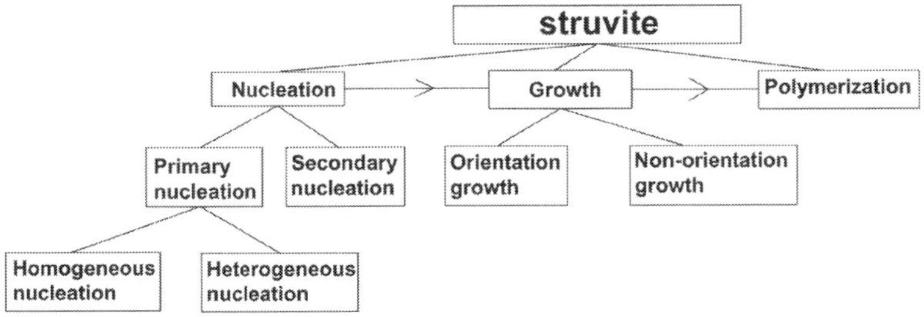

Figure 2. The crystal nucleation, growth and aggregation mechanism of magnesium ammonium phosphate.

After the crystal nucleus generates, the ions in the solution used to form the crystal begin to deposit on the crystal nucleus and the nucleus grow to the settling particles. During that time, there are two trends for the development of particles. One is orientation growth, which means the ions sequence in the crystal is arranged according to a certain lattice. The other one is nonorientation growth, which means these ions are too late to arrange in order. It is the fast growth rate that causes disorder. And two types of the crystal growth mechanisms lead to different trends. One is the integration mechanism, and the other is the mass transfer mechanism. The former is the integration of solute molecules into the surface; the latter is the transfer (by diffusion or convection) of solute molecules from the bulk solution to the crystal surface. When the effect of mass transfer is greater than the effect of integration, the crystal growth mainly depends on the diffusion effect and the growth of crystal would be orientable. However, if the effect of integration is greater, the integration on the surface of solute decides the crystal growth. And the relative sizes of the nucleation rate, aggregation rate, and directional array rate also decide the trend of crystal growth, which can be changed by precipitation conditions [31]. Abe et al. [34] showed that the growth rate of struvite was very slow. In the high concentration of phosphate (greater than 200 mg/L), the daily growth rate of struvite was 0.173 mm. In the low concentration of phosphate (30–100 mg/L), the daily growth rate of struvite was 0.061 mm. Therefore, increasing crystal growth rate and crystal size of struvite is not only beneficial to further removing phosphorus from livestock wastewater, but also to recycling phosphorus with a bigger size struvite. There is a metastable zone in industrial crystallization to make the crystal bigger and more even. The metastable zone is defined as a region bounded by the solubility curve in which the solution is supersaturated but the spontaneous nucleation cannot occur in such a short time [35]. In the metastable zone, the solute condenses on the nucleus as constantly as possible. As we known, the process of struvite constant growth is also the further recovery of phosphorus from livestock wastewater. So it is meaningful to study the metastable zone of struvite for the industrialized application.

4. The factors influencing struvite crystallization

4.1. Molar ratio of P and Mg

Generally speaking, livestock wastewater is rich in ammonium and phosphorus. So it is needed to add extra magnesium to form struvite. Therefore, the addition amount of magnesium affects the solubility product constant (K_{sp}) directly, which further affects the quantity of struvite crystal and the recovery rate of phosphorus in livestock wastewater. So the molar ratio of phosphate and magnesium is the key factor to control the yield of phosphorus recovery. The molar ratio of phosphate and magnesium is 1:1 in theory. However, the real molar of the added magnesium is larger than the total amount of phosphorus in the real livestock wastewater. As shown in **Table 2**, for a higher phosphorus removal rate, the molar ratio of phosphate and magnesium is about 1:1–1.2 from livestock wastewater, 1:1.4 from synthetic livestock wastewater, and 1:1–1.4 from anaerobic digesters of livestock wastewater. It is mainly based on the effect of coexisting ions in the livestock wastewater. The coexisting ions, such as OH^- and CO_3^{2-}, are apt to coprecipitate with Mg^{2+}, which prevent the Mg^{2+} from touching

Samples	Initial concentration of phosphate (mg/L)	Molar ratio of N, P, and Mg	pH	Reaction time	Removal rate of phosphate (%)	References
Animal manure wastewater	145	16.4:1:1.05	8	30 min	67	[40]
Animal manure wastewater	189.9	1:1:0.8–1	8.35	4 h	96	[41]
Animal manure wastewater	60.01	63.5:1:1	8.09	4 h	92.82	[42]
Animal manure wastewater	128 ± 13	1:1:1.2	9	1 h	98	[43]
Synthetic animal manure wastewater	80	8:1:1.4	9.5–10.5	2 h	97	[44]
Synthetic animal manure wastewater	130.2	1:1:5	7.9	–	92	[45]
Anaerobic digesters of manure wastewater	51.1	30.7:1:1.4	8.0–10.0	1 h	74–95	[46]
Anaerobic digesters of manure wastewater	55.4	9.6:1:1.2	9.0	20 min	85	[47]
Anaerobic digesters of manure wastewater	64.2	1:1.2:1.2	9.0	15 h	97.2	[48]

Table 2. The summary of parameters on magnesium ammonium phosphate crystallization.

with the NH_4^+ and PO_4^{3-}, so more magnesium is needed. Marit et al. [36] indicated that there would be many other kinds of magnesium phosphate precipitates except for struvite at different value of pH, such as $Mg_3(PO_4)_2$, $MgHPO_4$, and $Mg(H_2PO_4)_2$. So it does not mean that more magnesium means more struvite. Excessive amounts of magnesium would increase the pH value of the solution as well as the degree of saturation of the magnesium salts, resulting in the formation of other kinds of magnesium phosphate precipitates mentioned.

4.2. The wastewater pH value

The value of pH of livestock wastewater is an important parameter to the formation of struvite. It affects the quality and the purity of struvite at the same time. As shown in **Table 2**, the best value of pH to form struvite is between 8 and 10, while 8.0–9.0 is the best for livestock wastewater. However, Hao et al. [37] indicated that struvite could get the highest purity at pH = 7.0, and the purity seemed to have fallen with the increasing of the pH value of wastewater. When the value of pH is higher than 10, the formed precipitate mainly consists of $Mg_3(PO_4)_2$ ($K_{sp} = 9.8 \times 10^{-25}$). Song et al. [38] also found that the precipitate of $Mg(OH)_2$ would form at the pH of 11. It does not mean that it is better to form struvite at a lower value of pH. However, considering the phosphorous recovery, as long as the productions of phosphorus salts are harmless and nontoxic, the aim of recovering phosphorus from livestock wastewater is reached. Anyway, the pH value of livestock wastewater is generally between 7.5 and 8.5, which is more convenient to recover phosphorus without the need for adjusting the pH value. It is helpful to simplify the technology and reduce the cost of livestock wastewater treatment.

4.3. The coexisting ions

There are many kinds of coexisting ions interfering with the crystallization of struvite in livestock wastewater, such as calcium ion (Ca^{2+}), carbonate ion (CO_3^{2-}), suspended solids (SS), and heavy metal ions (HMI). Moerman et al. [39] found that Ca^{2+} could enhance the phosphorus removal with forming the precipitate of $Ca_3(PO_4)_2$. However, it reduces the size of struvite. Meanwhile, lots of $Ca_3(PO_4)_2$ powder flows out with effluent easily, declining the effluent water quality. Le Corre et al. [40] also declared that Ca^{2+} would compete with Mg^{2+} and form the precipitates of $Ca_3(PO_4)_2$ ($K_{sp} = 2.1 \times 10^{-33}$) and $CaHPO_4$ ($K_{sp} = 1.8 \times 10^{-7}$) at the pH value of 9. By performing batch experiments, Zhang et al. [41] found that the degree of the supersaturation would decrease with the increase of the concentration of CO_3^{2-}. The CO_3^{2-}, easily combining with Mg^{2+}, increases the ion saturation in the solution and decreases the concentration of Mg^{2+} forming struvite. Suzuki et al. [42] showed that negatively charged SS adsorbed NH_4^+ and Mg^{2+} easily in the alkaline environment, which retarded the struvite crystalline rate. And Muryanto et al. [43] studied on the influence of copper ions (Cu^{2+}) and zinc ions (Zn^{2+}) in struvite crystallization and showed that the existence of Cu^{2+} and Zn^{2+} would delay the nucleation rate and the growth rate of struvite. Although they had little impact on the crystal shape, the crystal would have some cracks on the surface.

All in all, in the process of recovering phosphorus from livestock wastewater, some pretreatments are necessary to implement for removing these coexisting ions before forming the struvite. Laridi et al. [44] tried to reduce the negative impacts of organics and SS by adding ferric chloride and flocculants into the livestock wastewater, and it worked with a higher phosphorus removal rate at the same time. Suzuki et al. [42] tried to separate the struvite from suspended solids containing heavy metals by the differences of their settlement characteristics. It improved the purity of struvite and reduced the negative impact of SS and heavy metal ions.

4.4. Seed crystal

Seed crystals have positive influence in the struvite growth. Adding seed crystals can reduce the saturation of struvite crystallization in need, shorten the nucleation time, and increase the rate of crystal growth. What is more, struvite crystallizes on the surface of seed crystals, which enhances the separation of crystals and water, prevents the tiny crystals from flowing out with the effluent, and improves the phosphorus removal efficiency. Ariyanto et al. [45] showed that the smaller the added crystal nucleus is, the faster is the rate of crystal growth. Kim et al. [18] emphasized that the excessive amount of seed crystals added could not improve the phosphorus removal efficiency, and the pH value of wastewater also influenced the efficiency at the same dosage of seed crystals. The phosphorus removal efficiency is more significant at the pH value of 9. So only adding proper amount of seed crystals with a proper average size can the phosphorus removal efficiency be higher.

5. Crystallizer of struvite

It is important to realize the phosphorus recycling in crystallizer, as the struvite crystallization equipment. The pros and cons of its design decide the shape and size of the struvite and the phosphorus removal efficiency from livestock wastewater. A series of struvite crystallizers had been developed and put into production successfully abroad previously, which had obtained environmental and economic benefits simultaneously. According to the mode of agitation, these crystallizers can be divided into air agitation type, water agitation type, and mechanical agitation type.

The air agitation type crystallizer is a kind of crystallizer that is studied widely. The special aerate system can not only mix the solution more efficiently, improving the collision chance of crystal forming ions, but also vent the gas carbon dioxide and insoluble ammonium from the solution, increasing the pH value of the solution, the ammonium removal efficiency, and the effluent water quality at the same time. The crystallizer used by the British slough sewage treatment plant (**Figure 3a**) reached the soluble phosphorus removal rate of 94% and the total phosphorus removal rate of 87.5% with drumming into air under the inner reaction zone [46]. Le Corre et al. designed two concentric meshes made of stainless steel as a substrate to grow struvite in the crystallizer, which can trap and then accumulate the struvite in the reactor as an adhesive (**Figure 3b**). With the help of crystallizers, the phosphorus removal rate can increase from 81 to 86%. However, because of the limitations of volume and growth time, the struvite crystal cannot grow large enough in the air agitation type crystallizer, which causes the loss of phosphorus recovery. Moreover, some kinds of air agitation type crystallizer have the problem of replacing padding or membranes frequently. What is more, the congestion problem becomes serious once the crystallizer broke down for some reasons, and it is hard to restart.

(a)

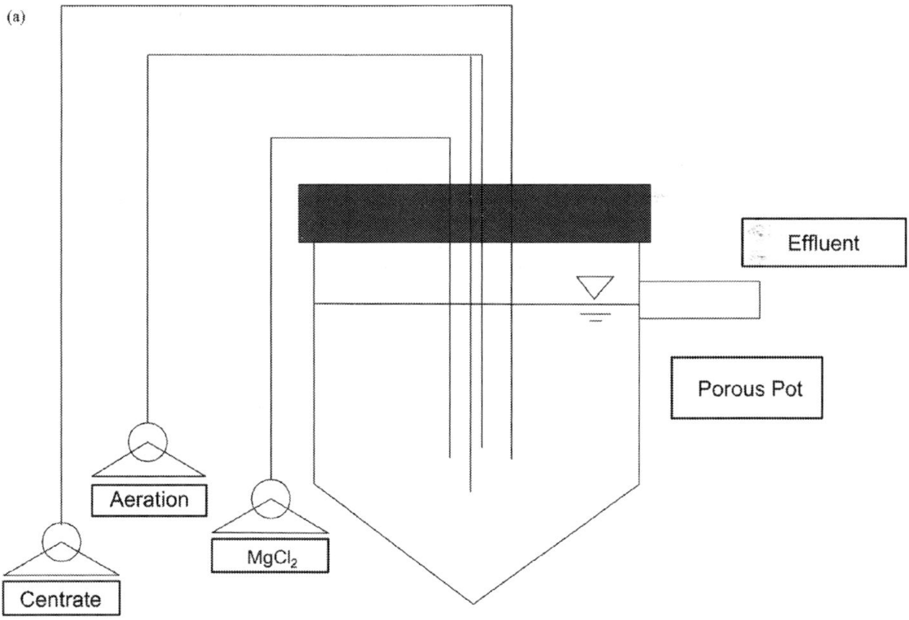

(b)

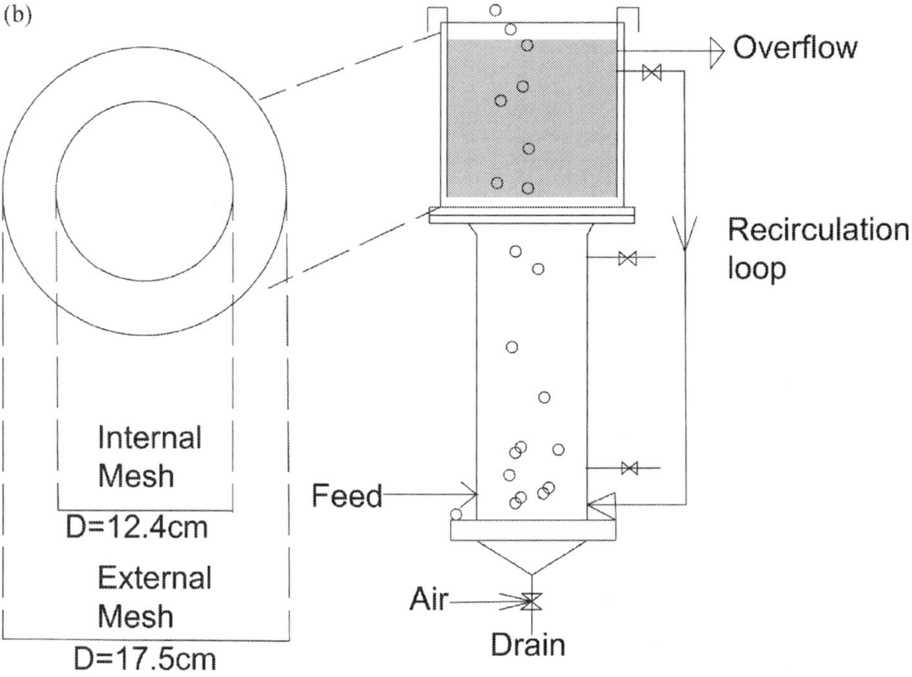
Overflow
Recirculation loop
Internal Mesh
D=12.4cm
External Mesh
D=17.5cm
Feed
Air
Drain

(c)

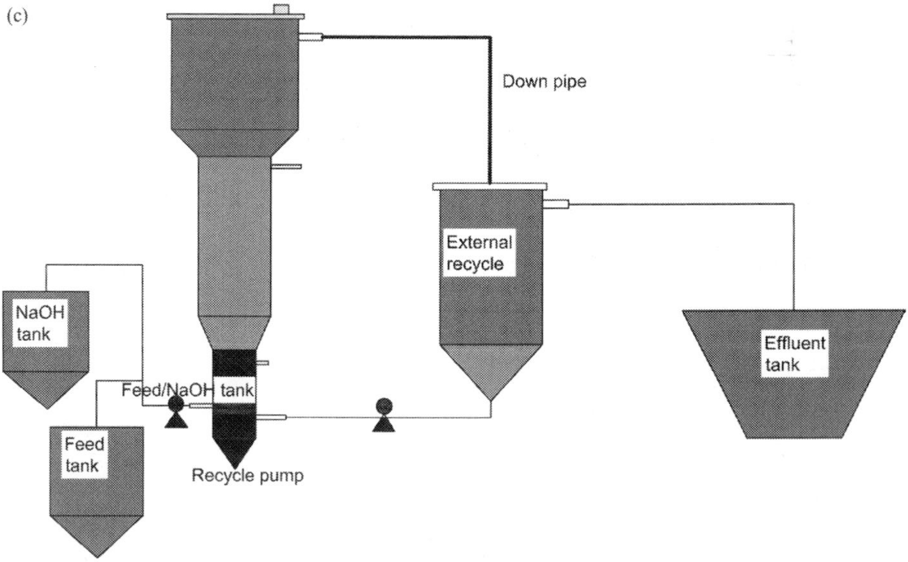
Down pipe
External recycle
NaOH tank
Feed/NaOH tank
Effluent tank
Feed tank
Recycle pump

http://dx.doi.org/10.5772/65692

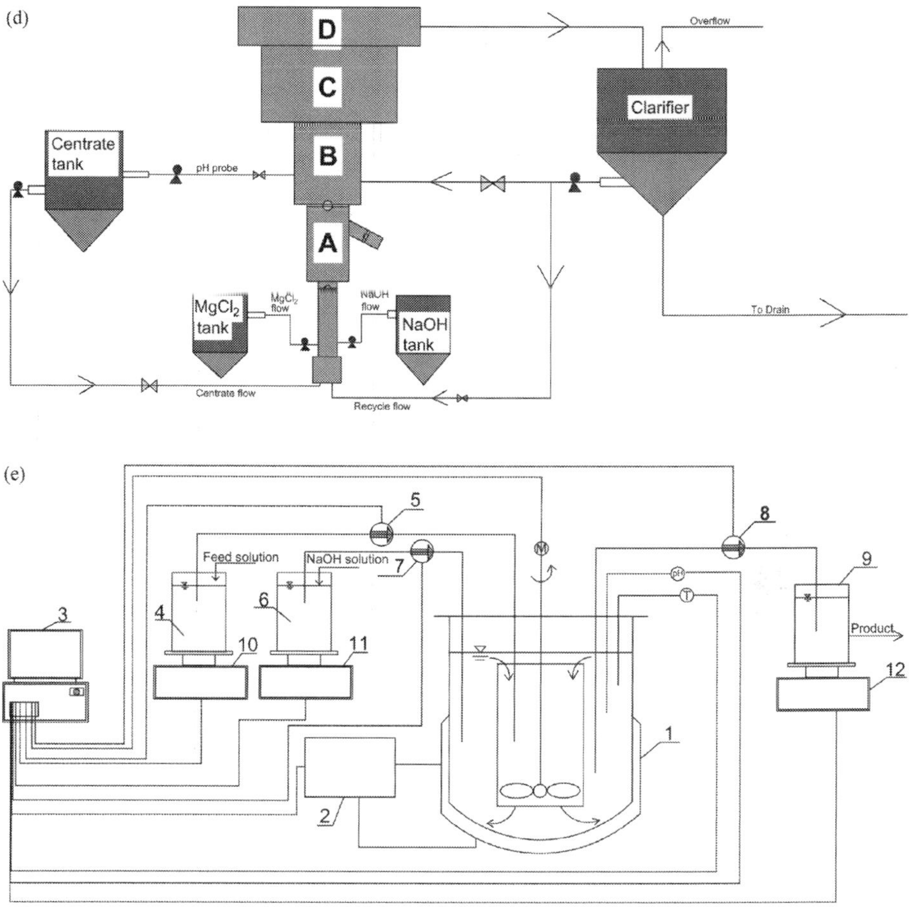

Figure 3. Different kinds of crystallizers forming struvite. (a) The air agitation crystallizer from the British slough sewage treatment plant. (b) The air agitation crystallizer from Le Corre et al. (c) The water agitation crystallizer from Guadia et al. (d) The water agitation crystallizer from Rahaman et al. (e) The MSMPR type crystallizer with 1—internal circulation of suspension, 2—thermostat, 3—computer, 4—rural wastewater (including aqueous solution of $MgCl_2$), 5—pump, 6—alkalinity agent tank: aqueous solution of NaOH, 7 and 8—pump, 9—storage tank of a product crystal suspension, 10, 11, and 12—electronic balances, M—stirrer speed control, T—temperature control, and pH—acid/alkaline reaction control.

The water agitation type crystallizer realizes uniform mixing by changing the solution flow direction, speed, or gravity changed the flow rate by increasing the diameters of the equipment from the bottom to the top, inserted cone-shape structures at an angle of 45° between every diameter-changed parts to reduce unwanted crystal loss at each junction, and recycled finer particles with the effluent through the external recycler (**Figure 3c**). This crystallizer can remove 92% phosphate, and the purity of struvite goes up to 99%. Rahaman et al. [47] designed four distinct zones at the same principle, added a settling zone (also called seed hopper) at the top, getting the phosphate removal rate of up to 90% and the size of particles up to 3.5 mm. Seed crystals are added into the crystallizer from the seed hopper and allowed the finer crystals to continue to grow up in the upper supersaturated solution.

The mechanical agitation type crystallizer, mixing solution by impellers, is simple in design and easy to operate. However, it causes greater energy consumption and uneven size of crystals distribution. Recently, a new crystallizer called mixed suspension, mixed product removal crystallizer (MSMPR for short) can solve the problem of uneven crystal size distribution (**Figure 3e**). With the mechanical agitation centered and the water agitation assisted, MSMPR can uniform the suspension density and particle size of the crystals, and remove productions evenly by controlling the speed and time of mixing. Hutnik et al. [48] and Kozik et al. [49] both got the phosphate removal rate up to 99% from industrial wastewater and wastewater with low concentration of phosphorus. And both of them confirmed that MSMPR could increase the size of crystals, improve the crystallization rate of the struvite, and enhance the phosphorus recovery rate.

6. Application of struvite as a fertilizer

It is reckoned that 100 m^3 wastewater could form 1 kg of struvite. If all the wastewater in the world is treated by struvite crystallization, 63,000 tons of P_2O_5 could be recovered, equaling to 16% of the phosphate rock production of the world [50]. And 171 g struvite can be recovered from livestock wastewater per square meter at most and the purity as high as 95% without washing. Therefore, recovery of struvite returning to the farmland is a developmental trend of struvite crystallization technology. Struvite, as a slightly soluble crystal, for containing the equal molar concentrations of magnesium (Mg), ammonium, and phosphate, has been successfully used on herbages [51], vegetables [14, 52], and grain crops [53] as a fertilizer, especially on the magnesium-fond crops, like sugar beet [54]. Especially, the presence of Mg in struvite makes it more attractive as an alternative to contemporary fertilizers for a few crops, which require magnesium [55]. Ryu et al. [52] found that the struvite source provided the essential crop nutrients of N, P, K, Ca, and Mg for Chinese cabbage as much as other commercial fertilizers. Moreover, it has a lasting positive function to roots and does not burn the seeding or roots due to its slow release characteristics. Besides, compared with other highly soluble fertilizers, struvite is more suitable for use in the vast areas of forest. Since the area of forest is too large to fertilize frequently, the use of struvite can decrease the frequency of fertilization and reduce the loss of nutrients [54]. However, as livestock wastewater is full of impurities, especially the heavy metal ions, the struvite recovered from livestock wastewater still contain more or less heavy metal ions. From livestock-based struvite, toxic substances may diffuse into the aquatic environment or accumulate in soils and have an adverse effect on the human health and environment [56, 57]. Although currently no specific threshold values are available for micropollutants in fertilizers, the introduction of potential hazardous substances into the environment should be avoided. The accumulation of heavy metal ions will be a serious concern for sustainability [58]. Ryu et al. [52] made a security evaluation for struvite as a fertilizer used in the soil. They affirmed the fertilizer efficiency of struvite and emphasized the negative effect of higher concentrations of copper and cadmium in struvite at the same time. Because copper and cadmium were tested in the cabbage fertilized by the struvite used as fertilizer, the struvite, especially recovered from livestock wastewater, needs to tested for the amount of toxic or harmful substances, followed by the security evaluation as a fertilizer.

7. Summary and outlook

Struvite crystallization represents a promising tool for recovering phosphorus from livestock wastewater. Based on this study, the conclusions are as follows.

Struvite is a white crystal which is formed in neural or mild alkali conditions. Nucleation and crystal growth are two steps for struvite crystallization from generation to growth. The molar ratio of magnesium and phosphate, and solution pH are the key factors to control. The coexisting substances, such as calcium, carbonate, suspended solids, and heavy metals, interfere the crystallization of struvite. Seed crystals have positive influence on struvite growth. Adding seed crystals can reduce the saturation of struvite crystallization, shorten the nucleation time, and improve the rate of crystal growth.

Crystallizer, its design decides the shape and size of struvite, is important to realize the phosphorus recycling. According to the agitation mode, it can be divided into air agitation, water agitation, and mechanical agitation.

The recovered struvite can be used on herbages, vegetables, and grain crops as a fertilizer, especially on the magnesium-fond crops, like sugar beet.

However, there are still some problems. Livestock wastewater belongs to the organic wastewater with high concentrations of ammonium, phosphorus, organics, and suspend solids. And the existing forms are complex, such as simple monoester phosphorus, phytate-like phosphorus, and polynucleotide-like phosphoric. So it is necessary to use some physical or chemical measures to transform different kinds of phosphorus to phosphate, as many as possible, before removing phosphorus from livestock wastewater.

The design for struvite crystallizer is still a key to struvite crystallization technology. Although enhancing the phosphorus removal rate has got a big breakthrough by the current crystallizer, the crystals are still too small to recover and block the crystallizer easily. It is the influence of negative zeta potential on the surface of crystals that makes further aggregating hard for small crystals. So finding a way to change the zeta potential on the surface of the crystals, enhancing the aggregation capability, and increasing the size of the crystals is required.

Struvite, a fertilizer with high concentrations of nutrients, might be difficult in application for the influence of other toxic or harmful impurities. Therefore, to reduce the environmental risk at source, it is necessary that estimating the potential effects of struvite on the ecosystem before use.

Acknowledgements

The work was supported by a grant from the National Natural Science Foundation of China (No. 31401944), the National key Research and Development Program of China (No. 2016YFD0501404), the Beijing Municipal Natural Science Foundation (No. 6144026), China Agricultural University Education Foundation "Da Bei Nong Group Education Foundation" (No. 1031-2415005), and the Specialized Research Fund for the Doctoral Program of Higher Education of China (No. 20120008120013).

Author details

Tao Zhang*, Rongfeng Jiang and Yaxin Deng

*Address all correspondence to: taozhang@cau.edu.cn

Key Laboratory of Plant-Soil Interactions of Ministry of Education, College of Resources and Environmental Sciences, China Agricultural University, Beijing, China

References

[1] U.S. Geological Survey. Mineral commodity summaries 2015[OL]. 2015-1. http://minerals.usgs.gov/minerals/pubs/mcs/2015/mcs2015.pdf

[2] U.S. Geological Survey. Mineral commodity summaries 2014[OL]. 2014-2. http://minerals.usgs.gov/minerals/pubs/mcs/2014/mcs2014.pdf

[3] Gale F, Marti D, Hu D. China's volatile pork industry. A report from the Economic Research Service. 2, 2012.

[4] Song Y H, Qiu G L, Yuan P, et al. Nutrients removal and recovery from anaerobically digested swine wastewater by struvite crystallization without chemical additions. Journal of Hazardous Materials, 2011, 190(1): 140–149.

[5] Bennett E, Elser J. A broken biogeochemical cycle. Nature, 2011, 478: 29–31.

[6] The Compilation Committee of the First National Sources of Pollution Survey. The first national sources of pollution survey data 4: the technical report about sources of pollution. Beijing: China Environmental Science Press, 2011: 436–442 (in Chinese).

[7] Zhang T, Ding L, Ren H, et al. Thermodynamic modeling of ferric phosphate precipitation for phosphorus removal and recovery from wastewater. Journal of Hazardous Materials, 2010, 176(1): 444–450.

[8] Rittmann B E, Mayer B, Westerhoff P, et al. Capturing the lost phosphorus. Chemosphere, 2011, 84(6): 846–853.

[9] Zhang T, Li P, Fang C, et al. Phosphate recovery from animal manure wastewater by struvite crystallization and CO_2 degasification reactor. Ecological Chemistry and Engineering S, 2014, 21(1): 89–99.

[10] Bektaş N, Akbulut H, Inan H, et al. Removal of phosphate from aqueous solutions by electro-coagulation. Journal of Hazardous Materials, 2004, 106(2): 101–105.

[11] Li J, Song C, Su Y, et al. A study on influential factors of high-phosphorus wastewater treated by electrocoagulation–ultrasound. Environmental Science and Pollution Research, 2013, 20(8): 5397–5404.

[12] Wang X H, Liu F F, Lu L, et al. Individual and competitive adsorption of Cr (VI) and phosphate onto synthetic Fe–Al hydroxides. Colloids and Surfaces A: Physicochemical and Engineering Aspects, 2013, 423: 42–49.

[13] Lahav O, Telzhensky M, Zewuhn A, et al. Struvite recovery from municipal-wastewater sludge centrifuge supernatant using seawater NF concentrate as a cheap Mg (II) source. Separation and Purification Technology, 2013, 108: 103–110.

[14] El Diwani G, El Rafie S, El Ibiari N N, et al. Recovery of ammonia nitrogen from industrial wastewater treatment as struvite slow releasing fertilizer. Desalination, 2007, 214(1): 200–214.

[15] Kim D, Kim J, Ryu H D, et al. Effect of mixing on spontaneous struvite precipitation from semiconductor wastewater. Bioresource Technology, 2009, 100(1): 74–78.

[16] Kim D, Ryu H D, Kim M S, et al. Enhancing struvite precipitation potential for ammonia nitrogen removal in municipal landfill leachate. Journal of Hazardous Materials, 2007, 146(1): 81–85.

[17] Cusick R D, Ullery M L, Dempsey B A, et al. Electrochemical struvite precipitation from digestate with a fluidized bed cathode microbial electrolysis cell. Water Research, 2014, 54: 297–306.

[18] Uysal A, Yilmazel Y D, Demirer G N. The determination of fertilizer quality of the formed struvite from effluent of a sewage sludge anaerobic digester. Journal of Hazardous Materials, 2010, 181(1): 248–254.

[19] Liu Y H, Kumar S, Kwag J H, et al. Magnesium ammonium phosphate formation, recovery and its application as valuable resources: a review. Journal of Chemical Technology and Biotechnology, 2013, 88(2): 181–189.

[20] Lee J E, Rahman M M, Ra C S. Dose effects of Mg and PO4 sources on the composting of swine manure. Journal of Hazardous Materials, 2009, 169(1): 801–807.

[21] Abbona F, Lundager Madsen H E, Boistelle R. Crystallization of two magnesium phosphates, struvite and newberyite: effect of pH and concentration. Journal of Crystal Growth, 1982, 57(1): 6–14.

[22] Lee S H, Yoo B H, Kim S K, et al. Enhancement of struvite purity by re-dissolution of calcium ions in synthetic wastewaters. Journal of Hazardous Materials, 2013, 261: 29–37.

[23] Zhang T, Ding L, Ren H. Pretreatment of ammonium removal from landfill leachate by chemical precipitation. Journal of Hazardous Materials, 2009, 166(2): 911–915.

[24] Rawn A M, Banta A P, Pomeroy R. Multiple-stage sewage sludge digestion. Transactions of the American Society of Civil Engineers, 1939, 104(1): 93–119.

[25] Snoeyink V L, Jenkins D. Water Chemistry. John Wiley and Sons, New York, 1980, 384.

[26] Ohlinger K N, PE, Young T M, et al. Kinetics effects on preferential struvite accumulation in wastewater. Journal of Environmental Engineering, 1999, 125(8): 730–737.

[27] Bhuiyan M I H, Mavinic D S, Beckie R D. A solubility and thermodynamic study of struvite. Environmental Technology, 2007, 28(9): 1015–1026.

[28] Adnan A, Koch F A, Mavinic D S. Pilot-scale study of phosphorus recovery through struvite crystallization-II: applying in-reactor supersaturation ratio as a process control parameter. Journal of Environmental Engineering and Science, 2003, 2(6): 473–483.

[29] Wang J, Song Y, Yuan P, et al. Modeling the crystallization of magnesium ammonium phosphate for phosphorus recovery. Chemosphere, 2006, 65(7): 1182–1187.

[30] Bouropoulos N C, Koutsoukos P G. Spontaneous precipitation of struvite from aqueous solutions. Journal of Crystal Growth, 2000, 213(3): 381–388.

[31] Bhuiyan M I H, Mavinic D S, Beckie R D. Nucleation and growth kinetics of struvite in afluidized bed reactor. Journal of Crystal Growth, 2008, 310(6): 1187–1194.

[32] Mehta C M, Batstone D J. Nucleation and growth kinetics of struvite crystallization. Water Research, 2013, 47(8): 2890–2900.

[33] Durrant A E, Scrimshaw M D, Stratful I, et al. Review of the feasibility of recovering phosphate from wastewater for use as a raw material by the phosphate industry. Environmental Technology, 1999, 20(7): 749–758.

[34] Abe S. Phosphate removal from dewatering filtrate by MAP process at Seibu treatment plant in Fukuoka City. Sewage Works in Japan, 1995, 43: 59–64.

[35] Nagy Z K, Fujiwara M, Woo X Y, et al. Determination of the kinetic parameters for the crystallization of paracetamol from water using metastable zone width experiments. Industrial & Engineering Chemistry Research, 2008, 47(4): 1245–1252.

[36] Marti N, Bouzas A, Seco A, et al. Struvite precipitation assessment in anaerobic digestion processes. Chemical Engineering Journal, 2008, 141(1): 67–74.

[37] Hao X, Wang C, van Loosdrecht M C M, et al. Looking beyond struvite for P-recovery. Environmental Science & Technology, 2013, 47(10): 4965–4966.

[38] Korchef A, Saidou H, Amor M B. Phosphate recovery through struvite precipitation by CO2 removal: effect of magnesium, phosphate and ammonium concentrations. Journal of Hazardous Materials, 2011, 186(1): 602–613.

[39] Moerman W, Carballa M, Vandekerckhove A, et al. Phosphate removal in agro-industry:pilot-and full-scale operational considerations of struvite crystallization. Water Research, 2009, 43(7): 1887–1892.

[40] Le Corre K S, Valsami-Jones E, Hobbs P, et al. Impact of calcium on struvite crystal size, shape and purity. Journal of Crystal Growth, 2005, 283(3): 514–522.

[41] Zhang T, Ding L, Ren H, et al. Ammonium nitrogen removal from coking wastewater by chemical precipitation recycle technology. Water Research, 2009, 43(20): 5209–5215.

[42] Suzuki K, Tanaka Y, Kuroda K, et al. Removal and recovery of phosphorous from swine wastewater by demonstration crystallization reactor and struvite accumulation device. Bioresource Technology, 2007, 98(8): 1573–1578.

[43] Muryanto S, Bayuseno A P. Influence of Cu^{2+} and Zn^{2+} as additives on crystallization kinetics and morphology of struvite. Powder Technology, 2014, 253: 602–607.

[44] Laridi R, Auclair J C, Benmoussa H. Laboratory and pilot-scale phosphate and ammonium removal by controlled struvite precipitation following coagulation and flocculation of swine wastewater. Environmental Technology, 2005, 26(5): 525–536.

[45] Ariyanto E, Sen T K, Ang H M. The influence of various physico-chemical process parameters on kinetics and growth mechanism of struvite crystallization. Advanced Powder Technology, 2014, 25(2): 682–694.

[46] Jaffer Y, Clark T A, Pearce P, et al. Potential phosphorus recovery by struvite formation. Water Research, 2002, 36(7): 1834–1842.

[47] Rahaman M S, Mavinic D S, Meikleham A, et al. Modeling phosphorus removal and recovery from anaerobic digester supernatant through struvite crystallization in a fluidized bed reactor. Water Research, 2014, 51: 1–10.

[48] Hutnik N, Kozik A, Mazienczuk A, et al. Phosphates (V) recovery from phosphorus mineral fertilizers industry wastewater by continuous struvite reaction crystallization process. Water Research, 2013, 47(11): 3635–3643.

[49] Kozik A, Hutnik N, Piotrowski K, et al. Continuous reaction crystallization of struvitefrom diluted aqueous solution of phosphate (V) ions in the presence of magnesium ionsexcess. Chemical Engineering Research and Design, 2014, 92(3): 481–490.

[50] Shu L, Schneider P, Jegatheesan V, et al. An economic evaluation of phosphorus recovery as struvite from digester supernatant. Bioresource Technology, 2006,97: 2211–2216.

[51] Yetilmezsoy K, Sapci-Zengin Z. Recovery of ammonium nitrogen from the effluent of UASB treating poultry manure wastewater by MAP precipitation as a slow release fertilizer. Journal of Hazardous Materials, 2009, 166(1): 260–269.

[52] Ryu H D, Lim C S, Kang M K, et al. Evaluation of struvite obtained from semiconductor wastewater as a fertilizer in cultivating Chinese cabbage. Journal of Hazardous Materials, 2012, 221: 248–255.

[53] Liu Y H, Rahman M M, Kwag J H, et al. Eco-friendly production of maize using struvite recovered from swine wastewater as a sustainable fertilizer source. Asian-Australasian Journal of Animal Sciences, 2011, 24: 1699–1705.

[54] De-Bashan L E, Bashan Y. Recent advances in removing phosphorus from wastewater and its future use as fertilizer (1997–2003). Water Research, 2004, 38(19): 4222–4246.

[55] Gaterell M R, Gay R, Wilson R, Gochin R J, Lester J N, An economic and environmental evaluation of the opportunities for substituting phosphorus recovered from wastewater treatment works in existing UK fertilizer markets. Environmental Technology, 2000, 21: 1067–1084.

[56] Sanderson H, Johnson D J, Wilson C J, Brain R A, Solomon K R. Probabilistic hazard assessment of environmentally occurring pharmaceuticals toxicity to fish, daphnids and algae by ECOSAR screening. Toxicology Letters, 2003, 144; 383–395.

[57] Halling-Sorensen B, Nors Nielsen S, Lanzky P F, Ingerslev F, Holten Lutzhoft H C, Jorgensen S E. Occurrence, fate and effects of pharmaceutical substances in the environment—a review. Chemosphere, 1998, 36(2); 357–393.

[58] Ronteltap M, Maurer M, Gujer W. The behaviour of pharmaceuticals and heavy metals during struvite precipitation in urine. Water Research, 2007, 41(9): 1859–1868.

Chapter 8

Slaughterhouse Wastewater: Treatment, Management and Resource Recovery

Ciro Bustillo-Lecompte and Mehrab Mehrvar

Additional information is available at the end of the chapter

http://dx.doi.org/10.5772/65499

Abstract

The meat processing industry is one of the largest consumers of total freshwater used in the agricultural and livestock industry worldwide. Meat processing plants (MPPs) produce large amounts of slaughterhouse wastewater (SWW) because of the slaughtering process and cleaning of facilities. SWWs need significant treatment for a sustainable and safe discharge to the environment due to the high content of organics and nutrients. Therefore, the treatment and final disposal of SWW are a public health necessity. In this chapter, the regulatory frameworks relevant to the SWW management, environmental impacts, health effects, and treatment methods are discussed. Although physical, chemical, and biological treatment can be used for SWW degradation, each treatment process has different advantages and drawbacks depending on the SWW characteristics, best available technology, jurisdictions, and regulations. SWWs are typically assessed using bulk parameters because of the various pollutant loads derived from the type and the number of animals slaughtered that fluctuate amid the meat industry. Thus, an on-site treatment using combined processes would be the best option to treat and disinfect the slaughterhouse effluents to be safely discharged into receiving waters.

Keywords: Anaerobic digestion, Activated sludge, Advanced oxidation processes, Combined processes, Slaughterhouse wastewater

1. Introduction

The meat processing industry consumes 29% of the total freshwater used by the agricultural sector worldwide [1, 2]. Moreover, the global production of beef, pork, and poultry meat has been doubled in the past decade and is projected to grow steadily until 2050. Thus, the number of slaughterhouse facilities is increasing, which results in an expected higher volume of slaughterhouse wastewater (SWW) to be treated [3]. SWWs are classified as one of the most detrimental industrial wastewaters to the environment by the United States Environmental Protection Agency (US EPA) because the inadequate disposal of SWW is one of the reasons for river deoxygenation and groundwater pollution [4]. Thus, SWWs require significant treatment for a safe and sustainable release to the environment, and the treatment and disposal of wastewater from slaughterhouses are an economic and public health necessity [5, 6].

The organic matter concentration in meat processing plant (MPP) effluents is usually high, and the residues are moderately solubilized, leading to a polluting effect due to the high levels of organics and pathogens present in SWW along with detergents used for cleaning purposes. SWWs are typically assessed using bulk parameters because of the various pollutant loads derived from the type and the number of animals slaughtered that fluctuate amid the meat industry [7].

Anaerobic treatment is the preferred biological treatment because of its effectiveness in treating high-strength wastewater such as SWW with less complex equipment requirements [8]. Although anaerobic treatment is efficient, anaerobically treated effluents require posttreatment to comply with required discharge limits where the complete stabilization of the organic matter is not possible by anaerobic treatment alone. Anaerobically treated effluents contain solubilized organic matters, which are more suited for treatment using aerobic processes. Therefore, aerobic treatment systems are more frequently used in wastewater treatment systems since they operate at higher rates than conventional anaerobic treatment methods. Taking into account that oxygen requirements and treatment time are directly proportional to an increase in wastewater strength, aerobic treatment is frequently applied as posttreatment of anaerobic effluents as well as for nutrient removal [9].

Nevertheless, biological processes alone do not produce effluents that comply with current effluent discharge limits when treating high-organic-strength wastewaters. The use of combined anaerobic and aerobic processes is beneficial for its potential resource recovery and high treatment efficiency [10].

On the other hand, some slaughterhouse effluents contain toxic, bioresistant, recalcitrant, and nonbiodegradable substances. Thus, advanced oxidation processes (AOPs) could be used to improve the biodegradability of SWW and inactivate pathogenic microorganisms and viruses, left after biological treatment of the wastewater. Consequently, AOPs are an attractive alternative and a complementary treatment method to biological processes for the treatment of slaughterhouse effluents, especially as a posttreatment method [5–7]. Adopting combined biological treatment and AOPs for the treatment of slaughterhouse effluents is considered operationally and economically advantageous. Combined processes incorporate advantages of diverse technologies to achieve high-quality effluents from industrial and high-strength wastewaters for water reuse and resource recovery purposes [9, 10].

In this chapter, the regulatory frameworks relevant to the SWW management, environmental impacts, and health effects are discussed along with common practices for SWW treatment. Significant progress in the combination of biological treatment and AOPs is emphasized. A

http://dx.doi.org/10.5772/65499

case study for the treatment of an actual SWW by integrated anaerobic baffled reactor (ABR)-aerobic activated sludge (AS)-UV/H_2O_2 processes is presented. The overall treatment efficiency of organics and nutrients, the potential energy recovery from CH_4 production, and the H_2O_2 residual are discussed. A cost-effectiveness analysis is used to minimize the treatment time as well as the overall incurred treatment costs required for the efficient treatment of slaughterhouse effluents. Finally, the chapter ends with a discussion on the need for an adequate SWW management, resource recovery, and the improvement actions required.

2. Characteristics of slaughterhouse wastewater

Meat processing effluents are considered harmful worldwide due to the SWW complex composition of fats, proteins, fibers, high organic content, pathogens, and pharmaceuticals for veterinary purposes. Slaughterhouse effluents are typically evaluated using bulk parameters because of the broad range of SWW and pollutant loads. SWW contains large amounts of biochemical oxygen demand (BOD), chemical oxygen demand (COD), total organic carbon (TOC), total nitrogen (TN), total phosphorus (TP), and total suspended solids (TSS) [7]. Typical characteristics of an actual SWW are summarized in **Table 1**.

Parameter	Range	Mean
BOD (mg/L)	150–8500	3000
COD (mg/L)	500–16,000	5000
TOC (mg/L)	50–1750	850
TN (mg/L)	50–850	450
TP (mg/L)	25–200	50
TSS (mg/L)	0.1–10,000	3000
K (mg/L)	0.01–100	50
Color (mg/L Pt scale)	175–400	300
Turbidity	200–300	275
pH	4.9–8.1	6.5

Table 1. Typical characteristics of the slaughterhouse wastewater.

As a result, due to the diverse characteristics of the SWW, it is appropriate to classify and minimize wastewater production at its source. Meat processing effluents are becoming one of the major agribusiness concerns due to the vast amount of water used during slaughtering, processing, and cleaning of the slaughtering facilities.

3. Regulations for slaughterhouse wastewater management

Regulations are necessary to mitigate the environmental impact of slaughterhouses, and the treatment methods are used as the main regulatory requirement [11]. The compliance with current environmental legislation and the state-of-the-art technologies may also provide some economic relief via resource recovery from biogas generation using high-rate anaerobic treatment.

Table 2 describes current regulations and discharge limits for organics and nutrients in SWW for an adequate release to the environment in different jurisdictions worldwide, including the World Bank Group [12], the Council of the European Communities [13], the US EPA [14], the Environment Canada [15, 16], the Colombian Ministry of Environment and Sustainable Development Colombia [17], the People's Republic of China Ministry of Environmental Protection [18], the Indian Central Pollution Control Board [19], and the Australian and New Zealand Environment and Conservation Council [20, 21].

Parameter	World Bank	EU	USA	Canada	Colombia	China	India	Australia
BOD (mg/L)	30	25	16–26	5–30	50	20–100	30–100	5–20
COD (mg/L)	125	125	n.a.	n.a.	150	100–300	250	40
TN (mg/L)	10	10–15	4–8	1.25	10	15–20	10–50	10–20
TOC (mg/L)	n.a.	n.a.	n.a.	n.a.	n.a.	20–60	n.a.	10
TP (mg/L)	2	1–2	n.a.	1.00	n.a.	0.1–1.0	5	2
TSS (mg/L)	50	35–60	20–30	5–30	50	20–30	100	5–20
pH	6–9	n.a.	6–9	6–9	6–9	6–9	5.5–9.0	5–9
Temperature (°C change)	n.a.	n.a.	n.a.	<1 °C	n.a.	n.a.	<5 °C	<2 °C

Table 2. Comparison of standard limits for slaughterhouse wastewater discharge in different jurisdictions worldwide.

Although it can be seen that Canadian standards are stricter than other international regulations such as those in the European Union (EU), Australia and New Zealand, or the USA, Canada does not have a specific regulation for the meat processing industry. Moreover, Australia and New Zealand and the USA have been incorporating an integrated approach to the regulation of the MPPs, where industry and regulatory sectors are working together to achieve a common goal of reducing the threats caused by the hazardous and high-strength wastewaters produced in slaughterhouses. Finally, emerging economies such as India, China, and Colombia have less strict standards, but their legislation is focused on specific industries to attain certain levels of treatment depending on the wastewater strength. Therefore, the selection of a specific treatment method depends on the characteristics of the SWW being treated, the best available technology economically achievable (BAT), and the compliance with regulations in different political jurisdictions.

4. Environmental impact and health effects of slaughterhouse wastewater

The commercialization of animal products for consumption leads to the production of a large volume of SWW. Although the environment can handle a certain amount of pollutants through natural degradation processes, as the SWW concentration increases, these mechanisms come to be overburdened, where contamination problems commence [22].

The discharge of raw SWW to water bodies affects the quality of water particularly by causing a reduction of dissolved oxygen (DO), which may lead to the death of aquatic life [23]. Moreover, macronutrients, such as nitrogen and phosphorus, may cause eutrophication events. The discharge of these nutrients triggers an excessive algae growth and subsequent decay. Thus, the mineralization of the algae may lead to the deterioration of aquatic life due to depletion of DO levels. Finally, SWW may contain compounds, such as chromium and un-ionized ammonia, which are directly toxic to aquatic life [24].

Another source of contamination of the meat processing industry is the addition of surfactants as a result of the cleaning process. Surfactants, major components in detergents, may enter the aquatic environment due to an inadequate SWW treatment, causing short-term and long-term changes in the ecosystem that affect humans, fish, and vegetation [25].

The environmental impact of SWW is not only characterized by pollution via surfactants, nitrate, and chloric anions but also pathogens, which persist in the soil and reproduce continuously. Pathogens from SWW can also be transmitted to humans who are exposed to the water body, making those areas nonsuitable for drinking, swimming, or irrigation purposes [5, 26].

The general public health effects of the meat processing industry are related to the direct interaction of human communities with the slaughterhouse activities and indirect interactions with the environment, which can be previously affected by the inadequate management of the liquid effluents, solid waste, and obnoxious odors [27]. According to Um et al. [28], conventional treatment processes have no major impact on the reduction of antibiotic-resistant *Escherichia coli* strains present in SWW, highlighting the public health risks associated with inadequately treated slaughterhouse effluents concerning the propagation of antibiotic-resistant and pathogenic bacteria into the environment.

The unsanitary conditions in some slaughterhouses allow the proliferation of pathogens to the final meat product to be consumed. People from developing countries in Africa, Asia, and South America have experienced serious gastrointestinal diseases, bloody diarrhea, liver malfunctions, and, in some cases, death associated with the presence of viruses, protozoa, helminthic eggs, and bacteria in SWW [5, 27]. Furthermore, the presence of hepatitis A and E viruses has been reported in the sewage of animal origin in Spain. Therefore, SWW must be treated efficiently before discharge into water bodies to avoid environmental pollution and human health effects [29].

5. Treatment methods for slaughterhouse wastewater

The freshwater consumption substantially varies in the meat processing sector, and a typical MPP generates a large amount of wastewater from the slaughtering process and cleaning of the facilities. Therefore, the water reuse and the recovery of valuable by-products from the meat processing effluents are the main focus in the agribusiness toward a cleaner production focused on high-quality effluents, biogas production and exploitation, and recovery of nutrients and fertilizers [7].

Treatment methods for SWW are comparable to those used in municipal wastewater treatment and include primary, secondary, and tertiary treatment. However, this does not eliminate the need for primary treatment. There are numerous SWW treatment methods after preliminary treatment, which can be divided into four main categories: physicochemical treatment, biological treatment, AOPs, and combined processes [2, 7]. Each method has advantages and disadvantages, which are discussed below.

5.1. Preliminary treatment

The purpose of the preliminary treatment is to separate solids and large particles from the liquid portion in SWW and remove up to 30% of the BOD. The most common unit operations for preliminary treatment of SWW include screeners, sieves, and strainers. Thus, large solids with a 10–30 mm diameter are retained while the SWW passes through. Other preliminary treatment methods include homogenization and equalization and flotation, among other systems such as catch basins and settlers [30].

5.2. Physicochemical treatment

After preliminary treatment, the effluent should be further treated using primary and secondary treatment. One of the most practical methods of primary treatment for SWW is dissolved air floatation (DAF) for the reduction of fat, oil, grease, TSS and BOD [31]. The most commonly used physicochemical treatment methods are presented below.

5.2.1. Coagulation-flocculation and sedimentation

In the coagulation process, colloidal particles in the SWW are grouped into larger particles, called flocs. The colloidal particles in SWW are nearly negatively charged which make them stable and resistant to aggregation. For this reason, coagulants with positively charged ions are added to destabilize the colloidal particles to form flocs and facilitate the sedimentation process. Various coagulant types can be found in the market, and the most widely used are inorganic metal based-coagulants such as aluminum sulfate, aluminum chlorohydrate, ferric chloride, ferric sulfate, and poly-aluminum chloride with removal efficiencies of up to 80% for BOD, COD, and TSS [32].

5.2.2. Dissolved air flotation

The DAF technology refers to the method of liquid-solid separation by air introduction. The fat and grease along with light solids are moved to the surface creating a sludge blanket. Thus, it can be continuously removed via scum scraping. Furthermore, flocculants and blood coagulants can be added to enhance the effectiveness of the DAF treatment for COD and BOD removals of up to 75%. Nevertheless, common DAF disadvantages include occasional malfunctioning, poor TSS elimination, and moderate nutrient removal [33].

5.2.3. Electrocoagulation

The electrocoagulation (EC) process has been employed as a cost-effective technology for the removal of organics, heavy metals, and pathogens from slaughterhouse effluents by inducing an electric current without chemical addition. The EC process generates M^{3+} ions, mainly Fe^{3+} and Al^{3+}, using different electrode materials. Other electrode types including Pt, SnO_2, and TiO_2 can interact with H^+ or OH^- ions in acidic or alkaline conditions, respectively. Thus, removal efficiencies of up to 80, 81, 84, 85, and 96% can be achieved for BOD, TSS, TN, COD, and color, respectively [34, 35].

5.2.4. Membrane processes

Membrane processes are becoming an alternative treatment method for meat processing effluents. Different membrane processes, including microfiltration (MF), ultrafiltration (UF), nanofiltration (NF), and reverse osmosis (RO), have been used for SWW treatment to remove particulates, colloids, macromolecules, organic matter, and pathogens with overall efficiencies of up to 90%. However, membrane processes are required to be coupled with conventional processes for nutrient removal in SWW. Another drawback of membrane processes refers to the membrane fouling when treating high-strength wastewater because of the formation of biofouling layers on the membranes, restricting the permeation rate [36].

5.3. Biological treatment

Primary treatment and physicochemical processes typically do not treat SWW completely, to a degree of satisfaction set by regulations. Thus, secondary treatment is used for the removal of the remaining soluble organic compounds from primary treatment. Biological processes include lagoons with anaerobic, aerobic, or facultative microorganisms, trickling filters, activated sludge (AS) bioreactors, and constructed wetlands (CWs) for organic and nutrient removal efficiencies of up to 90% [7].

5.3.1. Anaerobic treatment

Anaerobic digestion is the preferred method for SWW treatment due to its effectiveness in treating highly concentrated industrial effluents since organic compounds are degraded by anaerobic bacteria in the absence of oxygen into CO_2 and CH_4. Anaerobic systems have the advantage of achieving low sludge production, minimum energy requirements with potential resource recovery, and high COD removal. Typical anaerobic processes for the treatment of meat processing effluents comprise anaerobic baffled reactor (ABR), anaerobic digester (AD), anaerobic filter (AF), anaerobic lagoon (AL), septic tanks (ST), and up-flow anaerobic sludge blanket (UASB) [30].

Nevertheless, anaerobic treatment barely complies with current discharge limits. Complete stabilization of the organic compounds is difficult due to the high organic strength of SWW. Therefore, an additional treatment stage is recommended to remove the organics, nutrients, and pathogens that remain after anaerobic treatment. On the other hand, anaerobic treatment requires a higher space and a higher residence time to achieve high overall treatment efficiency, affecting the economic viability of anaerobic treatment alone. Accordingly, the combination of anaerobic and aerobic processes is necessary to achieve a maximum efficiency for the treatment of SWW [37].

5.3.2. *Aerobic treatment*

Aerobic processes are frequently employed for nutrient removal and further treatment after primary treatment. The required oxygen and treatment time are directly related to the strength of the SWW, which makes it inadequate as primary treatment of SWW but adequate after anaerobic treatment [38].

There are many advantages of using aerobic wastewater treatment processes, including low odor production, fast biological growth rate, and rapid adjustments to the temperature and loading rate changes. Conversely, the operating costs of aerobic systems are higher than those for anaerobic systems due to the maintenance and energy requirements for artificial oxygenation. There are different aerobic unit operations for SWW treatment, such as aerobic AS, rotating biological contactors (RBCs), and sequencing batch reactors (SBRs) [39].

5.3.3. *Constructed wetlands*

Constructed wetlands (CWs) emulate the degradation mechanisms of natural wetlands for water decontamination, integrating biological and physicochemical processes from the interaction of vegetation, soil, microorganisms, and atmosphere for the adsorption, biodegradation, filtration, photooxidation, and sedimentation of organics and nutrients.

The performance of CW systems for the treatment of SWW has been evaluated using both horizontal and vertical subsurface flow CWs. Results have shown a wide range of organic and nutrient removal for different vegetation with encouraging maximum removals of 99, 97, 85, and 78% for BOD, COD, TSS, and TN, respectively [40]. As a result, CWs are simple methods with low operation and maintenance costs and few negative impacts on the environment, which make them an attractive alternative to conventional treatment [41].

5.4. Advanced oxidation processes

AOPs are an interesting complementary treatment option for primary or secondary treatment of SWW, showing excellent overall treatment efficiencies for water reuse. AOPs are diverse and include gamma radiation, ozonation, ultrasound technology (UST), UV/H_2O_2, UV/O_3, and

http://dx.doi.org/10.5772/65499

photocatalysis, among others, for the oxidation and degradation of organic matter. The disinfection is another benefit of AOPs, which can inactivate pathogens without adding additional chemicals in comparison to other disinfection methods, such as chlorination, preventing the formation of hazardous by-products [5]. Another main advantage of the AOPs is the high reaction rates as well as very low treatment time.

Photocatalysis using photo-Fenton–based processes and photooxidation using UV/H_2O_2 are the most commonly used AOPs for SWW treatment. Although these processes are usually expensive if applied alone, removal efficiencies of over 90% can be achieved for SWW secondary effluents in terms of TOC and COD as a posttreatment method. Thus, the combination of biological processes and AOPs is recommended for SWW treatment [42, 43].

5.5. Combined processes

The implementation of combined processes is operationally and economically beneficial for SWW treatment since it couples the advantages of different technologies to treat high-strength industrial wastewaters. The combined ABR-AS-UV/H_2O_2 system is recognized as a cost-effective solution for SWW treatment with removal efficiencies of over 95% for organics and nutrients at optimum operating conditions [6, 9, 10].

An overview of the state-of-the-art technologies for SWW treatment, during the last two years, is presented in **Table 3**. Particular attention is given to organic and nutrient removal, in terms of bulk parameters such as BOD, COD, TOC, TN, and TP. As shown in **Table 3**, SWW treatment efficiencies vary extensively and depend on the SWW characteristics, the treatment time, and the influent concentration, as well as the type of treatment and BAT to comply with current regulations [7].

Method	HRT (h)	BOD_{in} (mg/L)	COD_{in} (mg/L)	TOC_{in} (mg/L)	TN_{in} (mg/L)	BOD_{rem} (%)	COD_{rem} (%)	TOC_{rem} (%)	TN_{rem} (%)	Ref.
EC	1	1123	2171	–	148	–	85	–	–	[44]
ABR-AS-AOP	41	4635	–	1200	841	100	–	100	82	[10]
SBR	96	–	6580	–	3321	–	81	–	95	[45]
AF	46	–	15800	–	–	–	60	–	–	[46]
AL	48	5088	9216	–	343	73	59	–	–	[47]
SBR	12	–	356	–	175	–	–	–	91	[48]
EC	1	1950	3337	–	–	–	78	–	–	[49]
SBR	161	4240	6057	1436	576	–	98	–	98	[50]
AF	24	–	88	–	–	–	80	–	90	[51]
AD	2640	–	18600	–	5200	–	–	–	66	[52]
MF	48	–	480	183	115	–	91	45	45	[53]
UST-AF-UF	144	–	3000	–	–	–	96	–	–	[54]
CW	28	–	468	–	61	–	60	–	46	[55]
AF-UF	48	–	1778	–	374	–	95	–	78	[8]

Method	HRT (h)	BOD_{in} (mg/L)	COD_{in} (mg/L)	TOC_{in} (mg/L)	TN_{in} (mg/L)	BOD_{rem} (%)	COD_{rem} (%)	TOC_{rem} (%)	TN_{rem} (%)	Ref.
SBR	12	4240	6057	1436	576	–	93	–	93	[56]
AOP	1	–	406	–	–	–	84	–	–	[57]
SBR	3	–	8604	–	1493	–	80	–	88	[58]
UST-AF-UF	343	–	5200	–	74	–	96	–	–	[59]
AOP	2	340	–	94	55	–	–	81	–	[42]
UF-RO	160	–	7970	–	–	–	80	–	–	[60]
AD-AOP	723	658	1494	513	181	98	98	97	31	[43]
ABR-AS-AOP	55	1635	2000	1200	841	100	99	100	85	[6]
ABR-AS-AOP	10	1831	2043	1691	866	100	99	100	90	[61]

Table 3. Comparison of different slaughterhouse wastewater treatment methods.

6. Case study

Actual SWW samples with average concentrations of 1950, 1400, 850, 750, 200, and 40 mg/L for COD, BOD, TOC, TSS, TN, and TP, respectively, were taken from selected licensed MPPs in Ontario, Canada [62]. Anaerobic and aerobic sludge inocula in concentrations of 40,000 and 3000 mg/L, respectively, were obtained from the Ashbridges Bay Municipal Wastewater Treatment Plant in Toronto, Canada. The inocula were acclimatized in a period of 60 days.

The combined ABR-AS-UV/H_2O_2 system consisted of a 36-L ABR with five equal-volume chambers and individual biogas collection, a 12.65-L aerobic AS reactor with controlled air flow to maintain DO concentrations of 2 mg/L, and a 1.35-L UV-C photoreactor with recycle, output power of 6 W, and uniform light distribution (**Figure 1**).

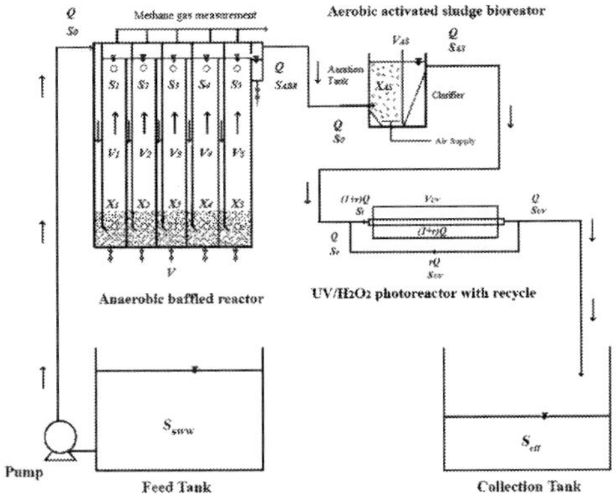

Figure 1. Schematic diagram of the combined ABR-AS-UV/H_2O_2 system for SWW treatment.

Bulk parameters including BOD, COD, TOC, TN, TP, and TSS were analyzed as the main parameters for the treatment of an actual SWW. **Figure 2** shows the obtained maximum removal values of more than 99% for COD (**Figure 2a**), BOD (**Figure 2b**), TOC (**Figure 2c**), TSS (**Figure 2d**), TN (**Figure 2e**), and TP (**Figure 2f**) from the SWW by the combined ABR-AS-UV/H_2O_2 processes, operated in continuous mode.

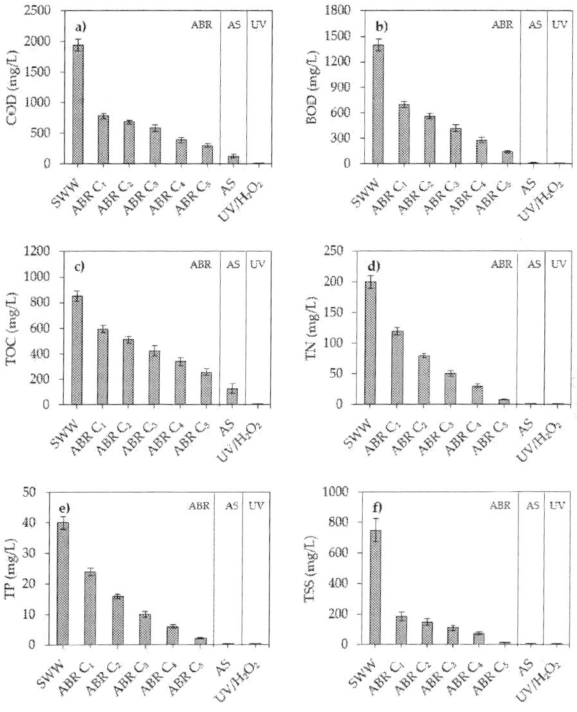

Figure 2. Maximum removal values of (a) COD, (b) BOD, (c) TOC, (d) TN, (e) TP, and (f) TSS from an actual slaughterhouse wastewater using combined ABR-AS-UV/H_2O_2 processes.

The ABR process alone achieved high TSS and TN removals, providing an effluent that complies with most of the current standards worldwide (**Table 2**), with concentrations of 15 and 8 mg/L, respectively. A further treatment with the aerobic AS bioreactor was required to achieve high BOD and TP removals reaching concentrations of 14 and 0.04 mg/L, respectively. However, the COD and TOC concentrations, which are not included broadly as standard parameters, remain with considerable concentrations of 132 and 128 mg/L, respectively. These concentrations are more related to nonbiodegradable organics that can be mineralized using AOPs as a posttreatment process. Thus, after the treatment by the UV/H_2O_2 process, the effluent concentrations for COD and TOC reached values of less than 0.4 and 0.1 mg/L, respectively, which could be used for water reuse.

The effects of the influent TOC concentration, flow rate, and pH on the TOC and TN removals, H_2O_2 residual, and CH_4 yield in the combined ABR-AS-UV/H_2O_2 system were also evaluated. **Figure 3** shows that the influent TOC concentration and the flow rate are inversely proportional to both TOC and TN removals. On the other hand, results indicate that an optimum TOC concentration with no pH adjustment and low flow rate are required to achieve a minimum H_2O_2 residual in the effluent. Finally, results also demonstrate that a high influent TOC concentration is needed to achieve a maximum CH_4 yield with an optimum flow rate and no pH adjustments.

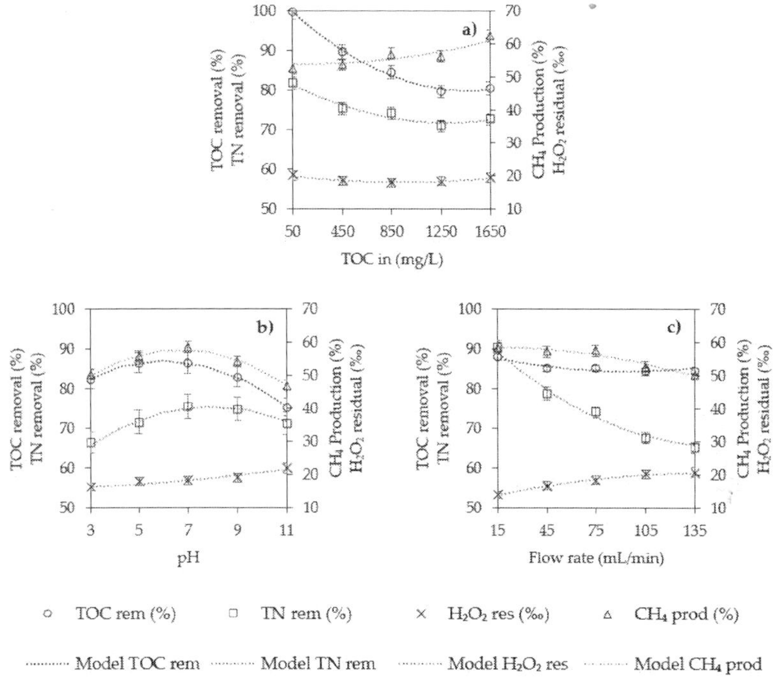

Figure 3. Individual effect of the (a) influent concentration of TOC, (b) flow rate, and (c) pH on the TOC removal, TN removal, H_2O_2 residual, and CH_4 yield. Dashed lines represent model predicted values, whereas marker points represent experimental values. The y-axis is in percentage (%) units, except H_2O_2 residual, which is expressed in per mil (%) for scaling purposes.

As a final point, the treatment costs per volume for the individual ABR, AS, and UV/H_2O_2 processes were compared with those of the combined ABR-AS-UV/H_2O_2 system for the treatment of an actual SWW and plotted versus the TOC removal for each configuration (**Figure 4**). Consequently, a minimum overall treatment cost of 0.12 $/m^3 for a maximum TOC removal of more than 90 % can be achieved at optimum operating conditions in the combined ABR-AS-UV/H_2O_2 system for the treatment of an actual SWW.

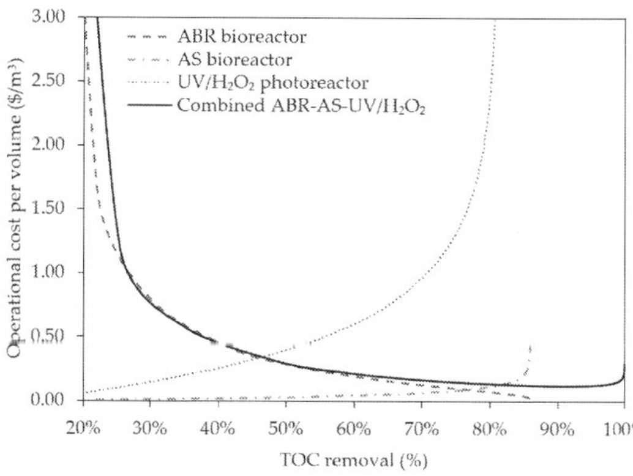

Figure 4. Comparison of operational costs per cubic meter of treated actual slaughterhouse wastewater using individual ABR, AS, and UV/H_2O_2 processes and the combined ABR-AS-UV/H_2O_2 system.

7. Slaughterhouse wastewater management and resource recovery

The meat processing industry needs to incorporate both waste minimization and resource recovery into SWW management strategies considering the portion of the industry's waste and by-products that have a potential of recovery for direct reuse, including nutrients and methane as biofuel. **Figure 5** presents a schematic illustration of the ideal operation of a meat processing plant and supply chain from the animal farming and raw materials to the final product, waste disposal, and recoverable resources [27, 63].

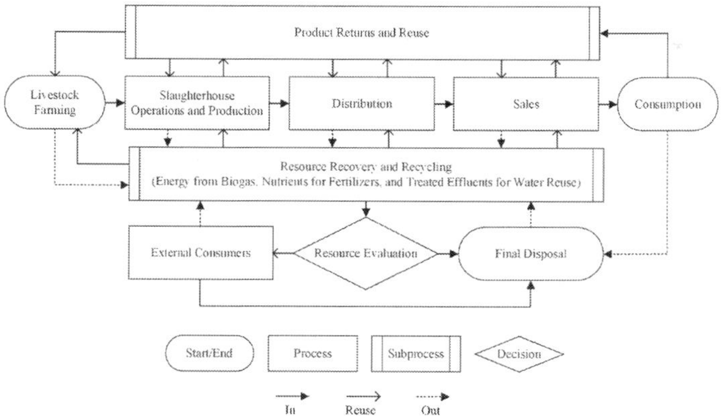

Figure 5. Schematic illustration of the ideal operation and supply chain of a meat processing plant.

A cleaner production should be the focus of meat processing plants due to the increasing interest in environmental initiatives and demands for green practices. Thus, it is appropriate to classify and minimize waste generation at the source, and on-site treatment is the preferred option for water reuse and potential energy recovery. As a result, there are some considerations to be made for the adequate treatment of SWW effluents. **Figure 6** presents a proposed layout of the pretreatment, treatment, and disinfection of slaughterhouse wastes for a typical meat processing plant, as well as the potential resource recovery for water reuse and products recycling [63, 64].

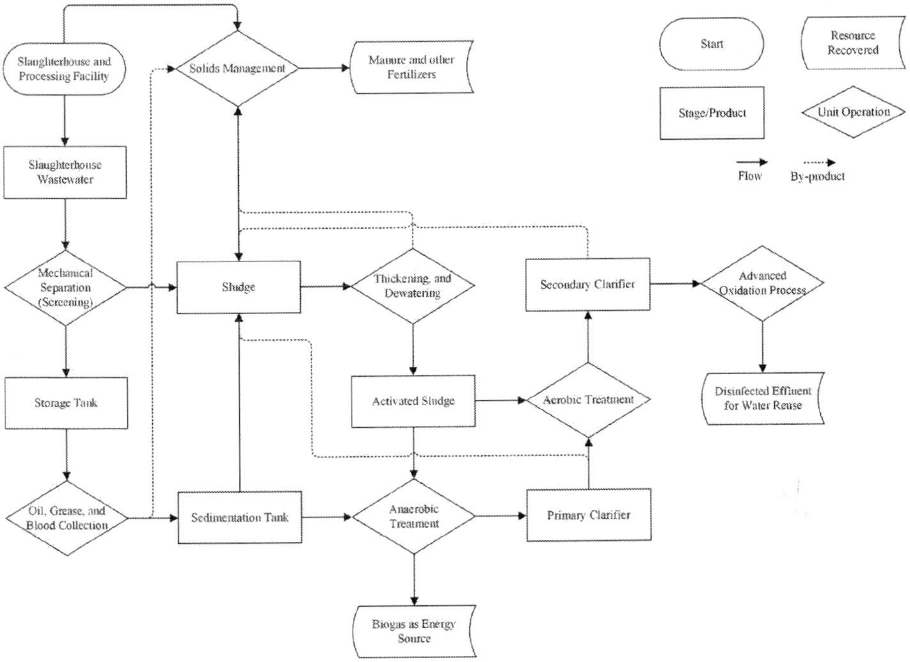

Figure 6. Proposed layout of the pretreatment, treatment, and disinfection of slaughterhouse wastes for a typical meat processing plant.

8. Conclusions

Meat processing effluents are usually pretreated using screeners, settlers, and blood collection systems, followed by physicochemical treatment methods, such as coagulation, flocculation, sedimentation, DAF, or secondary biological treatment. Although biological treatment is able to provide high organic and nutrient removal efficiencies, further treatment by AOPs, or other BAT, is required for a high-quality effluent.

The presented case study provided an example of the application of a combined ABR-AS-UV/H_2O_2 system for the treatment of an actual SWW. Maximum organic and nutrient removal reached over 90% in terms of TOC and TN, respectively. Moreover, a potential resource recovery achieved a maximum CH_4 yield of up to 56%, and minimization of residual by-products from disinfection was attained in terms of H_2O_2 residual of less than 2% at the effluent. Finally, the cost-effectiveness analysis found a minimum overall treatment cost of 0.12 $/m³ for the treatment of an actual SWW using the combined ABR-AS-UV/H_2O_2 system at optimum operating conditions.

All types of waste, liquid, solid, or gaseous, must be treated prior to their release into the environment. Whereas the use of recoverable resources is recommended as a feasible and practical alternative to conventional energy sources in the long term, costs associated to the application of these technologies will be offset by the reduction in local electricity consumption and by-product recycling and reuse. Thus, the potential of biogas production as an energy source, the use of fertilizers from nutrient recovery, and the SWW high-quality treated effluents for water reuse are to be considered toward a sustainable and cleaner production in the meat processing industry.

Consequently, the use of combined processes as an alternative to conventional methods has become a cost-effective approach for the treatment of meat processing effluents to comply with applicable current regulations worldwide.

Acknowledgements

The financial support of the Natural Sciences and Engineering Research Council of Canada (NSERC), the Ontario Trillium Scholarship (OTS) program, Colciencias (Colombian Research Funding Agency), and Ryerson University is greatly appreciated.

Nomenclature

ABR	anaerobic baffled reactor
AD	anaerobic digester
AF	anaerobic filter
AL	anaerobic lagoon
AOP	advanced oxidation process
AS	activated sludge
BAT	best available technology economically achievable
BOD	biochemical oxygen demand
COD	chemical oxygen demand
CW	constructed wetland

DAF	dissolved air floatation
DO	dissolved oxygen
EC	electrocoagulation
EU	European Union
MF	microfiltration
MPP	meat processing plant
NF	nanofiltration
RO	reverse osmosis
SBR	sequencing batch reactor
ST	septic tank
SWW	slaughterhouse wastewater
TOC	total organic carbon
TN	total nitrogen
TP	total phosphorus
TSS	total suspended solids
UF	ultrafiltration
UASB	up-flow anaerobic sludge blanket
US EPA	United States Environmental Protection Agency
UST	ultrasound technology

Author details

Ciro Bustillo-Lecompte and Mehrab Mehrvar*

*Address all correspondence to: mmehrvar@ryerson.ca

Ryerson University, Toronto, Canada

References

[1] Gerbens-Leenes P.W., Mekonnen M.M., Hoekstra A.Y. The water footprint of poultry, pork and beef: A comparative study in different countries and production systems. Water Resources and Industry. 2013;1–2:25–36. DOI: 10.1016/j.wri.2013.03.001

[2] Mekonnen M.M., Hoekstra A.Y. A global assessment of the water footprint of farm animal products. Ecosystems. 2012;15(3):401–415. DOI: 10.1007/s10021-011-9517-8

[3] Valta K., Kosanovic T., Malamis D., Moustakas K., Loizidou M. Overview of water usage and wastewater management in the food and beverage industry. Desalination and Water Treatment. 2015;53(12):3347. DOI: 10.1080/19443994.2014.934100

[4] United States Environmental Protection Agency (US EPA). Effluent limitations guidelines and new source performance standards for the meat and poultry products point source category [Internet]. 2004 [Updated: 2004-10-08]. Available from: https://federalregister.gov/a/04-12017 [Accessed: 2016-06-22]

[5] Barrera M., Mehrvar M., Gilbride K., McCarthy L., Laursen A., Bostan V., et al. Photolytic treatment of organic constituents and bacterial pathogens in secondary effluent of synthetic slaughterhouse wastewater. Chemical Engineering Research and Design. 2012;90(9):1335–1350. DOI: 10.1016/j.cherd.2011.11.018

[6] Bustillo-Lecompte C., Mehrvar M., Quiñones-Bolaños E. Slaughterhouse wastewater characterization and treatment: an economic and public health necessity of the meat processing industry in Ontario, Canada. Journal of Geoscience and Environment Protection. 2016;4:175–186. DOI: 10.4236/gep.2016.44021

[7] Bustillo-Lecompte C.F., Mehrvar M. Slaughterhouse wastewater characteristics, treatment, and management in the meat processing industry: a review on trends and advances. Journal of Environmental Management. 2015;161:287–302. DOI: 10.1016/j.jenvman.2015.07.008

[8] Jensen P.D., Yap S.D., Boyle-Gotla A., Janoschka J., Carney C., Pidou M., et al. Anaerobic membrane bioreactors enable high rate treatment of slaughterhouse wastewater. Biochemical Engineering Journal. 2015;97:132–141. DOI: 10.1016/j.bej.2015.02.009

[9] Bustillo-Lecompte C.F., Mehrvar M., Quiñones-Bolaños E. Combined anaerobic-aerobic and UV/H_2O_2 processes for the treatment of synthetic slaughterhouse wastewater. Journal of Environmental Science and Health, Part A: Toxic/Hazardous Substances and Environmental Engineering. 2013;48(9):1122–1135. DOI: 10.1080/10934529.2013.774662

[10] Bustillo-Lecompte C.F., Mehrvar M., Quiñones-Bolaños E. Cost-effectiveness analysis of TOC removal from slaughterhouse wastewater using combined anaerobic-aerobic and UV/H_2O_2 processes. Journal of Environmental Management. 2014;134:145–152. DOI: 10.1016/j.jenvman.2013.12.035

[11] Sneeringer S.E. Effects of environmental regulation on economic activity and pollution in commercial agriculture. The B.E. Journal of Economic Analysis and Policy . 2009;9(1): 1935–1682. DOI: 10.1016/S1642-3593(07)70117-9

[12] World Bank Group. Environmental, Health and Safety (EHS) Guidelines for Meat Processing. General EHS Guidelines: Environmental Wastewater and Ambient Water Quality. [Internet]. 2007 [Updated: 2013-11-15]. Available from: http://www.ifc.org/ehsguidelines [Accessed: 2016-06-24]

[13] Council of the European Communities (CEC). Urban Wastewater Treatment Directive 91/271/EEC. Official Journal of the European Communities. 1991;L 135(30.5):40–52.

[14] US Environmental Protection Agency (US EPA). Effluent limitations guidelines and new source performance standards for the meat and poultry products point source category; Final Rule; Federal Register. 2004;69(173):54476–54555.

[15] Environment Canada . Framework and recommendations concerning effluent quality of wastewater disposed by federal institutions, Final Report, Federal Committee on Environmental Management Systems (FCEMS)/Wastewater Working Group [Internet]. 2000 [Updated: 2003-06-11]. Available from: http://www.csc-scc.gc.ca/politiques-et-lois/318-6-gl-eng.shtml [Accessed: 2016-06-24]

[16] Environment Canada. Wastewater Systems Effluent Regulations. Fisheries Act. SOR/2012-139 [Internet]. 2012 [Updated: 2016-06-17]. Available from: http://laws-lois.justice.gc.ca/eng/regulations/sor-2012-139/FullText.html [Accessed: 2016-06-24]

[17] Colombian Ministry of Environment and Sustainable Development. Resolution 631 of 2015 Whereby the parameters and maximum permissible exposure limits in specific discharges to surface water and public sewer systems and other provisions are established [Internet]. 2015. Available from: https://www.minambiente.gov.co/images/normativa/app/resoluciones/d1-res_631_marz_2015.pdf [Accessed: 2016-06-24]

[18] Chinese Ministry of Environmental Protection. Discharge standard of water pollutants for meat packing industry GB 13457-92 [Internet]. 1992 [Updated: 1998?01?01]. Available from: http://english.mep.gov.cn/standards_reports/standards/water_environment/Discharge_standard/200710/t20071024_111799.htm [Accessed: 2016-06-24]

[19] Indian Central Pollution Control Board. General standards for discharge of environmental pollutants Part A: Effluents [Internet]. 1986 [Updated: 1993-12-31]. Available from: http://cpcb.nic.in/GeneralStandards.pdf [Accessed: 2016-06-24]

[20] Australian and New Zealand Environment and Conservation Council. Australian Guidelines for Sewerage Systems: Effluent Management [Internet]. 1997. Available from: https://www.environment.gov.au/system/files/resources/e52e452b-a821-4abe-9987-0e353/files/sewerage-systems-effluent-man-paper11.pdf [Accessed: 2016-06-24]

[21] Australian and New Zealand Environment and Conservation Council. Australian and New Zealand Guidelines for Fresh and Marine Water Quality. National Water Quality Management Strategy. 2000;1(1–7):4.

[22] Amorim A. K. B., De Nardi I. R., Del Nery V. Water conservation and effluent minimization: case study of a poultry slaughterhouse. Resources, Conservation and Recycling. 2007;51(1):93–100. DOI: 10.1016/j.resconrec.2006.08.005

[23] Torkian A., Eqbali A., Hashemian, S.J. The effect of organic loading rate on the performance of UASB reactor treating slaughterhouse effluent. Resources, Conservation and Recycling. 2003;40(1):1–11. DOI: 10.1016/S0921-3449(03)00021-1

[24] Belsky A.J., Matzke A., Uselman S. Survey of livestock influences on stream and riparian ecosystems in the western United States. Journal of Soil and Water Conservation. 1999;54(1): 419–431.

[25] Verheijen L.A.H.M., Wiersema D., Hulshoff Pol L.W. Management of Waste from Animal Product Processing [Internet]. 1996. Available from: http://www.fao.org/WAIRDOCS/LEAD/X6114E/X6114E00.HTM [Accessed: 2016-06-24]

[26] Cao W., Mehrvar M. Slaughterhouse wastewater treatment by combined anaerobic baffled reactor and UV/H_2O_2 processes. Chemical Engineering Research and Design. 2011;89(7):1136–1143. DOI: 10.1016/j.cherd.2010.12.001

[27] Mbuligwe S.E. Waste management and resource recovery in the food processing industry: the livestock industry. In: Bellinghouse V.C., editor. Food Processing: Methods, Techniques and Trends. New York, NY: Nova Science Publishers, Inc.; 2009. p. 77–113.

[28] Um M.M., Barraud O., Kérourédan M., Gaschet M., Stalder T., Oswald E., et al. Comparison of the incidence of pathogenic and antibiotic-resistant *Escherichia coli* strains in adult cattle and veal calf slaughterhouse effluents highlighted different risks for public health. Water Research. 2016;88:30–38. DOI: 10.1016/j.watres.2015.09.029

[29] Pina S., Buti M., Cotrina M., Piella J., Girones R. HEV identified in serum from human with acute hepatitis and sewage of animal origin in Spain. Journal of Hepatology. 2000;33(5):826–833. DOI: 10.1016/S0168-8278(00)80316-5

[30] Mittal G.S. Treatment of wastewater from abattoirs before land application—a review. Bioresource Technology. 2006;97(9):1119–1135. DOI: 10.1016/j.biortech.2004.11.021

[31] Al-Mutairi N.Z., Al-Sharifi F.A., Al-Shammari S.B. Evaluation study of a slaughterhouse wastewater treatment plant including contact-assisted activated sludge and DAF. Desalination. 2008;225(1–3):167–175. DOI: 10.1016/j.desal.2007.04.094

[32] De Sena R.F., Moreira R.F.P.M., José H.J. Comparison of coagulants and coagulation aids for treatment of meat processing wastewater by column flotation. Bioresource Technology. 2008;99(17):8221–8225. DOI: 10.1016/j.biortech.2008.03.014

[33] De Nardi I.R., Fuzi T.P., Del Nery V. Performance evaluation and operating strategies of dissolved-air flotation system treating poultry slaughterhouse wastewater. Resources, Conservation and Recycling. 2008;52(3):533–544. DOI: 10.1016/j.resconrec.2007.06.005

[34] Kobya M., Senturk E., Bayramoglu M. Treatment of poultry slaughterhouse wastewaters by electrocoagulation. Journal of Hazardous Materials. 2006;133(1–3):172–176. DOI: 10.1016/j.jhazmat.2005.10.007

[35] Bayramoglu M., Kobya M., Eyvaz M., Senturk E. Technical and economic analysis of electrocoagulation for the treatment of poultry slaughterhouse wastewater. Separation and Purification Technology. 2006;51(3):404–408. DOI: 10.1016/j.seppur.2006.03.003

[36] Gürel L., Büyükgüngör H. Treatment of slaughterhouse plant wastewater by using a membrane bioreactor. Water Science and Technology. 2011;64(1):214–219. DOI: 10.2166/wst.2011.677

[37] Chan Y. J., Chong M. F., Law C. L., Hassell D. A review on anaerobic-aerobic treatment of industrial and municipal wastewater. Chemical Engineering Journal. 2009;155(1–2): 1–18. DOI: 10.1016/j.cej.2009.06.041

[38] Al-Mutairi N.Z. Aerobic selectors in slaughterhouse activated sludge systems: a preliminary investigation. Bioresource Technology. 2009;100(1):50–58. DOI: 10.1016/j.biortech.2007.12.030

[39] Keskes S., Hmaied F., Gannoun H., Bouallagui H., Godon J.J., Hamdi M. Performance of a submerged membrane bioreactor for the aerobic treatment of abattoir wastewater. Bioresource Technology. 2012;103(1):28–34. DOI: 10.1016/j.biortech.2011.09.063

[40] Vymazal J. Constructed wetlands for treatment of industrial wastewaters: a review. Ecological Engineering. 2014;73:724–751. DOI: 10.1016/j.ecoleng.2014.09.034

[41] Gutiérrez-Sarabia A., Fernández-Villagómez G., Martínez-Pereda P., Rinderknecht-Seijas N., Poggi-Varaldo H.M. Slaughterhouse wastewater treatment in a full-scale system with constructed wetlands. Water Environment Research. 2004;76(4): 334–343. DOI: 10.2175/106143004X141924

[42] Bustillo-Lecompte C.F., Ghafoori S., Mehrvar M. Photochemical degradation of an actual slaughterhouse wastewater by continuous UV/H_2O_2 photoreactor with recycle. Journal of Environmental Chemical Engineering. 2016;4(1):719–732. DOI: 10.1016/j.jece.2015.12.009

[43] Vidal J., Huiliñir C., Salazar R. Removal of organic matter contained in slaughterhouse wastewater using a combination of anaerobic digestion and solar photoelectro-Fenton processes. Electrochimica Acta. 2016;210:163–170. DOI: 10.1016/j.electacta.2016.05.064

[44] Bayar S., Yildiz Y.S., Yilmaz A.E., Koparal A.S. The effect of initial pH on treatment of poultry slaughterhouse wastewater by electrocoagulation method. Desalination and Water Treatment. 2014;52(16–18):3047–3053. DOI: 10.1080/19443994.2013.800268

[45] Kundu P., Debsarkar A., Mukherjee S. Kinetic modeling for simultaneous organic carbon oxidation, nitrification, and denitrification of abattoir wastewater in sequencing batch reactor. Bioremediation Journal. 2014;18(4):267–286. DOI: 10.1080/10889868.2014.939134

[46] Martinez S.L., Torretta V., Minguela J.V., Siñeriz F., Raboni M., Copelli S., et al. Treatment of slaughterhouse wastewaters using anaerobic filters. Environmental Technology. 2014;35(3):322–332. DOI: 10.1080/09593330.2013.827729

[47] McCabe B.K., Hamawand I., Harris P., Baillie C., Yusaf T. A case study for biogas generation from covered anaerobic ponds treating abattoir wastewater: Investigation

of pond performance and potential biogas production. Applied Energy. 2014;114:798–808. DOI: 10.1016/j.apenergy.2013.10.020

[48] Mees J.B.R., Gomes S.D., Hasan S.D.M., Gomes B.M., Vilas Boas M.A. Nitrogen removal in a SBR operated with and without pre-denitrification: Effect of the carbon:nitrogen ratio and the cycle time. Environmental Technology. 2014;35(1):115–123. DOI: 10.1080/09593330.2013.816373

[49] Ozyonar F., Karagozoglu B. Investigation of technical and economic analysis of electrocoagulation process for the treatment of great and small cattle slaughterhouse wastewater. Desalination and Water Treatment. 2014;52(1–3):74–87. DOI: 10.1080/19443994.2013.787373

[50] Pan M., Henry L.G., Liu R., Huang X., Zhan X. Nitrogen removal from slaughterhouse wastewater through partial nitrification followed by denitrification in intermittently aerated sequencing batch reactors at 11°C. Environmental Technology. 2014;35(4):470–477. DOI: 10.1080/09593330.2013.832336

[51] Stets M.I., Etto R.M., Galvão C.W., Ayub R.A., Cruz L.M., Steffens M.B.R., et al. Microbial community and performance of slaughterhouse wastewater treatment filters. Genetics and Molecular Research. 2014;13(2):4444–4455. DOI: 10.4238/2014.June.16.3

[52] Yoon Y.M., Kim S.H., Oh S.Y., Kim C.H. Potential of anaerobic digestion for material recovery and energy production in waste biomass from a poultry slaughterhouse. Waste Management. 2014;34(1):204–209. DOI: 10.1016/j.wasman.2013.09.020

[53] Almandoz M.C., Pagliero C.L., Ochoa N.A., Marchese J. Composite ceramic membranes from natural aluminosilicates for microfiltration applications. Ceramics International. 2015;41(4):5621–5633. DOI: 10.1016/j.ceramint.2014.12.144

[54] Abdurahman N.H., Rosli Y.M., Azhari N.H., Bari H.A. The potential of ultrasonic membrane anaerobic systems in treating slaughterhouse wastewater. Journal of Water Reuse and Desalination. 2015;5(3):293–300. DOI: 10.2166/wrd.2015.107

[55] Odong R., Kansiime F., Omara J., Kyambadde J. Tertiary treatment of abattoir wastewater in a horizontal subsurface flow-constructed wetland under tropical conditions. International Journal of Environment and Waste Management. 2015;15(3):257–270. DOI: 10.1504/IJEWM.2015.069160

[56] Pan M., Hu Z., Liu R., Zhan X. Effects of loading rate and aeration on nitrogen removal and N2O emissions in intermittently aerated sequencing batch reactors treating slaughterhouse wastewater at 11°C. Bioprocess and Biosystems Engineering. 2015;38(4):681–689. DOI: 10.1007/s00449-014-1307-1

[57] Paramo-Vargas J., Camargo A.M.E., Gutierrez-Granados S., Godinez L.A., Peralta-Hernandez J.M. Applying electro-Fenton process as an alternative to a slaughterhouse effluent treatment. Journal of Electroanalytical Chemistry. 2015;754(2169):80–86. DOI: 10.1016/j.jelechem.2015.07.002

[58] Wosiack P.A., Lopes D.D., Rissato Zamariolli Damianovic M.H., Foresti E., Granato D., Barana A.C. Removal of COD and nitrogen from animal food plant wastewater in an intermittently-aerated structured-bed reactor. Journal of Environmental Management. 2015;154:145–150. DOI: 10.1016/j.jenvman.2015.02.026

[59] Abdurahman N.H., Rosli Y.M., Azhari N.H. The potential of ultrasonic membrane anaerobic system (UMAS) in treating slaughterhouse wastewater. ARPN Journal of Engineering and Applied Sciences. 2016;11(4):2653–2659.

[60] Coskun T., Debik E., Kabuk H.A., Manav Demir N., Basturk I., Yildirim B., et al. Treatment of poultry slaughterhouse wastewater using a membrane process, water reuse, and economic analysis. Desalination and Water Treatment. 2016;57(11):4944–4951. DOI: 10.1080/19443994.2014.999715

[61] Bustillo-Lecompte C.F., Mehrvar M Treatment of an actual slaughterhouse wastewater by integration of biological and advanced oxidation processes: Modeling, optimization, and cost-effectiveness analysis. Journal of Environmental Management. 2016;182: 651-666. DOI: 10.1016/j.jenvman.2016.07.044

[62] Ontario Ministry of Agricultural and Rural Affairs (OMAFRA). Provincially Licensed Meat Plants in Ontario [Internet]. 2016. Available from: http://www.omafra.gov.on.ca/english/food/inspection/maps/TblAllAbattoirs.htm [Accessed: 2016-06-24]

[63] Sgarbossa F., Russo I. A proactive model in sustainable food supply chain: insight from a case study. International Journal of Production Economics. 2016;1–11. DOI: 10.1016/j.ijpe.2016.07.022 (in press).

[64] Manios T., Gaki E., Banou S., Klimathianou A., Abramakis N., Sakkas N. Closed wastewater cycle in a meat producing and processing industry. Resources, Conservation and Recycling. 2003;38(4):335–345. DOI: 10.1016/S0921-3449(02)00169-6

Chapter 9

Treatment of Antibiotics in Wastewater Using Advanced Oxidation Processes (AOPs)

Ayşe Kurt, Berna Kiril Mert, Nihan Özengin,

Özge Sivrioğlu and Taner Yonar

Additional information is available at the end of the chapter

http://dx.doi.org/10.5772/67538

Abstract

Antibiotics are nonbiodegradable, can survive at aquatic environments for long periods and they have a big potential bio-accumulation in the environment. They are extensively metabolized by humans, animals and plants. After metabolization, antibiotics or their metabolites are excreted into the aquatic environment. Removal of these compounds from the aquatic environment is feasible by different processes. But antibiotics are not treated in conventional wastewater treatment plants efficiently. During the last years studies with advanced oxidation processes (AOPs) for removal of these pharmaceuticals from waters has shown that they can be useful for removing them fully. Advanced oxidation processes (AOPs) can work as alternatives or complementary method in traditional wastewater treatment, and highly reactive free radicals, especially hydroxyl radicals (OH) generated via chemical (O_3/H_2O_2, O_3/OH^-), photochemical (UV/O_3, O_3/H_2O_2) reactions, serve as the main oxidant. This study presents an overview of the literature on antibiotics and their removal from water by advanced oxidation processes. It includes almost all types of antibiotics which are consumed by human and veterinary processes. It was found that most of the investigated advanced oxidation treatment processes for the oxidation of antibiotics in water are direct and indirect photolysis with the combinations of H_2O_2, TiO_2, ozone and Fenton's reagent.

Keywords: wastewater, antibiotics, endocrine disrupter, advanced oxidation

1. Introduction

The "antibiotic" term qua generic is used to specify any class of organic molecule that blocks or ravage microbes by specific interactions with bacterial marks, without considering any compound or class [1]. Antibiotics are designed to act very effectively even at low doses and, in case of intracorporal administration, to be completely excreted from the body after a short time of residence [2]. They are nonbiodegradable and can survive in aquatic environments for long periods [3]. The entrance of these compounds into the environment owing to anthropogenic sources can result in a potential risk for organisms. Although antibiotics exist at residual levels, they can cause resistance in bacterial populations, making them inactive in the treatment of several diseases in the near future [4, 5]. And they cause endocrine-disrupting effects when they are consumed by living organisms. They interfere with the synthesis, secretion, transport, binding, action, and elimination of hormones in the human body [6].

The annual usages of antibiotics are determined between 100,000 and 200,000 t globally [7]. Traditionally, these compounds were not accepted as environmental contaminants, but their existence in the aquatic ecosystems has become an apprehension as biological impacts and potential threat to the environment [8–10]. Furthermore, it has been shown up that residual antibiotics are able to support the election of genetic variants of microorganisms concluding in the existence of antibiotic-resistant pathogens [11, 12].

Removal of these compounds from the aquatic environment is feasible by different processes. This can be carried out using biotic (biodegradation) or nonbiotic (chemical oxidation and advanced oxidation) ways. But antibiotics are not treated in conventional wastewater treatment plants efficiently. During the last years, studies with advanced oxidation processes (AOPs) for removal of these pharmaceuticals from waters have shown that they can be useful for removing them completely. In this chapter, we aim to introduce a review of literature on antibiotics and their removal from water by advanced oxidation processes. An effort to include as many studies as possible was made in order to highlight important findings and present the knowledge currently available on the removal efficiency of antibiotics from wastewater.

2. General description of antibiotics

Antibiotic as a word is reproduced from the Greek anti (=against) and biotikos (=living).

Most of the living organisms are able to compose matters that can influence other organisms' capacity for growth, endurance, and reproduction. Microorganisms have a versatile ability to inhibit the growth and purpose of other microorganisms and produce and release biologically effective substances at the appropriate moment. We denominate substances of this kind as *antibiotics* [13].

In addition, organisms' ability to compose antibiotics has been of great importance for the development of different life forms and their capability to accommodate to new circumambient. Nowadays, antibiotics are important components for the functions of various biological systems.

With the development of synthetic antibiotics, a large number of substances with specific areas of application have been given to access. Today, the level of usage of synthetic antibiotics and their effects to the environment are at a critical rate. Since antibiotics are bioavailable, they can show long-term biological effects in the environment.

Antibiotics can be grouped according to their chemical structure or mechanism of action. There are various groups of chemicals that can be arranged to different subgroups, such as ß-lactams, quinolones, tetracyclines, macrolides, sulphonamides, and others. They are complicated molecules, which may have different functionalities within the same molecule. Consequently, they act as neutral, cationic, anionic, or zwitterionic under different pH conditions. Owing to different functionalities in a single molecule, their physico-chemical and biological properties (like octanol-water partition coefficients (log P_{ow}), sorption behavior, photoreactivity and antibiotic activity, and toxicity) may change with pH [14].

b-Lactam antibiotics contain cefradine, amoxicillin ceftriaxone, sultamicillin, and penicillins G and VK. Actually, these antibiotics have been insulated from molds and have been adapted to obtain different physicochemical and pharmacological properties [15]. They suppress bacterial cell wall synthesis.

Sulfonamides are synthetic antibiotics, and they inhibit generation of bacteria by behaving as competitive inhibitors of p-aminobenzoic acid in the folic acid metabolism cycle [15]. A diversity of sulfonamides have been developed, consumed, and finally detected in the environment, and some of them have been studied for their degradation by ozonation and AOPs. These compounds include sulfadiazine, sulfadimethoxine, sulfachlopyridazine, sulfamethazine, sulfamethizole, sulfamethoxazole, sulfisoxazole, sulfapyridine, sulfathiazole, sulfamoxole, and sulfamerazine.

Most common quinolone antibiotics are enrofloxacin and ofloxacin. They have been examined in terms of their degradation by ozonation and AOPs. More particularly, both are fluoroquinolones. While enrofloxacin is utilized as a veterinary antibiotic, despite that ofloxacin is designed for human uptake. These compounds have a benefit to suppress the activity of bacterial DNA gyrase. It is known that the quinolones are metabolized in the liver and eliminated in the urine [15, 16].

When we examine other antibiotics, clarithromycin, azithromycin, erythromycin, and roxithromycin are macrolide antibiotics, and lincomycin is a lincosamide antibiotic. These antibiotics are described by a property to inhibit bacterial protein synthesis. These antibiotics are mostly eliminated in the bile [15, 16].

Antibiotic is a chemotherapeutic agent that inhibits the growth of microorganisms (bacteria, fungi, protozoa, or viruses) even at very low concentrations. They are nonbiodegradable and can survive in aquatic environments for long periods. So they can bio-accumulate in the environment [3]. Also antibiotics in the environment may contribute to the emergence of antibiotic-resistant bacteria [17]. And they cause endocrine-disrupting effects when they are consumed by living organisms. They interfere with the synthesis, secretion, transport, binding, action, and elimination of hormones in the human body [6].

Releasing of antibiotics into the aquatic environment by human beings and animals depends mainly on the consumption rates of antibiotics [14, 18]. According to the investigations, some antibiotics have toxic effects on humans, animals, and also microorganisms even at low concentrations. At the same time, they are nonbiodegradable and can survive in the environment even in the conventional wastewater treatments. So they cause bio-accumulation [3]. Therefore, the presence of antibiotics in the environment can cause the occurrence of antibiotic-resistant bacteria [17]. They are not treated in the conventional wastewater treatment plants completely. According to the recent studies, advanced oxidation processes (AOPs) are useful to remove these toxic compounds completely from waters.

2.1. Sources of antibiotics in the environment

In these last years, the use of antibiotics in veterinary and human medicine has been widespread, and consequently, the possibility of water contamination with such compounds has been increased [19]. These pollutants are continually discharged into the natural environment as parent compounds, metabolites/degradation products, or both forms by a diversity of input sources as shown in **Figure 1** [5].

Fertilizers present in the fields can contaminate soil and consequently surface and groundwater through runoff or filtration [20]. Likewise, human antibiotics which are present into the environment through discharge, entering in the sewage and reaching the Waste-water treatment plants (WWTP). Despite most of WWTPs are not projected to remove highly polar micropollutants [19], they can be transferred to surface waters and reach groundwater after leaching.

The sludge produced in WWTPs is utilized as soil manure and can cause problems when used as a fertilizer. Some other significant pollution source is the direct delivery of veterinary antibiotics through the implementation in aquaculture. Inappropriate elimination of unused/expired

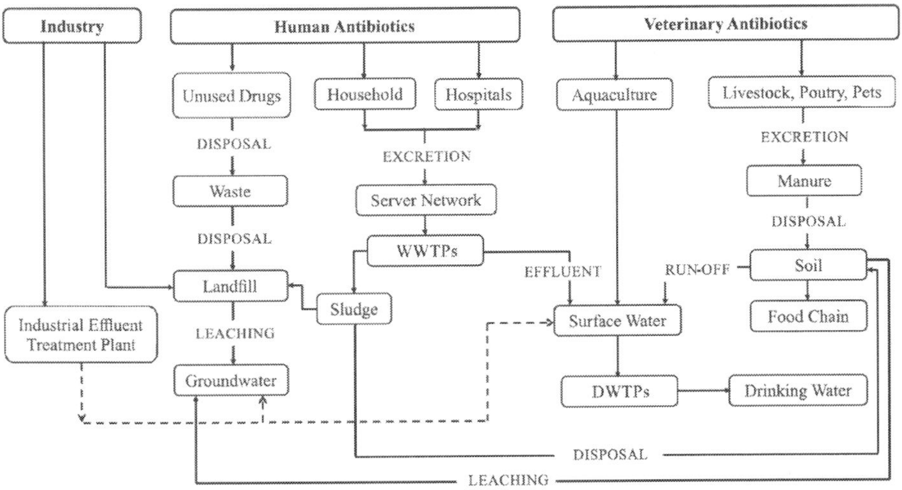

Figure 1. Pathways of antibiotics in the environment [5, 16].

drugs can also be considered as significant points of contamination. These are derived directly from sewage discharge or landfills deposition, waste effluents from manufacture, or accidental spills during manufacturing or distribution [5].

2.2. Occurrence

Research has quite extensively studied the presence of antibiotics in the environment (for a short overview, see **Table 1**). With respect to other pharmaceuticals, the concentrations of antibiotics measured in different countries were found in the same range of concentrations in the different compartments [14, 21–23]. The antibiotic groups that have been analyzed up to now include a number of different important classes of antibiotics. They include primarily macrolides, aminoglycosides, tetracyclines, sulfonamides, sulfanilamides, and quinolones to name just a few [14, 23–29].

Antibiotic	Sewage treatment plant effluent (ng L^{-1})	Surface water (ng L^{-1})	Ground water*/bank filtrate (ng L^{-1})	References
Penicillins				
Penicillin	up to 200	up to 3		[32]
Flucloxacillin		7		[33]
Piperacillin		48		[33]
Macrolides				
Macrolide	up to 700	up to 20	up to 2*	[32]
Azithromycin		up to 3		[33]
Erythromycin-H_2O	up to 287			[34, 35]
			up to 49	[36]
	up to 6000	up to 1700		[8]
		up to 190		[33]
		up to 15.9		[37]
		up to 220		[38]
	up to 400			[30]
Clarithromycin	up to 328	up to 65		[34, 35]
	up to 240	up to 260		[8]
		up to 37		[33]
		up to 20.3		[37]
		up to 20.3		[39]
	up to 38			[30]
Roxithromycin	up to 68			[30]
	up to 72			[34, 35]
			up to 26	[36]

Antibiotic	Sewage treatment plant effluent (ng L^{-1})	Surface water (ng L^{-1})	Ground water*/bank filtrate (ng L^{-1})	References
	up to 1000	up to 560		[8]
		up to 14		[33]
		up to 180		[38]
		up to 350		[31]
Chinolones				
Fluorchinolone	up to 100	up to 5		[32]
Fluorchinolone	up to 106	up to 19		[34, 35]
Ciprofloxacin		9		[33]
		up to 30		[38]
		up to 26.2		[37]
		up to 1300		[31]
		up to 26		[39]
Norfloxacin		up to 120		[38]
Ofloxacin	up to 82			[30]
		20		[33]
Sulfonamides				
Sulfamethoxazole	up to 370			[30]
	up to 2000	up to 480	up to 470	[8]
		up to 52		[33]
		up to 1900		[38]
		up to 2000		[31]
Sulfamethazin			up to 160	[8]
		up to 220		[38]
Sulfamethizole		up to 130		[38]
Sulfadiazine			up to 17	[36]
Sulfadimidine			up to 23	[36]
		up to 7		[33]
Tetracyclines				
Tetracycline (no more specified)	up to 20	up to 1		[32]
Tetracycline		up to 110		[38]
Chlortetracycline		up to 690		[38]
		up to 600		[31]
		up to 100		[38]
Oxytetracycline		up to 340		[38]
		up to 19.2		[37]

http://dx.doi.org/10.5772/67538

Antibiotic	Sewage treatment plant effluent (ng L^{-1})	Surface water (ng L^{-1})	Ground water*/bank filtrate (ng L^{-1})	References
Others				
Trimethoprim	up to 38			[30]
		up to 24		[36]
	up to 660	up to 200		[8]
		up to 12		[33]
		up to 710		[38]
Ronidazol			up to 10	[36]
Chloramphenicol	up to 68			[30]
	up to 560	up to 60		[8]
Clindamycin	up to 110			[30]
		up to 24		[33]
Lincomycin		up to 730		[38]
		up to 248.9		[39]
Spiramycin		up to 74.2		[37]
Oleandomycin		up to 2.8		[37]
Tylosin		up to 280		[38]
		up to 2.8		[37]

*Directly impacted by surface water.

Table 1. Examples of measured concentrations of antibiotics in the aquatic environment [14, 18, 30, 31].

3. Advanced oxidation processes (AOPs)

During the oxidation of organic contaminants, the ultimate goal is to produce simple, relatively harmless inorganic molecules [40]. Advanced oxidation processes are characterized by their production of the hydroxyl radical (·OH), a very strong oxidant, in sufficiently high concentrations to affect water quality. The symbol "·"represents the radical center, a single unpaired electron [41].

At optimum operation conditions, for instance sufficient contact time, it is possible to mineralize the target contaminant to CO_2 and H_2O, the most stable end products of chemical oxidation. For this reason, the extraordinary definition of AOPs on chemical processes is that they are completely described as "environmentally friendly" [42].

The basic treatment of AOPs can be explained in two steps: one is the generation of hydroxyl radicals and the other is the oxidative reaction of these radicals with molecules [43]. The dissolved organic pollutants can be converted into CO_2 and H_2O by AOPs. The generation of hydroxyl radical might be by the use of UV, UV/H_2O_2, UV/O3, TiO_2/H_2O_2, Fe^{+2}/H_2O_2 and one or two processes [44].

AOPs can be classified in two groups: (1) nonphotochemical AOPs and (2) photochemical AOPs. Nonphotochemical AOPs include cavitation, ozonation, Fenton and Fenton-like processes, wet air oxidation, ozone/hydrogen peroxide, etc. Photochemical oxidation processes include homogeneous and heterogeneous processes [45].

3.1. Nonphotochemical oxidation processes

Nonphotochemical oxidation processes can be classified as follows: ozonation, peroxide, Fenton process, ozone/hydrogen, supercritical water oxidation, electrochemical oxidation, cavitation, gamma-ray, X-ray, electrical discharge-based nonthermal plasma, and electron beam.

3.1.1. Ozonation

Ozone is a powerful oxidizer and has been increasingly used for the treatment of wastewater [46]. High pH values (>11.0) causes high efficiency and ozone behaves randomly with all organic and inorganic compositions present in the reacting medium [45]. Ozone reacts with substances in two different ways: indirect and direct. These two reaction pathways are managed by different type of kinetics and lead to different oxidation products [47].

Simplified reaction mechanism of ozone at high pH is given in below:

$$3O_3 + H_2O \xrightarrow{OH-} 2OH\bullet + 4O_2 \tag{1}$$

3.1.2. Ozone/hydrogen peroxide (peroxone) process (O_3/H_2O_2)

The principle of peroxonation is based on the coupling between ozone (O_3) and H_2O_2, resulting in the generation of oxidizing radicals. As pointed out by Zaviska et al. [47], the peroxonation mechanism could be more productive than ozonation alone, and H_2O_2 impacts on increasing the decomposition percentage of O_3 in water, which generates a larger number of very reactive $\bullet$OH radicals [49]. Because of the high cost of ozone generation, this combination makes the process economically feasible [50]. Several factors limit the usefulness of the peroxonation process such as important energetic consumption, low water solubility of ozone, and its sensitivity to several factors [51]. A general mechanism of peroxon process is given below:

$$H_2O_2 + 2O_3 \rightarrow 2OH\bullet + 3O_2 \tag{2}$$

Solution pH is critical as well for the process output like other AOPs. Higher production rates of hydroxyl radicals will be obtained by the addition of hydrogen peroxide to the aqueous O_3 solution at high pH conditions. Independence of peroxone process from any light source or UV delivers a certain benefits to this operation [44].

3.1.3. Fenton process

Fenton's reaction is known as the dark reaction of ferrous iron (Fe(II)) with H_2O_2 (Eq.(15)) [6]. $\bullet$OH radical is generated through the agency of reaction between H_2O_2 and Fe^{+2} salts as described below.

$$Fe^{+2} + H_2O_2 \rightarrow Fe^{+3} + OH^- + \bullet OH \tag{3}$$

Thus, composed hydroxyl radical can react with Fe(II) to develop ferric ion (Fe(III)) (Eq. (16))

$$\cdot OH + Fe^{+2} \rightarrow Fe^{+3} + OH^- \tag{4}$$

As an alternative, hydroxyl radicals are able to react with organic pollutants and start oxidation in a waste stream,

$$RH + \cdot OH \rightarrow R \cdot + H_2O \tag{5}$$

Reactions can result into the degradation of Fe^{+3} to Fe^{+2} at a value of pH between 2.7 and 2.8.

$$Fe^{2+} + H_2O_2 \rightarrow Fe^{3+} + \bullet OH + OH^- \tag{6}$$

$$Fe^{3+} + H_2O_2 \rightarrow Fe^{2+} + \bullet OOH + H^+ \tag{7}$$

In these circumstances, iron can be considered as a true catalyst [156].

Process efficiency is closely related to the solution pH whose optimal values are between 2 and 4 as well as the COD:H_2O_2:catalyst ratio in the feed [52].

Basically, the Fenton process possesses several important advantages for water/wastewater treatment [48, 53]:

- A plain and adaptable operation permitting easy execution in existing plants
- Easy-to-use and relatively cheap chemicals
- No need of energy input

3.2. Photochemical oxidation processes

3.2.1. Homogeneous photochemical oxidation processes

3.2.1.1. Vacuum UV (VUV) photolysis

The vacuum ultraviolet (UV) is absorbed by all the materials from water to air, therefore can only be transferred in a vacuum. The absorption of a VUV photon causes breaking of one or more bond. As an example, water is decomposed by

$$H_2O + h\nu(< 190\ \text{nm}) \rightarrow H\bullet + HO\bullet \tag{8}$$

$$H_2O + h\nu(< 190\ \text{nm}) \rightarrow H + +e - +HO\bullet \tag{9}$$

VUV photolysis has a high feasibility for the oxidative degradation of organic pollutants in water. In spectral domain (approx. 140–200 nm), it produces hydrogen atoms and hydroxyl radicals. Due to the high absorption cross-section of water and quantum yields of water homolysis of 0.45–0.3 at stimulation wavelengths (between 140 and 185 nm) provide productive local concentrations of hydroxyl and hydrogen radicals. VUV photolysis is a new technique for water treatment and suggests the benefit to generate unusually high local concentrations of oxidative reactive intermediates without the addition of supplementary oxidant [44, 54].

3.2.1.2. Hydrogen peroxide/UV (H_2O_2/UV) process

Hydrogen peroxide can be photolyzed by UV radiations by producing the homolytic scission of the O–O bond of the H_2O_2 and resulting the formation of •OH radicals which can also be supplied to the decomposition of H_2O_2 by secondary reactions [48]. The main reaction is given below:

$$H_2O_2 + hv \rightarrow 2HO\bullet \quad (10)$$

UV/H_2O_2 process is effective in mineralizing organic pollutants. As an disadvantage the process cannot use solar light as the source of UV light owing to the fact that the required UV energy is not available in the solar spectrum [55]. Over and above, H_2O_2 has poor UV absorption characteristics. At last, special reactors designed for UV illumination are required [56].

The major factors influencing the process are the amount of H_2O_2 used, presence of bicarbonate, wastewater pH, the initial concentration of the object compound, and reaction time [57].

3.2.1.3. Ozone/UV (O_3/UV) process

The advanced oxidation process with ozone and UV radiation is initiated by the photolysis of ozone. Hydroxyl radicals can be composed by those in hydrogen peroxide under UV and/or ozone. The equations are given below:

$$O_3 + hv + H_2O \rightarrow H_2O_2 + O_2 \quad (11)$$

$$H_2O_2 + hv \rightarrow 2 \cdot OH \quad (12)$$

$$2O_3 + H_2O_2 \rightarrow 2 \cdot OH + 3O_2 \quad (13)$$

All kinds of UV light origins can be utilized for this process, especially low-pressure mercury vapor lamps. The O_3/UV process does not have the same limitations as that of H_2O_2/UV process. Many variables (temperature, pH, UV intensity, tubidity, lamp spectral characteristics, and pollutant type, etc.) affect the performance of the system [42, 44].

3.2.1.4. Ozone/hydrogen peroxide/UV (O_3/H_2O_2/UV) process

This method is considered to be the most effective and powerful method, which provides a fast and complete mineralization of pollutants [42, 50]. The addition of H_2O_2 to the O_3/UV process accelerates the decomposition of ozone, which results in an increased rate of ·OH generation.

The main short mechanism of the O_3/H_2O_2/UV process is given below:

$$2O_3 + H_2O_2 \overset{UV}{\rightarrow} 2\,HO\cdot + 3O_2 \quad (14)$$

The capital and operating costs for the system vary widely depending on the wastewater flow rate, types, and concentrations of contaminants present and the degree of removal required [58].

http://dx.doi.org/10.5772/67538

3.2.1.5. *Photo-Fenton process*

The Photo-Fenton process occurs by the combination of H_2O_2 and UV radiation with Fe(II) or Fe(III). The main factor of the mechanism is that iron salts act as photocatalysts and H_2O_2 as an oxidizing agent. It offers a productive and cheap method for wastewater treatment and produces hydroxyl radicals to a greater extent [59].

The reaction is given below:

$$Fe(OH)^{+2} + hv \rightarrow Fe^{+3} + HO\cdot \quad (15)$$

A highly low reaction time is required for the photo-Fenton process, and depending on the operating pH value, the concentrations of H_2O_2 and iron are added.

3.2.2. *Heterogeneous photochemical oxidation processes*

Widely investigated and applied heterogeneous photochemical oxidation processes are semiconductor-sensitized photochemical oxidation processes.

A semiconductor consists of two energy bands: one is high energy conduction and the other is low energy valence band. This kind of photolytic chemical oxidation is used for the generation of OH radical in heterogeneous processes. Zinc oxide, strontium titanium trioxide, and TiO_2 have been used for commercial implementation. Valance and conduction bands of a semiconductor material are distinguished by energy gap/band gap [60].

Moreover, the photocatalyst TiO_2 is a wide band gap semiconductor (3.2 eV) and is successfully used as a photocatalyst for the treatment of organic pollutants [61, 62]. To summarize, in the TiO_2 process, the photon energy given to achieve the band gap energy and to induce an electron into the transmission band from the valence band can be fed with a wavelength shorter than 387.5 nm. Clarified reaction mechanisms of TiO_2/UV process are given below [Eq. (16)–(19)].

$$TiO_2 + hv \rightarrow e^-{}_{CB} + h^+{}_{VB} \quad (16)$$

$$H_2O + h^+{}_{VB} \rightarrow OH\bullet + H^+ \quad (17)$$

$$O_2 + e^-{}_{CB} \rightarrow O_2\bullet^- \quad (18)$$

$$O_2\bullet^- + H_2O \rightarrow OH\bullet + OH^- + O_2 + HO_2^- \quad (19)$$

The basic reason of this reversal is the production of photons. The reversal mechanism importantly decreases the photocatalytic efficiency of a semiconductor. Main benefit of TiO_2/UV process is low energy consumption thus sunlight can be utilized as a light source [44].

AOPs have been examined in terms of limitations and summarized below.

As an example, UV oxidation process with H_2O_2 is just effective at low wavelengths (especially under 200 nm).The treated aqueous flux must supply good transmission of UV light. Scavengers and high doses of chemical subscriptions may limit the process. Insoluble oil and grease, heavy metal ions, insoluble oil and grease, carbonates, and high alkalinity may cause clogging of the UV quartz handle. Air emission problems with O_3 may arise. The cost of the AOPs is expensive when compared to rival technologies [42].

3.3. Assessment of AOPs performance for antibiotic removal

Ozonation and AOPs are required for efficient degradation of antibiotics in water and wastewater. These treatment processes have an advantage of elimination of such pollutants through mineralization or conversion to the products that are less harmful to human health and the aquatic environment.

Various studies have published the effective AOPs treatment for the removal of antibiotics from wastewater [63–67]. High-quality and effective publications relevant with the AOP studies on the mechanisms and applications of water and wastewater treatments have been pronounced for last two decades. From the theoretical, environmental, and economical point of view, they demonstrate a great and increasing interest. As shown in **Table 2**, several studies have been conducted on the applicability of AOPs on different antibiotic classes.

Ozone is a potent oxidant and has been progressively applied for the treatment of wastewater. Ozone and/or hydroxyl radicals passivate bactericidal characteristics of antibiotics by disrupting or modulating their pharmaceutically active functional groups, such as N-etheroxime and dimethylamino groups of macrolides [68, 69], aniline moieties of sulfonamides [70], thioether groups of penicillins, unsaturated bonds of cephalosporin, and the phenol ring of trimethoprim [69]. High removal rates (>90%) were achieved by ozonation of the compounds with electron-rich aromatic systems, such as hydroxyl, amino (e.g., sulfamethoxazole), acylamino, alkoxy, and alkyl aromatic compounds, as well as those compounds with deprotonated amine (e.g., erythromycin, ofloxacin, and trimethoprim) and nonaromatic alkene groups, since these structural moieties are highly amendable to oxidative attack [1]. Ozonation process was found to be effective for the removal of b-lactams, macrolides, sulfonamides, trimethoprim, quinolones, tetracyclines, and lincosamides [5].

The performance of ozone treatment can be improved providing ozone is combined with UV irradiation, hydrogen peroxide, or catalysts such as iron or copper complexes [52]. Regardless, optimum process and operational circumstances have still been determined for the different water and wastewater types together with various types of antibiotics [152].

In general, Fenton process has been widely used successfully for the oxidation of many groups of antibiotics, including b-lactams, quinolones, trimethoprim, and tetracyclines. Fenton's oxidation is a homogeneous oxidation process and considered to be a metal-catalyzed oxidation reaction, in which iron acts as a catalyst [65, 153]. The main handicap of the process is the low pH value. It is required to avoid iron precipitation that takes place at high pH [154, 155].

Heterogeneous photocatalysis with TiO_2 semiconductor is generally accomplished by the illumination of a suspension of TiO_2 in aqueous solution with light energy which is greater than its bandgap energy. This causes the generation of high energy electron-hole pairs (e^-/h^+), which may migrate to the surface of the catalyst and may either reunite producing thermal

http://dx.doi.org/10.5772/67538

Compound name	AOP	Concentration	Reaction conditions	References
Amoxicillin, sulfamethoxazole, and ciprofloxacin	Direct photolysis with UV	1 mgL^{-1}	250 W lamp (254 nm), UV doses: 0–2.5 × 10^4 μW s cm^{-2}, urban wastewater	[71]
β-Lactam antibiotics (amoxicillin and ampicillin)	Ferrate (VI)	0.1 mM	Fe(VI): 0.1–10 mM, pH 7.0, synthetic wastewater	[72]
Amoxicillin and cloxacillin	Photo-Fenton	150 mgL^{-1}	Solar intensity 0.85 kWm^{-2}, pH 3, synthetic wastewater	[73]
Enrofloxacin	Wet air oxidation and ozonation	0.2 mM	*Wet air oxidation*: 50 mL Teflon-lined stainless steel autoclave, 0.5 MPa,150°C stirring speed: 300 rps *Ozonation*: Pyrex glass tubular photoreactor, flow rate: 7.3 L h^{-1}, OGV-500 catalyst, synthetic wastewater	[74]
Amoxicillin, ampicillin, and cloxacillin	Fenton	AMX, AMP, CLX:104, 105, 103 mgL^{-1}	pH 3, COD: 520 mgL^{-1}, synthetic wastewater	[75]
Amoxicillin, ampicillin, and cloxacillin	Photo-Fenton	AMX, AMP, CLX: 104, 105, 103 mgL^{-1}	UV lamp, 230 V, 0.17 A, 6 W, 365 nm, synthetic wastewater	[76]
Amoxicillin, ampicillin and cloxacillin	UV/TiO_2 and $UV/H_2O_2/TiO_2$	AMX, AMP, CLX: 104, 105, 103 mgL^{-1}	pH ~ 5, COD: 520 mgL^{-1}, BOD_5/COD ~0 and DOC 145 mgL^{-1}, synthetic wastewater	[77]
Amoxicillin and cloxacillin	$UV/TiO_2/H_2O_2$	AMX: 138 mgL^{-1} CLX: 84 mgL^{-1}	6-W lamp, wavelength ≈ 365 nm, pharmaceutical industry wastewater	[67]
Amoxicillin	UV-A/TiO_2	2.5–30 mgL^{-1}	Degussa P25 TiO_2, TiO_2: 100–750 mgL^{-1}, pH 5 or 7.5, photon flux of 8 × 10^{-4} E/(L min), 9 W lamp, 350–400 nm, 25°C, synthetic wastewater	[78]
Amoxicillin, ampicillin, and cloxacillin	Fenton, photo-Fenton, TiO_2 photocatalytic and UV/ZnO	AMX, AMP, CLX: 104, 105, 103 mgL^{-1}	UV lamp, 230 V, 0.17 A, 6 W, 365 nm, synthetic wastewater	[77]
Amoxicillin	O_3/OH^-, H_2O_2/UV, Fe^{2+}/H_2O_2, Fe^{3+}/H_2O_2, Fe^{2+}/H_2O_2/UV and Fe^{3+}/H_2O_2/UV	AMX: 400 mgL^{-1}	O_3 generated from O_2, 21W Hg lamp (253.7 nm), flow rate: 1.3 L/min, light intensity: 3.65 WL^{-1} (1.73 × 10^{-4} $EinsteinL^{-1}s^{-1}$), effective pathlength: 1.72 cm, pharmaceutical wastewater	[3]
Amoxicillin and cloxacillin	Fenton	AMX and CLX: 150 mgL^{-1}	pH 3.0, H_2O_2/COD: 1.0–3.0, H_2O_2/Fe^{2+}: 2–150 and reaction time: 60–120 min, synthetic wastewater	[79]
Amoxicillin, oxacillin, and ampicillin	Nonthermal plasma	OX, AMX, AMP: 100 mgL^{-1}	Discharge was generated at the gas-liquid interface at room temperature and atmospheric pressure, in	[80]

Compound name	AOP	Concentration	Reaction conditions	References
			oxygen flow rate: 600 sccm, power: 2 W, pH 8, synthetic wastewater	
Amoxicillin	UV, O_3, Fenton, Fenton-like, photo-Fenton, UV/O_3/H_2O_2, TiO_2, Fe(II), and Fe(III)	1 μM	Temperature: 20°C, 15 W Hg lamp (254 nm), light intensity: 1.81 micro Einstein s^{-1}, optical path: 5.09 cm, $O_3$was generated from O_2.flow rate of O_3: 16 mgh^{-1}photo-Fenton and photo-Fenton: pH 3 and other experiments were carried out at natural pH for Fenton and photo-Fenton Fe(II) and H_2O_2 concentration:10 μM, ultra-pure water, reservoir water, groundwater, secondary effluents from municipal WWTP	[81]
Amoxicillin	Photo-Fenton	0.1 mM	15-W black-light fluorescent lamp (365 nm), pharmaceutical solution flow rate: 80 mL min^{-1}, pH 2.5, ferric nitrate or FeOx conc.: 0.2 mmol L^{-1}, H_2O_2 conc.: 1.0–10.0 mmol L^{-1}, sewage treatment plant effluent	[82]
Amoxicillin	Microwave assisted Fenton	450 μg L^{-1}	H_2O_2 conc.: 2 g L^{-1}, $FeSO_4.7H_2O$ conc.: 0.2 g L^{-1}, pH 3.5, microwave-assisted oxidation reactions were performed with a modified version of the domestic electric oven: power of 1200 W, of 2450 MHz, synthetic wastewater	[83]
Amoxicillin	Ozonation	5.0×10^{-4} M	Ozonation were performed in a semicontinuous stirred gas–liquid reactor. 25°C, flow rate: 361 h^{-1}, synthetic wastewater	[84]
Amoxicillin	Photo-Fenton	50 mg L^{-1}	Solar simulator: 1100-W xenon arc lamp (290 nm), minimum intensity (250 W m^{-2}) pH 6.2, TOC: 26.3 mg C L^{-1}, $FeSO_4.7H_2O$ or FeO*x* conc.: 0.05 mM, H_2O_2 concentration used was 120 mg L^{-1}, pH 2.5–2.8, synthetic wastewater	[85]
Amoxicillin	Sulfate radicals under ultrasound irradiation	0.095 $mmolL^{-1}$	Ultrasonic generator: 20 kHz, Ti probe, synthetic wastewater	[86]
Amoxicillin	UV and UV/H_2O_2	00 μM	Low-pressure Hg arc-UV (254 nm) Photon fluence rate: 8×10^{-7} Einstein L^{-1} s^{-1}, effective light path: 5.5 cm, *T*: 20 ± 2°C pH 7	[87]

http://dx.doi.org/10.5772/67538

Compound name	AOP	Concentration	Reaction conditions	References
			H_2O_2: 0.4, 2, 3, 4, 5, and 10 mM, synthetic wastewater	
Trimethoprim, sulfamethoxazole, clarithromycin, erythromycin, and roxithromycin	Ozonation and UV	TMP:0.34 μg L^{-1} SMX: 0.62 μg L^{-1}CMI: 0.21 μg L^{-1}EMC: 0.62 μg L^{-1} RXM:0.54 μg L^{-1}	pH: 7.2, DOC: 23.0 mg L^{-1}, COD: 30.0 mg L^{-1} AOX: 100 mg L^{-1}, BOD_5: 2.8 mg L^{-1}, low-pressure UV unit (254 + 185 nm, 110 W power rating, 400 J m^{-2} by a flowrate: 2 m^3 h^{-1}) municipal wastewater	[88]
Beta lactam antibiotics	Sulfate radical oxidation	-	Linear accelerator (LINAC) electron pulse radiolysis system was used, T: 20–22°C, 4–6 ns pulses of 8.0 MeV electrons generating sulfate Radical concentrations of 5–10 μM per pulse were used, synthetic wastewater	[89]
Flumequine, ofloxacin, and sulfamethoxazole	Photo-Fenton	100 μg L^{-1}	pH 5, Fe^{2+} conc.: 5 mg L^{-1}, natural water	[90]
Cefalexin	Electro Fenton	50, 100, 200, and 300 mgL^{-1}	Cathode: activated carbon fiber (ACF), resistivity: 18.2 MΩ cm, *T*: 25°C, $FeSO_4{\cdot}7H_2O$ conc.: 0.5–1 mM, pH 2–5, wastewater	[91]
Cefazolin	TiO_2/UV and sunlight	1.0 × 10^{-2} mol L^{-1}	TiO_2 Degussa P25 and the N-doped TiO_2 were used; 5 × 8 W blacklight Fluorescent lamps (max. 365 nm), photonic fluence: 3.1 × 10^{-7} Einstein s^{-1}, *T*: 23 ± 2°C, pH 6.4 ± 0.1, synthetic wastewater	[92]
Ceftriaxone, cephalosporine, penicillin VK, penicillin group, enrofloxacin, and quinolone	O_3 and O_3/H_2O_2	COD: 450 mg L^{-1}	Ozone generated from O_2, pH 3, 7, and 10.6, oxygen flow rate: 100 Lh^{-1}, *T*: 20°C ± 2, synthetic wastewater	[93]
Chloramphenicol	Photo-Fenton	200 mg L^{-1},	400 W high-pressure Hg vapor lamp (295–390 and 295–710 nm), photonic flux: 6.0 × 10^{-7} and 3.3 × 10^{-6} Einstein s^{-1}, *T*: 25–30°C, synthetic wastewater	[94]
Chloramphenicol	Solar photoelectro-Fenton	245 mg L^{-1}	pH 3.0, *T*:35°C, synthetic wastewater	[95]
Chloramphenicol	UV/H_2O_2	100 mg L^{-1}	*T*: 20°C ± 2, 6-W low-pressure Hg lamp (254 nm), pH 5.5 ± 0.1, synthetic wastewater	[96]
Chloramphenicol	Direct photolysis (UVC), hydrogen peroxide/UVC and solar radiation	20 mg L^{-1}	30W three UVC lamps, illuminance: 2500 lux, 53 μW cm^{-2} (290 and 390 nm) and 18.6 μW cm^{-2} (254 nm), synthetic wastewater	[97]

Compound name	AOP	Concentration	Reaction conditions	References
Chlortetracycline, doxycycline, oxytetracycline	Ozone	5×10^{-6} M	Flow rate: 80 cm^3/min, *T*: 20–21°C, synthetic wastewater	[98]
Chlortetracycline, sulfamethoxazole	UV, electron beam, ozone	30 mgL^{-1}	pH 4.63 and 4.33, atmospheric pressure T: 22 ± 2°C, 6-W single UV-C lamp (254 nm), ozone was produced from O_2, electron accelerator (1 MeV and 40 kW), synthetic wastewater	[99]
Chlortetracycline	Photocatalytic ozonation	0.15 mM	Ozone generated from O_2 (air pressure: 5 bar, air flow rate: 1200 L h^{-1}), ozone input: 20 g m^{-3} and flow rate of the ozone/air: 20 L h^{-1}, T: 25°C, 15-W UV low-pressure lamp (254 nm), synthetic wastewater	[100]
Chlortetracycline	Photocatalytic ozonation	0.5 mM	Ozone generated from pure oxygen. flow rate: 20 mg min^{-1}, T: 20 ± 2°C, 125-W high-pressure UV lamp (260, 275, 290, 302, 307, 315, 336, 366, 406, and 434 nm), TiO_2: Degussa P25 and 0.1 g L^{-1}, synthetic wastewater	[101]
Ciprofloxacin and sulfamethoxazole	persulfate	0.15 mM	Initial pH 6 and decreased to 3–4.$K_2S_2O_8$ and Fe (II)/Fe(II)-chelate: 4.8 and 4.8 mM, river water	[102]
Ciprofloxacin, moxifloxacin	UV and TiO_2/UV	CIP:45.3 μM and MOX: 37.4 μM	Photocatalyst: TiO_2-P25, TiO_2: 0.5 g L^{-1},*T*: 298 ± 1°K, synthetic wastewater	[103]
Ciprofloxacin	Electron ionization	100 mg L^{-1}	10 MeV, 10-kW electron ionizing energy unit, synthetic wastewater	[104]
Ciprofloxacin	O_3/H_2O_2	45.27 μM	A bubble reactor was used for ozonation. Ozone generated from O_2. *T*: 6.0–62°C, ozone conc.: 2500 ppm, gas flow rate: 120 mL min^{-1}, *T*: 27.5°C, H_2O_2: 2–990 μmol L^{-1}, synthetic wastewater	[105]
Ciprofloxacin	UV, TiO_2/UV, O_3 and H_2O_2	200 μgL^{-1}	Photocatalyst: TiO_2-P25, 125W medium-pressure Hg lamp, pH 3 (UV and TiO_2/UV), TiO_2 conc.: 571 ppm, ozone generated from O_2,flowrate: 8 $Lmin^{-1}$, pH 9 (O_3 and H_2O_2), H_2O_2 conc.: 500 and 1000 mgL^{-1}, hospital wastewater	[106]
Ciprofloxacin	Photo-Fenton	0.15 mM	*T*: 298 K, 125-W high-pressure lamp, photonic flux (9×10^4 μEs m^{-2} s^{-1}), synthetic wastewater	[100]

http://dx.doi.org/10.5772/67538

Compound name	AOP	Concentration	Reaction conditions	References
Ciprofloxacin	Pulsed radiolysis, UV	100 mM	Electron pulse radiolysis 8-MeV TBS-8/16–1S linear accelerator, pulse lengths: 2.5–10 ns, $\lambda = 472$ nm, $((G\epsilon) = 5.2 \times 10^{-4}$ $m^2 J^{-1}$,3–5 Gy per 2–3 ns pulse, 8–12 replicate pulses, T: 25 ± 1°C, 125-W high-pressure Hg lamp ($E_{max} = 365$ nm) Light intensity: 0.38 $mWcm^{-2}$, TiO_2: 1.5 gL^{-1} (Degussa P25), synthetic wastewater	[107]
Amoxicillin and cloxacillin	Photo-Fenton	AMX: 138 ± 5 mgL^{-1} CLX: 84 ± 4 mgL^{-1}	230 V, 0.17 A, 6-W UV lamp (365 nm), antibiotic wastewater	[67]
Dicloxacillin and ceftazidime	Ozonation	1.5 mg L^{-1}	Ozone gas-phase concentration (mg L^{-1}): 5 ± 0.5–30 ± 0.5, volumetric ozone-gas flow rate (mL min^{-1}): 40 ± 0.5, ozone inlet pressure (bar): 2.5 ± 0.1, transmembrane pressure (TMP) (bar): 2.1 ± 0.1, volumetric cross-flow rate ($Lmin^{-1}$): 0.55 ± 0.05, temp. (°C): 24 ± 1, surface water	[108]
Doxycycline and norfloxacin	UV C, ozonation	5×10^{-5} M	Ozone was produced from pure oxygen, gas flow rate: 30 L h^{-1}, 15-W low-pressure Hg vapor lamp (254 nm), commercial activated carbon Hydraffin P-110 was used in granular form. Titanium dioxide Degussa P-25 was also used. Synthetic wastewater.	[109]
Enrofloxacin	Anodic oxidation, electro-Fenton (EF), photoelectro-Fenton (PEF) and solar photo electro-Fenton	158 mgL^{-1}	Fluorescent lamp (360 nm, 1.4 Wm^{-2}), pH 3.0, T: 35°C, synthetic wastewater	[110]
Enrofloxacin, ciprofloxacin	UV/H_2O_2, $UV/H_2O_2/Fe(II)$, O_3, O_3/UV, $O_3/UV/H_2O_2$ and $O_3/UV/H_2O_2$	0.15 mM	Ozone, generated from pure oxygen air Pressure: 5 bar, air flow rate 1200 L h^{-1} Flow rate of the ozone/air mixture: 20 L h^{-1}, T: 25°C, 15W UV low-pressure lamp (254 nm), synthetic wastewater	[100]
Ciprofloxacin, erythromycin, ofloxacin, sulfamethoxazole, trimethoprim	Ozonation	ERYC: 346 ngL^{-1} CIP: 5524 ngL^{-1} OFX: 2275ngL^{-1} SMX: 279ngL^{-1} TMP: 104 ngL^{-1}	pH 7.54, COD (mgL^{-1}) 269, BOD_5 (mgL^{-1}) 42,T: 25°C, ozone was produced by a corona discharge ozonator (Ozomatic, 119 SWO100) fed by an AirSep AS-12 PSA oxygen generation unit.gas flow rate: 0.36 $Nm^{-3}h^{-1}$, pH value of 8.5 ± 0.1 urban	[111]

Compound name	AOP	Concentration	Reaction conditions	References
			wastewaters were taken from Alcala de Henares (Madrid)	
Flumequine, ofloxacin, sulfamethoxazole	Modified photo-Fenton	100 μgL^{-1}	5 mgL^{-1} Fe, 35 mgL^{-1}, 50 mgL^{-1} H_2O_2, oxalic acid, initial pH ≈ 7, λ < 400 nm Solar UV power: 30 Wm^{-2}, municipal wastewater treatment plant effluent was taken downstream of the MWTP secondary biological treatment in El Ejido (province of Almería, Spain). pilot compound parabolic collector (CPC) was used for photo-Fenton experiment	[112]
Flumequine	Fenton and photo-Fenton	500 μg L^{-1}	Low-pressure mercury lamp, 15 W, λ_{max} = 254 nm, irradiance: 8.3 mW cm^{-2}, H_2O_2: 0.5–10.0 mmol L^{-1}, Fe(II): 0.25–1.0 mmol L^{-1}. $NaHSO_4/H_2O_2$: 1	[97]
Levofloxacin	OzonationandTiO_2/UV	20 mgL^{-1}	Ozone flow rate: 3.3 gh^{-1}, oxygen was used as a feed gas, commercial TiO_2 Degussa P25 was used as catalyst, *T*: 17°C, pH 6.5, synthetic wastewater	[113]
Metronidazole	Electro-Fenton	80 mg L^{-1}	*T*: 20°C, synthetic wastewater	[114]
Moxifloxacin	TiO_2/UV	37.4 and 124.6 μM	*T*: 25°C, pH: 3.0, 7.0 and 10.0, stirring speed: 13.2 rps, reactor volume: 200 mL, catalyst loading: 1.0 g L^{-1}, air flow: 60 mLmin^{-1}, phosphate buffer conc.: 10 mM, light intensity UV-A 104 mW, synthetic wastewater	[115]
Moxifloxacin	TiO_2/UV	12.5, 24.9, 37.4, 49.9, 62.3 and 124.6 μM	*T*: 5, 15, 25, 35, 45, and 65°C, pH 7, stirring speed 2.3, 7.9 and 13.2 rps, reactor volume 200 mL, catalyst loading 0.25, 0.5, 1, 3, 5, and 8 g L^{-1}, oxygen, air, nitrogen flow: 60 mL min^{-1}, buffer concentration 10 mM, light intensity UV-A at 3 cm, 485 W cm^{-2}, ISO concentration μmol L^{-1} 37.4, 374, 3740, 37.4 × 10^3, 74.8 × 10^3, and 18.7 × 10^4, KI concentration 3.74, 37.4, 374, 3740, and 7480 mol L^{-1}, synthetic wastewater	[116]
Ofloxacin and trimethoprim	Solar photo-Fenton process	100 μg L^{-1}	*T*: 25°C, UV power: 30 W m^{-2}, secondary treated domestic effluents	[117]

http://dx.doi.org/10.5772/67538

Compound name	AOP	Concentration	Reaction conditions	References
Ofloxacin	Solar Fenton	10 mg L^{-1}	*T*: 20°C, H_2O_2 conc.: 2.5 mg L^{-1} and Fe^{2+} conc.: 2 mg L^{-1},demineralized water, simulated natural freshwater, simulated effluent from municipal wastewater treatment plant and pre-treated real effluent from municipal wastewater treatment plant	[1]
Oxolinic acid	TiO_2/UV	20 mg L^{-1}	Titanium dioxide Degussa P-25 with a surface area of 50 m^2 g^{-1} (size ~20–30 nm) was used as provided. 14 Wm^{-2}, emission maximum at 365 nm, synthetic wastewater	[118]
Oxytetracycline	Photo-Fenton	20 mg L^{-1}	*T*: 25°C, $I = 500$ Wm^{-2},wastewater	[119]
Oxytetracycline	TiO_2/UV	20 mg L^{-1}	*T*: 25°C, photocatalyst: Titanium dioxide Degussa P-25, 1000-W Xe-OP lamp, radiant power: 3.55 J s^{-1}, synthetic wastewater	[120]
Roxithromycin, sulfamethoxazole, and trimethoprim	H_2O_2/UV, Fenton, photo-Fenton, UV, ozon	50–100 μgL^{-1}	Membrane bioreactor Hollow-fiber ultrafiltration (UF) membranes, nominal pore size: 0.04 mm, pH: 7.2 municipal wastewater UV radiation, O_3 and AOP *T*: 20°C, pH: 3.0, synthetic wastewater and MBR permeate	[121]
Sulfachlorpyridazine, sulfapyridine, and sulfisoxazole	TiO_2/UV	50–200 μM	Xe arc lamp, 172 nm, power: 125 W, *T*: 20°C, photocatalyst: degussa P25, synthetic wastewater	[122]
Sulfamethazine	Electrochemical incineration	193 mg dm^{-3}	Synthetic wastewater	[123]
Sulfamethazine	Gamma irradiation/H_2O_2	20 mgL^{-1}	Dose rate: 339 Gymin^{-1}, pH: 6.0–7.5, H_2O_2 concentration: 0, 10, and 30 mgL^{-1}, synthetic wastewater	[124]
Sulfamethoxazole and acetaminophen	Ozone, Fenton-like	30 mg L^{-1}	Ozone was produced from pure oxygen, gas flow rate: 20 Lh^{-1},15-W black light lamps, 365-nm radiation, synthetic wastewater	[125]
Sulfamethoxazole, ciprofloxacin, clarithromycin, erythromycin, sulfamethoxazole	UV	763.31 and 2.32 μgL^{-1}	Two different UV lamps: medium pressure (MP) lamp with power of 2–10 kW and low-pressure (LP) UV lamp with power of 0.25 kW LP lamp wavelengths: 254 nm and 185 nm. MP	[126]

Compound name	AOP	Concentration	Reaction conditions	References
			lamp has polychromatic emission, hospital wastewater	
Sulfamethoxazole, roxithromycin, erythromycin, ciprofloxacin and sulfathiazole	UV	1 μgmL^{-1}	Mercury vapor lamp (UV 254 nm) or black light phosphor bulb (UV 350 nm), xenon lamp (750 Wcm^{-2}, 250 Wcm^{-2}, pH: 5.5–8.1, *T*: 20°C, synthetic wastewater	[127]
Sulfamethoxazole, sulfamethazine, sulfadiazine, trimethoprim	UV and UV/H_2O_2	4 μM	Low-pressure UV lamps, fluence = 540 mJ cm^{-2}, H_2O_2 dose = 6 mg L^{-1}, synthetic wastewater, surface water, wastewater treatment plant effluent	[128]
Sulfamethoxazole	Anodic oxidation and electro-Fenton	1.3 mM	Catalyst: 0.2 mM Fe^{2+} and/or 0.2 mM Cu^{2+}, pH 3.0 and T: 23 ± 2°C, current: 30–450 mA, synthetic wastewater	[129]
Sulfamethoxazole	Photoelectro-Fenton	200–300 mg L^{-1}	*T*: 20°C, anode: RuO2/Ti, cathode: RuO_2/Ti, UV Lamp SLUV-8, 254/365 nm, energy input: 1407 Wcm^{-}2, current: 0.36 A, synthetic wastewater	[130]
Sulfamethoxazole	Ozone	0.150 mM	Ozone generated from pure oxygen, pH: 2 and 8, H_2O_2: 0.013 M, flow: 3.0 ml min^{-1}, *T*: 25°C, gas flow: 8.5 gNm^{-3}, synthetic wastewater	[131]
Sulfamethoxazole	Ozone	200 μg L^{-1}	Ozone generated from pure oxygen, 15-W black light lamps (365 nm), flux of radiation: 7.05 ± 0.05 × 10^{-5} Einstein min^{-1}, primary wastewater effluent	[132]
Sulfamethoxazole	Photo-Fenton	200 mg L^{-1}	Black-light blue lamps with power of 8W (350 and 400 nm), photon flow: 6.85–5.67 Einstein s^{-1}. *T*: 25 ± 0.8°C; TOC = 94.5 mgL^{-1} and COD = 290 mgO_2 L^{-1}, synthetic wastewater	[133]
Sulfamethoxazole	Solar photo-Fenton	10 mg L^{-1}	1100-W xenon arc lamp (below 290 nm), intensity: 250 Wm^2, *T*: 25°C, synthetic wastewater, seawater	[134]
Sulfamethoxazole	TiO_2/UV	100 mg L^{-1}	Catalyst: TiO_2 Degussa P25, T: 25°C, xenon lamp (1000 W), wavelength: below 290 nm, synthetic wastewater	[135]
Sulfamethoxazole	TiO_2/UV	2.5–30 mgL^{-1}	9W lamp (350–400 nm), photon flux: 2.81 × 10^{-4} Einstein min^{-1}, *T*: 25°C, synthetic wastewater	[136]

http://dx.doi.org/10.5772/67538

Compound name	AOP	Concentration	Reaction conditions	References
Sulfasalazine	Fenton-like	100 mgL^{-1}	Initial pH:3.0, industrial wastewater	[137]
Sulfadiazine	Gamma irradiation	10–30 mgL^{-1}	The dose rate of gamma-ray: 103 $Gymin^{-1}$, pH: 5.5–6.5, wastewater	[138]
Sulfamethoxazole	Catalytic ozonation	50 ppm	Catalysts: commercial activated carbon and commercial multi-walled carbon nanotubes, pH: 4.8, flow rate: 150 cm^3min^{-1}, ozone concentration: 50 g m^{-3}, *T*: 20°C, synthetic wastewater	[139]
Sulfamethoxazole	UV/H_2O_2	1 mg L^{-1}	0.45-kW polychromatic (200–300 nm) medium-pressure (MP) Hg vapor lamp, H_2O_2 concentrations: 0–4.41 mM, synthetic wastewater	[140]
Sulfamethoxazole	UV, O_3, O_3/TiO_2, O_3/UVA, O_2/TiO_2/UVA, O_3/TiO_2/UVA	30–80 mg L^{-1}	Ozone was generated from pure oxygen. catalyst: TiO_2 Degussa P25, high-pressure mercury lamp (700 W, 238–579 nm) Radiation intensity: 0.111 Einstein h^{-1}, synthetic wastewater	[141]
Sulfamethazine	Gamma irradiation	20 mgL^{-1}	Dose rate: 320 $Gymin^{-1}$, G (Fe^{3+}): 15.6 (per 100eV), Fe^{2+} concentrations: 0, 0.1, 0.2, 0.4, and 0.6 mM, pH: 6.0–7.5, irradiation: 200, 400, 600, 800, and 1000 Gy, synthetic wastewater	[142]
Sulfamethazine	Sonophotolytic goethite/oxalate Fenton-like	25 mg L^{-1}	*T*: 20°C, 9-W UVA lamp (λ_{max} = 365 nm), light intensity: 7.7 mW cm^{-2}, ultrasonic shockwave Frequency: 20 kHz, purified air flow:1.0 L min^{-1}, synthetic wastewater	[143]
Sulfanilamide	Electro-Fenton and UVA photoelectro-Fenton	239–2511 mg L^{-1}	6-W fluorescent black light blue tube (320–400 nm), photoionization energy: 5 Wm^{-2}, synthetic wastewater	[123]
Tetracycline, chlortetracycline, oxytetracycline, doxycycline	Electron pulse radiolysis, gamma radiolysis	0.5 mM	Pulse radiolysis *k* = 472 nm, dose of radiolysis:3–5 Gy per 2–3 ns pulse, pH 7, *T*: 22°C, Xe arc lamp (172 nm), synthetic wastewater γ-radiolysis pH 7, *T*: 22°C, synthetic wastewater	[12]

Compound name	AOP	Concentration	Reaction conditions	References
Tetracycline	Electrochemical oxidation	TC = 200 mg L^{-1}	The process was performed using a DSA (mixed metal oxide, Ti/RuO_2-IrO_2) anode carbon-felt from cathode. synthetic wastewater	[144]
Tetracycline	Electrochemical oxidation, electro Fenton	100 mg L^{-1}	Three electrodes as anode: commercial pure Pt, boron-doped diamond (BDD, thin-film deposited on a niobium substrate), and commercial DSA (mixed metal oxide Ti/RuO_2–IrO_2), and a tridimensional, carbon-felt electrodes as cathode were used. *T*: 23°C, synthetic wastewater	[145]
Tetracycline	Ozonation	20–100 mgL^{-1}	O_3 was generated from oxygen. *T*: 25°C, synthetic wastewater	[146]
Tetracycline	Photo-Fenton	TOC: 13 mg L^{-1}	15-W black-light lamp (365 nm), irradiance: 19 Wm^{-2}, flow rate: 80 mL min^{-1}, synthetic wastewater, surface water and a sewage treatment plant effluent	[147]
Tetracycline	Photocatalysis	67 mgL^{-1}	Medium mercury lamp, synthetic wastewater	[148]
Tinidazole	Ozone	30 mgL^{-1}	*T*: 25°C, synthetic wastewater, surface water and a sewage treatment plant effluent	[149]
Tinidazole	Sonolysis	45, 80, and 100 ppm	pH: 3, 5, 7, 9; H_2O_2 conc.: 83, 167, 250, 333, and 417 mML^{-1}, frequency: 40, 80, 120, and 160 kHz, input power:750 W, pharmaceutical wastewater (Tehran, Iran), synthetic wastewater	[150]
Trimethoprim	Anodic oxidation, electro-Fenton, photoelectro-Fenton, solar photoelectro-Fenton	20.0 mg L^{-1}	6-W fluorescent blacklight blue lamp (350–410 nm), synthetic wastewater, wastewater	[151]
Trimethoprim	BDD, electrochemical oxidation	1.72×10^{-4} mol L^{-1}	*T*: 25°C, synthetic wastewater	[120]

Table 2. Summary of reaction conditions for antibiotic removal from water by AOP.

http://dx.doi.org/10.5772/67538

energy or participate in redox reactions with the compounds that are adsorbed on the catalyst's surface [1].

Due to some disadvantages of the heterogeneous photocatalysis (e.g., rather small quantum efficiency of the process; comparatively narrow light-response reach of TiO_2; the requirement of postseparation and recovery of the catalyst particles from the reaction mixture in aqueous slurry systems), TiO_2 appear to have some interesting properties, such as high chemical stability in a wide pH range, strong resistance to chemical breakdown and photocorrosion, and high efficiency. The catalyst is also inexpensive and can be reprocessed [134, 156]. The characteristics of antibiotics to be treated like pKa and molecular structure will identify not only the performance of their photocatalytic breakdown but also the mechanisms of the oxidation products formation.

Ultraviolet (UV) disinfection is progressively discovering practices in UWTPs. Photolytic breakdown can be either direct or indirect. In direct photolysis, the target pollutant assimilates a solar photon, which causes to a breakup of the molecule. In an indirect photolysis mechanism, as a matter of course occurring molecules in the system such as dissolved organic matter (DOM) behave as sensitizing species, which creates strong reactive agents such as hydroxyl radicals, singlet oxygen, and hydrate electrons under solar radiation [1, 117].

Ultraviolet irradiation has been greatly used for the treatment of waters and wastewaters worldwide. Different studies have stated the effective treatment of UV irradiation for the removal of antibiotics in wastewater effluents [63]. It has been lately stated that at high UV doses of almost 11,000–30,000 mJ cm^2, a nearly complete removal of tetracyclines and ciprofloxacin was obtained. Kim et al. [99] noticed that sulfonamides and quinolones demonstrate high removal efficiency in the reach of 86–100% throughout the UV process [1].

Many of the antibiotics have aromatic rings, structural moieties (such as phenol and nitro groups) heteroatoms, and other functional chromophore groups that can either absorb solar radiation or react with photogenerated transient species in natural. The organic material, UV dosage, contact time, and the chemical construction of the compound are significant agents ruling the removal performance of antibiotics throughout direct photolysis. This technique is only practicable to wastewater-containing photosensitive compounds and waters with low COD concentrations [5].

Most traditional operation performed in WWTPs and DWTPs (such as coagulation, flocculation, sedimentation, and filtration) were ineffective in the removal of these compounds [63], taking the improvement of new effective methodologies. Owing to the recalcitrant nature of the effluents including antibiotics residues, the implementation of the AOPs arises as an alternate. In fact, ozonation and Fenton's oxidation are the most tried methodologies. Although ozonation has the benefit of being used to fluctuate flow rates and compositions, the high cost of material and the energy required to provide the process constitute a primary disadvantage. Oxyhydroxides produce precipitate (if the pH range is not controlled well) when a homogeneous process is used and the necessity of recovering dissolved catalyst is a disadvantage. This is another process that is applied often to the group of beta-lactam antibiotics, combined with UV irradiation (photo-Fenton).

4. Conclusions

The consumption of antibiotics worldwide by human and veterinary uses has been increasing significantly. This is an important public concern because they have endocrine-disrupting properties even in trace concentrations and can cause microorganism resistance in aquatic environments. According to the researches made in recent years, advanced oxidation processes are promising treatment methods for the removal of the antibiotic compounds from water.

In the event of the photochemical technologies, we can determine that the photochemical AOPs are usually easy, clean, comparatively inexpensive, and productive against the classical, chemical AOPs. Four basic types of photochemical AOPs (H_2O_2/UV, O_3/UV, H_2O_2/Fe^{2+}/UV, and TiO_2/UV) have been enforced to reduce and/or mineralize organic pollutants. We have defined that, within these photochemical processes, the photocatalytic ones had mainly a better performance.

Furthermore, it is significant to point that heterogeneous photocatalysis has been the aim of an enormous improvement in the last decade. In fact, TiO_2 is a semiconducter approach that exists, for example, as a chemically very stable, biologically inefficient, very easy to manufacture, cheaper than the photocatalytic viewpoint, active and several important photocalysts with an energy vacancy comparable to solar photons.

The economic robust of AOPs for full-scale wastewater treatment needs to be extensively investigated. These technologies should be modified to achieve both technical efficiency and cost effectiveness so that water industries could afford the adaptation of such technologies.

Author details

Ayse Kurt, Berna Kiril Mert, Nihan Özengin, Özge Sivrioğlu and Taner Yonar*

*Address all correspondence to: yonar@uludag.edu.tr

Department of Environmental Engineering, Faculty of Engineering, Uludağ University, Görükle Campus, Bursa, Turkey

References

[1] Michael I, Rizzo L, McArdell CS, Manaia CM, Merlin C, Schwartz T, Dagot C, Fatta-Kassinos D. Urban wastewater treatment plants as hotspots for the release of antibiotics in the environment: A review. Water Res. 2013;**47**:957–995. DOI:10.1016/j.watres.2012.11.027

[2] Thiele-Brühn S. Pharmaceutical antibiotic compounds in soils: A review. J. Plant Nutr. Soil Sci. 2003;**166**:145–167. DOI: 10.1002/jpln.200390023

[3] Alaton I, Dogruel S. Pre-treatment of penicillin formulation effluent by advanced oxidation processes. J. Hazard. Mater. 2004;**B112**:105–113. DOI:10.1016/j.jhazmat.2004.04.009

[4] Schwartz T, Volkmann H, Kirchen S, Kohnen W, Schon-Holz K, Jansen B, Obst U. Real-time PCR detection of *Pseudomonas aeruginosa* in clinical and municipal wastewater and genotyping of the ciprofloxacin-resistant isolates. FEMS Microbiol. Ecol. 2006;**57**:158–167. DOI: 10.1111/j.1574-6941.2006.00100.x 1

[5] Homem V, Santos L. Degradation and removal methods of antibiotics from aqueous matrices: A review. J. Environ. Manage. 2011;**92**:2304–2347. DOI:10.1016/j.jenvman.2011.05.023

[6] EPA. Handbook on Advanced Non-Photochemical Oxidation Process, US. EPA, Washington, DC; 2001.

[7] Kümmerer K. Significance of antibiotics in the environment. J. Antimicrob. Chemother. 2003;**52**:5–7. DOI: 10.1093/jac/dkg293

[8] Hirsch R, Ternes T, Haberer K, Kratz KL. Occurrence of antibiotics in the aquatic environment. Sci. Total Environ. 1999;**225**:109–118. DOI:10.1016/S0048-9697(98)00337-4

[9] Boxall ABA, Kolpin DW, Halling-Sorensen B, Tolls J. Peer reviewed: Are veterinary medicines causing environmental risks? Environ. Sci. Technol. 2003;**37**(15):286A–294A. DOI: 10.1021/es032519b

[10] Banik KK, Hossain S. Pharmaceuticals in drinking water: A future water quality threat. Indian J. Environ. Prot. 2006;**26**:926–932.

[11] Chee-Sanford JC, Aminov RI, Krapac IJ, Garrigues-Jeanjean N, Mackie RI. Occurrence and diversity of tetracycline resistance genes in lagoons and groundwater underlying two swine production facilities. Appl. Environ. Microbiol. 2001;**67**:1494–1502. DOI: 10.1128/AEM.67.4.1494-1502.2001

[12] Jeong J, Song W, Cooper WJ, Jung J, Greaves J. Degradation of tetracycline antibiotics: Mechanisms and kinetic studies for advanced oxidation/reduction processes. Chemosphere. 2010;**78**:533–540. DOI:10.1016/j.chemosphere.2009.11.024

[13] Johansson N. Antibiotics in the environment. Environment and pharmaceuticals. Swedish Environmental Protection Agency, and Roland Möllby, Karolinska Institute; 2006.

[14] Kümmerer K. Antibiotics in the aquatic environment-a review-Part I. Chemosphere. 2009;**75**:417–434. DOI:10.1016/j.chemosphere.2008.11.086

[15] Merck and Co. The Merck Manual of Diagnosis and Therapy. 17th edition. John Wiley & Sons, Indianapolis, IN; 1999.

[16] Ikehata K, Naghashkar NJ, El-Din MG. Degradation of aqueous pharmaceuticals by ozonation and advanced oxidation processes: A review. Ozone Sci. Eng. 2006;**28**:353–414.

[17] Rozas O, Contreras D, Mondaca MA, Pérez-Moya M, Mansilla HD. Experimental design of Fenton and photo-Fenton reactions for the treatment of ampicillin solutions. J. Hazard. Mater. 2010;177:1025–1030. DOI:10.1016/j.jhazmat.2010.01.023

[18] Alexy R, Kümpel T, Kümmerer K. Assessment of degradation of 18 antibiotics in the closed bottle test. Chemosphere. 2004;**57**:505–512. DOI:10.1016/j.chemosphere.2004.06.024

[19] Xu WH, Zhang G, Zou SC, Li XD, Liu YC. Determination of selected antibiotics in the Victoria Harbour and the Pearl River, South China using high performance liquid chromatography electrospray ionization tandem mass spectrometry. Environ. Pollut. 2007;**145**:672–679. DOI:10.1016/j.envpol.2006.05.038

[20] Kemper N. Veterinary antibiotics in the aquatic and terrestrial environment. Ecol. Indic. 2008;**8**:1–13. DOI:10.1016/j.ecolind.2007.06.002

[21] Botitsi E, Frosyni C, Tsipi D. Determination of pharmaceuticals from different therapeutic classes in wastewaters by liquid chromatography electrospray ionization-tandem mass spectrometry. Anal. Bioanal. Chem. 2007;**387**:1317–1327. DOI: 10.1007/s00216-006-0804-8

[22] Chang H, Hu J, Asami M, Kunikane S. Simultaneous analysis of 16 sulfonamide and trimethoprim antibiotics in environmental waters by liquid chromatography-electrospray tandem mass spectrometry. J. Chromatogr. A. 2008;**1190**:390–393. DOI:10.1016/j.chroma.2008.03.057

[23] Duong HA, Pham NH, Nguyen HT, Hoang TT, Pham HV, Pham VC, Berg M, Giger W, Alder AC. Occurrence, fate and antibiotic resistance of fluoroquinolone antibacterials in hospital wastewaters in Hanoi, Vietnam. Chemosphere. 2008;**72**:968–973. DOI:10.1016/j.chemosphere.2008.03.009

[24] Hartmann A, Alder AC, Koller T. Widmer RM. Identification of fluoroquinolone antibiotics as the main source of umuC genotoxicity in native hospital wastewater. Environ. Toxicol. Chem. 1998;**17**:377–382. DOI: 10.1002/etc.5620170305

[25] Lindberg R, Jarnheimer PA, Olsen B, Johansson M, Tysklind M. Determination of antibiotic substances in hospital sewage water using solid phase extraction and liquid chromatography/mass spectrometry and group analogue internal standards. Chemosphere. 2004;57:1479–1488. DOI:10.1016/j.chemosphere.2004.09.015

[26] Turiel E, Bordin G, Rodríguez AR. Study of the evolution and degradation products of ciprofloxacin and oxolinic acid in river water samples by HPLC–UV/MS/MS–MS. J. Environ. Monitor. 2005;**7**:189–195. DOI: 10.1039/B413506G

[27] Brown KD, Kulis J, Thomson B, Chapman TH, Mawhinney DB. Occurrence of 24 antibiotics in hospital, residential, and dairy effluent, municipal wastewater, and the Rio Grande in New Mexico. Sci. Total Environ. 2006;**366**:772–783. DOI:10.1016/j.scitotenv.2005.10.007

[28] Thomas KV, Dye C, Schlabach M, Langford KH. Treatment works. J. Environ. Monitor. 2007;**9**:1410–1418. DOI: 10.1039/B709745J

[29] Martins AF, Vasconcelos TG, Henriques DM, Frank CD, König A, Kümmerer K. Concentration of ciprofloxacin in Brazilian hospital effluent and preliminary risk assessment: A case study. Clean. 2008;**36**:264–269. DOI: 10.1002/clen.200700171

[30] Alexy R, Sommer A, Lange FT, Kümmerer K. Local use of antibiotics and their input and fate in a small sewage treatment plant—Significance of balancing and analysis on a local scale vs. nationwide scale. Acta Hydroch. Hydrob. 2006;**34**:587–592. DOI: 10.1002/aheh.200400657

[31] Watkinson AJ, Murby EJ, Kolpin DW, Constanzo SD. The occurrence of antibiotics in an urban watershed: From wastewater to drinking water. Sci. Total Environ. 2009;**407**:2711–2723. DOI: 10.1016/j.scitotenv.2008.11.059

[32] Färber H. Antibiotika im Krankenhausabwasser (Antibiotics in the hospital sewage). Hyg. Med. 2002;**27**:35.

[33] Christian T, Schneider RJ, Färber HA, Skutlarek D, Meyer MT, Goldbach HE. Determination of antibiotic residues in manure, soil, and surface waters. Acta Hydroch. Hydrob. 2003;**31**:36–44. DOI: 10.1002/aheh.200390014

[34] Giger W, Alder AC, Golet EM, Kohler HPE, McArdell CS, Molnar E, Siegrist H, Suter MJF. Occurrence and fate of antibiotics as trace contaminants in wastewaters, sewage sludges, and surface waters. Chimia. 2003a;**57**:485–491. DOI:10.2533/000942903777679064

[35] Giger W, Alder AC, Golet EM, Kohler HPE, McArdell CS, Molnar E, Pham Thi NA, Siegrist H. Antibiotikaspuren auf dem Weg von Spital-und Gemeindeabwasser in die Fliessgewässer: Umweltanalytische Untersuchungen über Einträge und Verhalten (Traces of antibiotics on the way from hospital and local sewage treatment plant in flowing bodies of water: Environmental analytical investigations on their entry and behavior). In: Track, T., Kreysa, G. (Eds.), Spurenstoffe in Gewässern. Pharmazeutische Reststoffe und endokrinwirksame Substanzen (Trace materials in bodies of water. Pharmaceutical trace materials and endocrine active substances). Wiley-VCH GmbH &Co, New Jersey, USA. 2003b p. 21–33.

[36] Sacher F, Gabriel S, Metzinger M, Stretz A, Wenz M, Lange FT, Brauch HJ, Blankenhorn I. Arzneimittelwirkstoffe im Grundwasser – Ergebnisse eines Monitoring-Programms in Baden-Württemberg (Active pharmaceutical ingredients in ground water – the results of a monitoring program in Baden-Württemberg (Germany)). Vom Wasser. 2002;**99**:183–196. ISSN 0083-6915

[37] Calamari D, Zuccato E, Castiglioni S, Bagnati R, Fanelli R. Strategic of therapeutic drugs in the rivers Po and Lambro in Northern Italy. Environ. Sci. Technol. 2003;**37**:1241–1248. DOI: 10.1021/es020158e

[38] Kolpin DW, Furlong ET, Meyer MT, Thurman EM, Zaugg SD, Barber LB, Buxton HT. Pharmaceuticals, hormones, and others organic wastewater contaminants in US streams, 1999–2000: A national reconnaissance. Environ. Sci. Technol. 2002;**36**:1202–1211. DOI: 10.1021/es011055j

[39] Zuccato E, Chiabrando C, Castiglioni S, Calamari D, Bagnati R, Schiarea S, Fanelli R. Cocaine in surface waters: A new evidence-based tool to monitor community drug abuse. Environ. Health. 2005;**5**:4–14. DOI: 10.1186/1476-069X-4-14

[40] Parsons SA, Williams M. Introduction. In: Parsons, S. (Ed.). Advanced Oxidation Processes for Water and Wastewater Treatment. IWA Publishing, London, UK. 2004, p. 1–6.

[41] Ince NH, Apikyan IG. Combination of activated carbon adsorption with light enhanced chemical oxidation via hydrogen peroxide. Water Res. 2000;**34**:4169–4176. DOI:10.1016/S0043-1354(00)00194-9

[42] Azbar N, Kestioğlu K, Yonar T. Application of Advanced Oxidation Processes (AOPs) to Wastewater Treatment. Case Studies: Decolourization of Textile Effluents, Detoxification of Olive Mill Effluent, Treatment of Domestic Wastewater, Ed. A.R. BURK. Water Pollution: New Research, Nova Science Publishers, New York, 2005 p. 99–118. ISBN-1-59454-393-3

[43] Mandal A, Ojha K, De AK, Bhattacharjee S. Removal of catechol from aqueous solution by advanced photo-oxidation process. Chem. Eng. J. 2004;**102**:203–208. DOI:10.1016/j.cej.2004.05.007

[44] Yonar T. Decolorisation of Textile Dyeing Effluents Using Advanced Oxidation Processes. Eds. Hauser P.J. In: Advances in Treating Textile Effluent. Intech Publisher, Rijeka, Croatia, 2011.

[45] Litter MI. Introduction to photochemical advanced oxidation processes for water treatment. Environ. Photochem. Part II. 2005;**2**:325–326. DOI:10.1007/b138188

[46] Gottschalk C, Libra JA, Saupe A. Ozonation of Water and Wastewater. Wiley-Vch, Weinheim, Deutschland, 2009 ISBN:978-3-527-31962-6.

[47] Zaviska F, Drogui P, Mercier G, Blais JF. Proc'ed'es d'oxydation avanc'ee dans le traitement des eaux et des effluents industriels: Application 'a la d'egradation des polluants r'efractaires. Rev. Sci. Eau. 2009;**22**(4):535–564. DOI: 10.7202/038330ar

[48] Oturan MA, Aaron JJ. Advanced oxidation Processes in water/wastewater treatment: Principles and Applications: A review. Crit. Rev. Environ. Sci. Technol. 2014;**44**:2577–2641. DOI:10.1080/10643389.2013.829765

[49] Buxton GU, Greenstock CL, Helman WC, Ross AB. Critical review of rate constant for reactions of hydrated electrons, hydrogen atoms and hydroxyl radicals (HO/O^-) in aqueous solution. J. Phys. Chem. Ref. Data. 1988;**17**(2):513–886. DOI: 10.1063/1.555805

[50] Mokrini M, Oussi D, Esplugas S. Oxidation of aromatic compounds with UV radiation/ozone/hydrogen peroxide. Water Sci. Technol. 1997;**35**:95–102

[51] Hernandez R, Zappi M, Colluci J, Jones R. Comparing the performance of various advanced oxidation process for treatment of acetone contaminated water. J. Hazard. Mater. 2002;**92**(1): 33–50. DOI:10.1016/S0304-3894(01)00371-5

[52] Klavarioti M, Mantzavinos D, Kassinos D. Removal of residual pharmaceuticals from aqueous systems by advanced oxidation process. Environ. Int. 2009;**35**:402–417. DOI:10.1016/j.envint.2008.07.009

[53] Bautista P, Mohedano AF, Casas JA, Zazo JA, Rodriguez JJ. An overview of the application of Fenton oxidation to industrial wastewaters treatment. J. Chem. Technol. Biotechnol. 2008;**83**(10):1323–1338. DOI: 10.1002/jctb.1988

[54] Gonzalez MG, Oliveros E, Worner M, Braun AM. Vacuum-ultraviolet photolysis of aqueous reaction systems. J. Photochem. Photobiol. C Photochem. Rev. 2004;**5**:225–246. DOI: 10.1016/j.jphotochemrev.2004.10.002

[55] Niaounakis M, Halvadakis CP. Olive processing waste management—Literature review and patent survey, 2nd ed., Elsevier, Amsterdam, The Netherlands, 2006. ISSN:1478-7482

[56] Crittenden JC, Trussell RR, Hand DW, Howe KJ, Tchobanoglous G. Water treatment: Principles and Design, 2nd ed., Wiley, Oxford, UK, Estonian Academy Publishers, Talinn, Estonia, 2005.

[57] Stasinakis AT. Use of selected advanced oxidation processes (AOPs) for wastewater treatment—A mini review. Global Nest J. 2008;**10**(3):376–385.

[58] Munter R. Advanced oxidation processes-currents status and prospects. Proceedings of the Estonian Academy of Sciences. Aben, H., editor; 2001 ISSN: 1406-0124.

[59] Ghaly MY, Härtel G, Mayer R, Haseneder R. Photochemical oxidation of *p*-chlorophenol by UV/H_2O_2 and photo-Fenton process. A comparative study. Waste Manage. 2001;**21**:41–47. DOI: 10.1016/S0956-053X(00)00070-2

[60] Gosavi VD, Sharma S. A general review on various treatment methods for textile wastewater. JECET. 2014;3(1):29–39. E-ISSN: 2278–179X

[61] Kormann C, Bahnemann DF, Hoffmann MR. Photolysis of chloroform and other organic molecules in aqueous TiO_2 suspensions. Environ. Sci. Technol. 1991;**25**:494–500. DOI: 10.1021/es00015a018

[62] Zahraa O, Chen HY, Bouchy M. Adsorption and photocatalytic degradation of 1,2-dichloroethane on suspended TiO_2. J. Adv. Oxid. Technol. 1999;**4:**167–173.

[63] Adams C, Asce M, Wang Y, Loftin K, Meyer M. Removal of antibiotics from surface and distilled water in conventional water treatment processes. J. Environ. Eng. 2002;**128**:253–260. DOI: 10.1061/(ASCE)0733-9372

[64] Arslan Alaton I. Dogruel S, Baykal E, Gerone G. Combined chemical and biological oxidation of penicillin formulation effluent. J. Environ. Manag. 2004;**73**:155–163. DOI: 10.1016/j.jenvman.2004.06.007

[65] Saritha P, Aparna C, Himabindu V, Anjaneyulu Y. Comparison of various advanced oxidation processes for the degradation of 4-chloro-2nitrophenol. J. Hazard. Mater. 2007;**149**:609–614. DOI:10.1016/j.jhazmat.2007.06.111

[66] Naddeo V, Meric S, Kassinos D, Belgiorno V, Guida M. Fate of pharmaceuticals in contaminated urban wastewater effluent under ultrasonic irradiation. Water Res. 2009;**43**:4019–4027. DOI:10.1016/j.watres.2009.05.027

[67] Elmolla ES, Chaudhuri M. The feasibility of using combined TiO_2 photocatalysis-SBR process for antibiotic wastewater treatment. Desalination. 2011;**272**:218–224. DOI: 10.1016/j.desal.2011.01.020

[68] Lange F, Cornelissen S, Kubac D, Sein MM, Von Sonntag J, Hannich CB, Golloch A, Heipieper HJ, Moder M, Von Sonntag C. Degradation of macrolide antibiotics by ozone: A mechanistic case study with clarithromycin. Chemosphere. 2006;**65**(1):17–23. DOI: 10.1016/j.chemosphere.2006.03.014

[69] Dodd MC. Potential impacts of disinfection processes on elimination and deactivation of antibiotic resistance genes during water and wastewater treatment. J. Environ. Monit. 2012;**14**:1754–1771. DOI: 10.1039/C2EM00006G

[70] Huber MM, Göbel A, Joss A, Hermann N, Loffler D, McArdell CS, Ried A, Siegrist H, Ternes TA, Von Gunten U. Oxidation of pharmaceuticals during ozonation of municipal wastewater effluents: A pilot study. Environ. Sci. Technol. 2005;**39**:4290–4299.DOI: 10.1021/es048396s

[71] Rizzo L, Fiorentino A, Anselmo A. Advanced treatment of urban wastewater by UV radiation: Effect on antibiotics and antibiotic-resistant *E. coli* strains. Chemosphere. 2013;**92**:171–176. DOI: 10.1016/j.chemosphere.2013.03.021

[72] Sharma VK, Liu F, Tolan S, Sohn M, Kim H, Oturan MA. Oxidation of b-lactam antibiotics by ferrate(VI). Chem. Eng. J. 2013;**221**:446–451. DOI: 10.1016/j.cej.2013.02.024

[73] Chaudhuri M, Wahap M, Affam AC. Treatment of aqueous solution of antibiotics amoxicillin and cloxacillin by modified photo Fenton process. Desalination and Water Treatment. 2013;**51**:7255–7268. DOI:10.1080/19443994.2013.773565

[74] Li Y, Zhang F, Liang X, Yediler A. Chemical and toxicological evaluation of an emerging pollutant (enrofloxacin) by catalytic wet air oxidation and ozonation in aqueous solution. Chemosphere. 2013;**90**:284–291. DOI:10.1016/j.chemosphere.2012.06.068

[75] Elmolla ES, Chaudhuri M. Photocatalytic degradation of amoxicillin, ampicillin and cloxacillin antibiotics in aqueous solution using UV/TiO_2 and UV/H_2O_2/TiO_2 photocatalysis. Desalination. 2010a;**252**:46–52. DOI:10.1016/j.desal.2009.11.003

[76] Elmolla ES, Chaudhuri M. Degradation of the antibiotics amoxicillin, ampicillin and cloxacillin in aqueous solution by the photo-Fenton process. J. Hazard. Mater. 2009;**172**:1476–1481. DOI: 10.1016/j.jhazmat.2009.08.015

[77] Elmolla ES, Chaudhuri M. Comparison of different advanced oxidation processes for treatment of antibiotic aqueous solution. Desalination. 2010b;**256**:43–47. DOI: 10.1016/j.desal.2010.02.019

[78] Dimitrakopoulou D, Rethemiotaki I, Frontistis Z, Xekoukoulotakis N, Venieri D, Mantzavino, D. Degradation, mineralization and antibiotic inactivation of amoxicillin by UV-A/TiO_2 photocatalysis. J. Environ. Manag. 2012;**98**:168–174. DOI: 10.1016/j.jenvman.2012.01.010

[79] Affamand AC, Chaudhuri M. Optimization of Fenton treatment of amoxicillin and cloxacillin antibiotic aqueous solution. Desalination Water Treatment. 2013;**52**:1878–1884. DOI: 10.1080/19443994.2013.794015

[80] Magureanu M, Piroi D, Mandache NB, David V, Medvedovici A, Bradu C, Parvulescu VI. Degradation of antibiotics in water by non-thermal plasma treatment. Water Res. 2011;**45**:3407–3416. DOI: 10.1016/j.watres.2011.03.057

[81] Benitez FJ, Acero JL, Real FJ, Roldan G, Casas F. Comparison of different chemical oxidation treatments for the removal of selected pharmaceuticals in water matrices. Chem. Eng. J. 2011;**168**:1149–1156. DOI: 10.1016/j.cej.2011.02.001

[82] Trovo AG, Melo SAS, Nogueira RFP. Photodegradation of the pharmaceuticals amoxicillin, bezafibrate and paracetamol by the photo-Fenton process-Application to sewage treatment plant effluent. J. Photochem. Photobiol. A Chem. 2008;**198**:215–220. DOI: 10.1016/j.jphotochem.2008.03.011

[83] Homem V, Alves A, Santos L. Microwave-assisted Fenton's oxidation of amoxicillin. Chem. Eng. J. 2013;**220**:35–44. DOI: 10.1016/j.cej.2013.01.047

[84] Andreozzi R, Canterino M, Marotta R, Paxeus N. Antibiotic removal from wastewaters: The ozonation of amoxicillin. J. Hazard. Mater. 2005;**122**:243–250. DOI:10.1016/j.jhazmat.2005.03.004

[85] Trovo AG, Nogueira RFP, Aguera A, Fernandez-Alba AR, Malato S. Degradation of the antibiotic amoxicillin by photo-Fenton process, chemical and toxicological assessment water research. 2011;**45**:1394–1402. doi:10.1016/j.watres.2010.10.029

[86] Su S, Guo W, Yi C, Leng Y, Ma Z. Degradation of amoxicillin in aqueous solution using sulphate radicals under ultrasound irradiation. Ultrasonics Sonochem. 2012;**19**:469–474. DOI: 10.1016/j.ultsonch.2011.10.005

[87] Jung YJ, Kim WG, Yoon Y, Kang JW, Hong YM, Kim HW. Removal of amoxicillin by UV and UV/H_2O_2 processes. Sci. Total Environ. 2012;**420**:160–167. DOI: 10.1016/j.scitotenv.2011.12.011

[88] Ternes TA, Stuber J, Herrmanna N, McDowella D, Ried A, Kampmann M, Teiser B. Ozonation: A tool for removal of pharmaceuticals, contrast media and musk fragrances from wastewater? Water Res. 2003;**37**:1976–1982. DOI:10.1016/S0043-1354(02)00570-5

[89] Biń AK, Madej SS. Comparison of the Advanced Oxidation Processes (UV, UV/H_2O_2 and O_3) for the Removal of Antibiotic Substances during Wastewater Treatment. Ozone Sci. Eng. 2012;**34**:136–139. DOI: 10.1080/01919512.2012.650130

[90] Cuevas SM, Arqués A, Maldonado MI, Pérez JAS, Rodríguez SM. Combined nanofiltration and photo-Fenton treatment of water containing micropollutants. Chem. Eng. J. 2013;**224:**89–95. DOI: 10.1016/j.cej.2012.09.068

[91] Estrada AL, Li YY, Wanga A. Biodegradability enhancement of wastewater containing cefalexin by means of the electro-fenton oxidation process. J. Hazard. Mater. 2012;**227–228**:41–48. DOI: 10.1016/j.jhazmat.2012.04.079

[92] Gurkan YY, Turkten N, Hatipoglu A, Cinar Z. Photocatalytic degradation of cefazolin over N-doped TiO_2 under UV and sunlight irradiation: Prediction of the reaction paths via conceptual DFT. Chem. Eng. J. 2012;**184**:113–124. DOI: 10.1016/j.cej.2012.01.011

[93] Balcıoğlu IA, Ötker M. Treatment of pharmaceutical wastewater containing antibiotics by O_3 and O_3/H_2O_2 processes. Chemosphere. 2003;**50**:85–95. DOI:10.1016/S0045-6535(02)00534-9

[94] Trovo AG, Paiva VAB, Machado AEH, Oliveira CA, Santos RO. Degradation of the antibiotic chloramphenicol by photo-fenton process at lab-scale and solar pilot plant: Kinetic, toxicity and inactivation assessment. Solar Energy. 2013;**97**:596–604. DOI: 10.1016/j.solener.2013.09.017

[95] Segura SG, Cavalcanti EB, Brillasa E. Mineralization of the antibiotic chloramphenicol by solar photoelectro-Fenton. From stirred tank reactor to solar pre-pilot plant. Appl. Catal. B Environ. 2014;**144**:588–598. DOI:10.1016/j.apcatb.2013.07.071

[96] Zuorro A, Fidaleo M, Fidaleo M, Lavecchia R. Degradation and antibiotic activity reduction of chloramphenicol in aqueous solution by UV/H_2O_2 process. J. Environ. Manag. 2014;**133**:302–308. DOI: 10.1016/j.jenvman.2013.12.012

[97] Rocha ORS, Pinheiro RB, Duarte MMB, Dantas RF, Ferreira AP, Benachour M, Silva VL. Degradation of the antibiotic chloramphenicol using photolysis and advanced oxidation process with UV C and solar radiation, Desalination Water Treatment. 2013;**51**:7269–7275. DOI: 10.1080/19443994.2013.792148

[98] Hopkins ZR, Blaney L. A novel approach to modeling the reaction kinetics of tetracycline antibiotics with aqueous ozone. Sci. Total Environ. 2014;**468–469**:337–344. DOI: 10.1016/j.scitotenv.2013.08.032

[99] Kim TH, Kim SD, Kim HY, Lim SJ, Lee M, Yu S. Degradation and toxicity assessment of sulfamethoxazole and chlortetracycline using electron beam, ozone and UV. J. Hazard. Mater. 2012;**227–228**:237–242. DOI: 10.1016/j.jhazmat.2012.05.038

[100] Bobu M, Yediler A, Siminiceanu I, Zhang F, Hostede S. Comparison of different advanced oxidation processes for the degradation of two fluoroquinolone antibiotics in aqueous solutions. J. Environ. Sci. Health, Part A. 2013;**48**:251–262. DOI: 10.1080/10934529.2013.726805

[101] Khanet MH, Jung HS, Lee W, Jung JY. Chlortetracycline degradation by photocatalytic ozonation in the aqueous phase: Mineralization and the effects on biodegradability. Environ. Technol. 2013;**34**:495–502. DOI:10.1080/09593330.2012.701332

[102] Jiet Y, Ferronato C, Salvador A, Yanga X, Chovelon J. Degradation of ciprofloxacin and sulfamethoxazole by ferrous-activated persulfate: Implications for remediation of groundwater contaminated by antibiotics. Sci. Total Environ. 2014;**472**:800–808. DOI: 10.1016/j.scitotenv.2013.11.008

[103] Doorslaer XV, Demeestere K, Heynderickx PM, Langenhove HV, Dewulf J. UV-A and UV-C induced photolytic and photocatalytic degradation of aqueous ciprofloxacin and moxifloxacin: Reaction kinetics and role of adsorption. Appl. Catal. B Environ. 2011;**101**:540–547. DOI: 10.1016/j.apcatb.2010.10.027

[104] Cho JY, Chung BY, Lee K, Geon-Hwi Lee Hwang SA. Decomposition reaction of the veterinary antibiotic ciprofloxacin using electron ionizing energy. Chemosphere. 2014;**117**:158–163. DOI: 10.1016/j.chemosphere.2014.06.039

[105] Witte BD, Dewulf J, Demeestere K, Langenhove HV. Ozonation and advanced oxidation by the peroxone process of ciprofloxacin in water. J. Hazard. Mater. 2009;**161**:701–708. DOI: 10.1016/j.jhazmat.2008.04.021

[106] Vasconcelos TG, Kümmerer K, Henriques DM, Martins AF. Ciprofloxacin in hospital effluent: Degradation by ozone and photoprocesses. J. Hazard. Mater. 2009;**169**:1154–1158. DOI: 10.1016/j.jhazmat.2009.03.143

[107] An T, Yang H, Li G, Song W, Cooper WJ, Nie X. Kinetics and mechanism of advanced oxidation processes (AOPs) in degradation of ciprofloxacin in water. Appl. Catal. B Environ. 2010;**94**:288–294. DOI: 10.1016/j.apcatb.2009.12.002

[108] Alpatova AL, Davies SH, Masten SJ. Hybrid ozonation-ceramic membrane filtration of surface waters: The effect of water characteristics on permeate flux and the removal of DBP precursors, dicloxacillin and ceftazidime. Sep. Purif. Technol. 2013;**107**:179–186. DOI: 10.1016/j.seppur.2013.01.013

[109] Rivas J, Encinas Á, Beltrán F, Graham N. Application of advanced oxidation processes to doxycycline and norfloxacin removal from water. J. Environ. Sci. Health A Tox. Hazard. Subst. Environ. Eng. 2011;**46**:9:944–951. DOI: 10.1080/10934529.2011.586249

[110] Guinea E, Garrido JA, Rodríguez RM, Cabot P, Arias C, Centellas F, Brillas E. Degradation of the fluoroquinolone enrofloxacin by electrochemical advanced oxidation processes based on hydrogen peroxide electrogeneration. Electrochim. Acta. 2010;**55**,:2101–2115. DOI: 10.1016/j.electacta.2009.11.040

[111] Rosal R, Rodrıguez A, Perdigon-Melon JA, Petre A, Calvo EG, Gomez MJ, Aguera A, Fernandez-Alba AR. Occurrence of emerging pollutants in urban wastewater and their removal through biological treatment followed by ozonation. Water Res. 2010;**44**:578–588. DOI:10.1016/j.watres.2009.07.004

[112] Klamerth N, Malato S, Maldonado MI, Agüera A, Fernández-Alba A. Modified photo-Fenton for degradation of emerging contaminants in municipal wastewater effluents. Catal. Today. 2011;**161**:241–246. DOI:10.1016/j.cattod.2010.10.074

[113] Nasuhoglu D, Rodayan A, Berk D, Yargeau V. Removal of the antibiotic levofloxacin (LEVO) in water by ozonation and TiO_2 photocatalysis. Chem. Eng. J. 2012;**189–190**:41–48. DOI: 10.1016/j.cej.2012.02.016

[114] Cheng W, Yang M, Xie Y, Liang B, Fang Z, Tsang EP. Enhancement of mineralization of metronidazole by the electro-Fenton process with a Ce/SnO_2–Sb coated titanium anode. Chem. Eng. J. 2013;**220**:214–220. DOI: 10.1016/j.cej.2013.01.055

[115] Doorslaer XV, Demeestere K, Heynderickx PM, Caussyn M, Langenhove HV, Devlieghere F, Vermeulen A, Dewulf J. Heterogeneous photocatalysis of moxifloxacin: Identification of degradation products and determination of residual antibacterial activity. Appl. Catal. B Environ. 2013;**138– 139**:333– 341. DOI: 10.1016/j.apcatb.2013.03.011

[116] Doorslaer XV, Heynderickx PM, Demeestere K, Debevere K, Langenhove HV, Dewulf J. TiO_2, mediated heterogeneous photocatalytic degradation of moxifloxacin: Operational variables and scavenger study. Appl. Catal. B Environ. 2012;**111–112**:150–156. DOI:10.1016/j.apcatb.2011.09.029

[117] Michael I, Hapeshi E, Michael C, Varela AR, Kyriakou S, Manaia CM, Fatta-Kassinos D. Solar photo-Fenton process on the abatement of antibiotics ata pilot scale: Degradation kinetics, ecotoxicity and phytotoxicity assessment and removal of antibiotic resistant enterococci. Water Res. 2012;**46**:5621–5634. DOI: 10.1016/j.watres.2012.07.049

[118] Giraldo A, Penuela G, Torres-Palma R, Pino N, Palominos R, Mansilla H. Degradation of the antibiotic oxolinic acid by photocatalysis with TiO_2 in suspension. Water Res. 2010;**44**:5158–5167. DOI: 10.1016/j.watres.2010.05.011

[119] Pereira J, Queirós D, Reis A, Nunes O, Borges M, Boaventura R, Vilar V. Process enhancement at near neutral pH of a homogeneous photo-Fenton reaction using ferricarboxylate complexes: Application to oxytetracycline degradation. Chem. Eng. J. 2014;**253**:217–228. DOI: 10.1016/j.cej.2014.05.037

[120] Pereira J, Vilar V, Borges M, Gonzalez O, Esplugas S, Boaventura R. Photocatalytic degradation of oxytetracycline using TiO_2 under natural and simulated solar radiation. Solar Energy. 2011;**85**:2732–2740. DOI: 10.1016/j.solener.2011.08.012

[121] Tambosi JL, de Sena RF, Gebhardt WRFPM, Moreira, José HJ, Schröder HF. Physico-chemical and advanced oxidation processes—A comparison of elimination results of antibiotic compounds following an MBR treatment. Ozone Sci. Eng. J. Int. Ozone Assoc. 2009;**31**:428–435. DOI: 10.1080/01919510903324420

[122] Yang H, Li G, An T, Gao Y, Fu J. Photocatalytic degradation kinetics and mechanism of environmental pharmaceuticals in aqueous suspension of TiO_2: A case of sulfa drugs. Catal. Today. 2010;**153**:200–207. DOI: 10.1016/j.jhazmat.2010.03.079

[123] El-Ghenymy A, Oturan N, Oturan MA, Garrido JA, Cabot PL, Centellas F, Rodríguez RM, Brillas E. Comparative electro-Fenton and UVA photoelectro-Fenton degradation of the antibiotic sulfanilamide using a stirred BDD/air-diffusion tank reactor. Chem. Eng. J. 2013;**234**:115–123.

[124] Liu Y, Wang J. Degradation of sulfamethazine by gamma irradiation in the presence of hydrogen peroxide. J. Hazard. Mater. 2013;**250–251**:99–105. DOI: 10.1016/j.jhazmat.2013.01.050

[125] Aguinaco A, Beltrán FJ, Sagasti JJP, Gimeno O. In situ generation of hydrogen peroxide from pharmaceuticals single ozonation: A comparative study of its application on Fenton like systems. Chem. Eng. J. 2014;**235**:46–51. DOI: 10.1016/j.cej.2013.09.015

[126] Köhler C, Venditti S, Igos E, Klepiszewski K, Benetto E, Cornelissen A. Elimination of pharmaceutical residues in biologically pre-treated hospital wastewater using advanced UV irradiation technology: A comparative assessment. J. Hazard. Mater. 2012;**239–240**:70–77. DOI:10.1016/j.jhazmat.2012.06.006

[127] Batchu SR, Panditi VR, O'Shea KE, Gardinali PR. Photodegradation of antibiotics under simulated solar radiation: Implications for their environmental fate. Sci. Total Environ. 2014;**470–471**:299–310. DOI: 10.1016/j.scitotenv.2013.09.057

[128] Baeza C, Knappe DRU. Transformation kinetics of biochemically active compounds in low-pressure UV Photolysis and UV/H_2O_2 advanced oxidation processes. Water Res. 2011;**45**:4531–4543. DOI: 10.1016/j.watres.2011.05.039

[129] Dirany A, Sirés I, Oturan N, Oturan MA. Electrochemical abatement of the antibiotic sulfamethoxazole from water. Chemosphere. 2010;**81**:594–602. DOI: 10.1016/j.chemosphere.2010.08.032

[130] Wang A, Li Y, Estrada AL. Mineralization of antibiotic sulfamethoxazole by photoelectro-Fenton treatment using activated carbon fiber cathode and under UVA irradiation. Applied Catal. B Environ. 2011;**102**:378–386. DOI:10.1016/j.apcatb.2010.12.007

[131] Ramos MMG, Mezcua M, Agüera A, Alba ARF, Gonzalo S, Rodríguez A, Rosal R. Chemical and toxicological evolution of the antibiotic sulfamethoxazole under ozone treatment in water solution. J. Hazard. Mater. 2011;**192**:18–25. DOI: 10.1016/j.jhazmat.2011.04.072

[132] Espejo A, Aguinaco A, Amat AM, Beltrán FJ. Some ozone advanced oxidation processes to improve the biological removal of selected pharmaceutical contaminants from urban wastewater. J. Environ. Sci. Health A Tox. Hazard. Subst. Environ. Eng. 2014;**49**:410–421. DOI: 10.1080/10934529.2014.854652

[133] Gonzalez O, Sans C, Esplugas S. Sulfamethoxazole abatement by photo-Fenton toxicity, inhibition and biodegradability assessment of intermediates. J. Hazard. Mater. 2007;**146:**459–464. DOI: 10.1016/j.jhazmat.2007.04.055

[134] Trovo AG, Nogueira RFP, Aguera A, Alba ARF, Sirtori C, Malato S. Degradation of sulfamethoxazole in water by solar photo-Fenton: Chemical and toxicological evaluation. Water Res. 2009;**43**:3922 – 3931. DOI: 10.1016/j.watres.2009.04.006

[135] Abellan MN, Bayarri B, Gimenez J, Costa J. Photocatalytic degradation of sulfamethoxazole in aqueous suspension of TiO_2. Appl. Catal. B Environ. 2007;**74**:233–241. DOI: 10.1016/j.apcatb.2007.02.017

[136] Xekoukoulotakis NP, Drosou C, Brebou C, Chatzisymeon E, Hapeshi E, Kassinos DF, Mantzavinos D. Kinetics of UV-A/TiO_2 photocatalytic degradation and mineralization

of the antibiotic sulfamethoxazole in aqueous matrices. Catal. Today. 2011;**161**:163–168. DOI: 10.1016/j.cattod.2010.09.027

[137] Fan X, Hao H, Shen X, Chen F, Zhang J. Removal and degradation pathway study of sulfasalazine with Fenton-like reaction. J. Hazard. Mater. 2011;**190**:493–500. DOI: 10.1016/j.jhazmat.2011.03.069

[138] Liu Y, Hu J, Wang J. Radiation-induced removal of sulphadiazine antibiotics from wastewater. Environ. Technol. 2014a;**35**(16):2028–2034. DOI: 10.1080/09593330.2014.889761

[139] Goncalves AG, Órfão JJM, Pereira MFR. Catalytic ozonation of sulphamethoxazole in the presence of carbon materials: Catalytic performance and reaction pathways. J. Hazard. Mater. 2012;**239–240**:167–174. DOI: 10.1016/j.jhazmat.2012.08.057

[140] Lester Y, Avisar D, Mamane H. Photodegradation of the antibiotic sulphamethoxazole in water with UV/H_2O_2 advanced oxidation process. Environ. Technol. 2010;**31**(2):175–183. DOI: 10.1080/09593330903414238

[141] Beltran FJ, Aguinaco A, Araya JFG, Oropesa A. Ozone and photocatalytic processes to remove the antibiotic sulfamethoxazole from water. Water Res. 2008;**42**:3799–3808. DOI: 10.1016/j.watres.2008.07.019

[142] Liu Y, Hu J, Wang J. Fe^{2+} enhancing sulfamethazine degradation in aqueous solution by gamma irradiation. Radiation Phys. Chem. 2014b;**96**:81–87. DOI: 10.1016/j.radphyschem.2013.08.018

[143] Zhou T, Wu X, Zhang Y, Li J, Lim T. Synergistic catalytic degradation of antibiotic sulfamethazine in a heterogeneous sonophotolytic goethite/oxalate Fenton-like system. Appl. Catal. B Environ. 2013;**136–137**:294–301. DOI: 10.1016/j.apcatb.2013.02.004

[144] Wu J, Zhang H, Oturan N, Wang Y, Chen L, Oturan MA. Application of response surface methodology to the removal of the antibiotic tetracycline by electrochemical process using carbon-felt cathode and DSA(Ti/RuO_2–IrO_2) anode. Chemosphere. 2012;**87**:614–620. DOI: 10.1016/j.chemosphere.2012.01.036

[145] Oturan N, Wu J, Zhang H, Sharma VK, Oturan MA. Electrocatalytic destruction of the antibiotic tetracycline in aqueous medium by electrochemical advanced oxidation processes: Effect of electrode materials. Appl. Catal. B: Environ. 2013;**140–141**:92–97. DOI: 10.1016/j.apcatb.2013.03.035

[146] Pacheco CV, Polo MSJ, Utrilla R, Penalver JL. Tetracycline removal from waters by integrated technologies based on ozonation and biodegradation. Chem. Eng. J. 2011;**178**:115–121. DOI: 10.1016/j.cej.2011.10.023

[147] Bautitz IR, Nogueira RFP. Degradation of tetracycline by photo-fenton process-solar irradiation and matrix effects. J. Photochem. Photobiol. A Chem. 2007;**187**:33–39. DOI: 10.1016/j.jphotochem.2006.09.009

[148] Mboula VM, Héquet V, Gru Y, Colin R, Andrès Y. Assessment of the efficiency of photocatalysis on tetracycline biodegradation. J. Hazard. Mater. 2012;**209–210**:355–364. DOI: 10.1016/j.jhazmat.2012.01.032

[149] Utrilla JR, Polo MS, Joya GP, García MAF, Toledo IB. Removal of tinidazole from waters by using ozone and activated carbon in dynamic regime. J. Hazard. Mater. 2010;**174**:880–886. DOI: 10.1016/j.jhazmat.2009.09.059

[150] Rahmani H, Gholami M, Mahvi AH, Alimohammadi M, Azarian G, Esrafili A, Rahmani K, Farzadkia M. Tinidazole removal from aqueous solution by sonolysis in the presence of hydrogen peroxide. Bull. Environ. Contam. Toxicol. 2014;**92**:341–346. DOI: 10.1007/s00128-013-1193-2

[151] Moreira FC, Segura S, Boaventura RAR, Brillas E, Vilar VJP. Degradation of the antibiotic trimethoprim by electrochemical advanced oxidation processes using a carbon-PTFE air-diffusion cathode and a boron-doped diamond or platinum anode. Appl. Catal. B Environ. 2014;**160–161**:492–505. DOI:10.1016/j.apcatb.2014.05.052

[152] Yargeau V, Leclair C. Impact of operating conditions on decomposition of antibiotics during ozonation: A review. Ozone Sci. Eng. 2008;**30:**175–188. DOI: 10.1080/01919510701878387

[153] Tekin H, Bilkay O, Ataberk SS, Balta TH, Ceribasi IH, Sanin FD. Use of Fenton oxidation to improve the biodegradability of a pharmaceutical wastewater. J. Hazard. Mater. 2006;**136:**258–265.DOI: 10.1016/j.jhazmat.2005.12.012

[154] Melero JA, Calleja G, Martinez F, Molina R, Pariente MI. Nanocomposite Fe2O3/SBA-15: An efficient and stable catalyst for the catalytic wet peroxidation of phenolic aqueous solutions. Chem. Eng. J. 2007;**131**:245–256. DOI: 10.1016/j.cej.2006.12.007

[155] Santos A, Yustos P, Rodriguez S, Simon E, Garcia-Ochoa F. Abatement of phenolic mixtures by catalytic wet oxidation enhanced by Fenton's pretreatment: Effect of H_2O_2 dosage and temperature. J. Hazard. Mater. 2007;**146**:595–601. DOI: 10.1016/j.jhazmat.2007.04.061

[156] Andreozzi R, Insola A, Caprio V, Amore MGD. Ozonation of pyridine aqueous solution: Mechanistic and kinetic aspects. Water Res. 1991;**25**:655–659. DOI: 10.1016/0043-1354(91)90040-W

Chapter 10

Wastewater Treatment through Low Cost Adsorption Technologies

Gonzalo Montes-Atenas and Fernando Valenzuela

Additional information is available at the end of the chapter

http://dx.doi.org/10.5772/67097

Abstract

This chapter addresses the wastewater treatment of mining residues through adsorption methodologies. It preferentially focuses its attention on (but not limited to) the removal of heavy metals. It begins with a brief description of the most used wastewater treatment pathways highlighting both their advantages and disadvantages and focusing on adsorption industrial practice. Classic models of adsorption thermodynamics and kinetics are presented. It finalises with a more detailed description of two methodologies of low cost sorbents: (i) inorganic nanostructured silicates and (ii) organic-based sorbents—pine bark.

Keywords: adsorption, adsorbent, wastewater treatment, sorbent, mathematical modelling, mining

1. Wastewaters containing heavy metals and other contaminants: mining applications

1.1. Introduction

One of the two major environmental problems concerning the management of liquids in mining and mining-related activities, particularly relevant to a country like Chile, is the natural generation of acidic mine drainages (AMD) or acidic rock drainages (ARD) produced by chemical and/or microbial oxidation of sulphides in the presence of air and water [1, 2]. Obviously, the other problem is the impact of industrial residual aqueous solutions originated in the metallurgical and mining processing plants. All these solutions exhibit an important amount of chemical contaminants, either dissolved or suspended, at concentrations that normally surpass the limit fixed by the national discharge regulations and their discard in an acceptable manner into surface and groundwater bodies is an imperious need [3]. The complexity behind the latter issue is even higher considering that the water requirements in the mining industry are becoming a real problem, especially in regions such as the Atacama Desert or The Andes Mountains, where this vital liquid is scarce. Mining activities must then share the few water resources with the needs of the local communities for human life and agriculture. Therefore, it is absolutely necessary to optimise the use of all the available water resources, by controlling the waters entering mining and metallurgical plants and by treating the liquid residues that the processes such as leaching, solvent extraction and froth flotation plants produce.

For instance, AMD occurs when sulphide ores, mainly iron- and copper-bearing sulphides minerals, are exposed to water and air, resulting in the formation of sulphuric acid and metal hydroxides that release toxic heavy metal ions, and acidic hydrogen into surface and ground waters [4, 5]. They are also generated at waste disposal sites and around abandoned mine sites. Its formation is primarily a function of the local geology and hydrology. The AMD usually consists of an aqueous solution that exhibits a variable acidity and contains many and variable toxic or valuable dissolved metals, with a high content of sulphate (SO_4^{2-}) and other anions such as phosphate (PO_4^{3-}), nitrate (NO_3^-), molybdate (MoO_3^{2-}) and chloride (Cl^-), and a high amount of colloidal-type suspended fine solids difficult to settle, becoming a water pollution problem difficult to remediate. The metals remain in solution until the pH increases to a level where precipitation occurs. However, precipitation presents many problems, such as redissolving of precipitates, the need of a large amount of chemicals and the generation of large volumes of sludge whose disposal is quite complex. Just as an exemplification, some chemical reactions that represent the chemistry of AMD formation are as follows [6, 7]:

$$4\,FeS_{2(s)} + 15\,O_{2(g)} + 14\,H_2O_{(l)} \rightarrow 4Fe\,(OH)_{3(s)} + 8\,H_2SO_{4(aq)} \tag{1}$$

$$4\,Fe^{2+}_{(aq)} + O_{2(g)} + 4\,H^{+}_{(aq)} \leftrightarrow 4\,Fe^{3+}_{(aq)} + 2\,H_2O_{(l)} \tag{2}$$

$$4\,Fe^{3+}_{(aq)} + 12\,H_2O_{(l)} \leftrightarrow 4Fe\,(OH)_{3(s)} + 12\,H^{+}_{(aq)} \tag{3}$$

$$FeS_{2(s)} + 14\,Fe^{3+}_{(aq)} + 8\,H_2O_{(l)} \rightarrow 15\,Fe^{2+}_{(aq)} + 2\,SO^{2-}_{4(aq)} + 16\,H^{+}_{(aq)} \tag{4}$$

Equation (1) represents the oxidation of pyrite by oxygen. This reaction generates two moles of acid per mole of oxidised pyrite. Reaction (2) involves the conversion of Fe(II) into Fe(III), reaction that is enhanced by the bacteria *Acidithiobacillus ferrooxidans* (whenever these microorganisms are present in the system). Reaction (3) represents the hydrolysis of iron with the formation of the corresponding hydroxide and the production of new acid molecules. Finally, reaction (4) represents the oxidation of additional pyrite by ferric iron generated in the previous reactions. These reactions are based on the dissolution of pyrite (FeS_2); nevertheless, it is expected that they also occur simultaneously with other metal sulphides present in the ore. An overall chemical reaction cycle takes place very rapidly and continues until either the ferric iron or the metal sulphide is depleted. The latter implies that a variety of metals and anions may dissolve in the process both conferring a significant toxicity to the streams containing them and turning AMD into a potential source of many valuable and, sometimes, scarce metals.

1.2. Wastewater treatment paths and their comparison to adsorption operations

Solubility driven treatments: Precipitation operations widely used in the removal of heavy metals are implemented based on pH or the addition of counter-ions leading the formation of sparingly soluble compounds (Eqs. (5) and (6), respectively)

$$M^{n+}_{(aq)} + n\,OH^{-}_{(aq)} \leftrightarrow M\,(OH)_{n\,(s)} \tag{5}$$

$$m\,M^{n+}_{(aq)} + n\,X^{m-}_{(aq)} \leftrightarrow M_{m}X_{n\,(s)} \tag{6}$$

In these cases, the solubility product constants and the solubility of the compounds are written as in Eqs. (7) and (8).

$$K_{ps} = a_{M^{n+}}\,a^{n}_{OH^{-}} \quad s = \sqrt[n+1]{\frac{K_{ps}}{n}} \tag{7}$$

$$K_{ps} = a^{m}_{X^{n-}}\,a^{n}_{X^{m-}} \quad s = \sqrt[m+n]{\frac{K_{ps}}{m+n}} \tag{8}$$

where K_{ps} and a_i represent the solubility product constant and the activity or real (effective) concentration of the ion i in the aqueous solution.

The remaining amount of the pollutant in the solution depends on the species solubility which is not only a function of the solubility product constant but also of the number and equilibrium constants associated with the formation of anionic and others species related to counterions forming other molecular complexes [8]. The formation of metal hydroxides may leave large amounts of pollutants in aqueous solutions. In this case, counterions with lower K_{ps} values are used, for instance sulphide ions (S^{2-}) are used to improve the wastewater treatment of waters containing mercury or cadmium. Most of the treatments employed in our country only consider a alkaline chemical treatment by contacting the acidic stream with $CaCO_3$ or CaO or $Ca\,(OH)_2$ raising the pH up to a point where metals are partially precipitated.

Cementation and other electrochemically driven treatments: Cementation operations correspond to spontaneous electrochemical reactions largely displaced towards the formation of the products. They have identical problems to those found in classic precipitation operations and, additionally, the process have to deal with the release of the reduction agent (R) to the aqueous phase (Eq. (9)).

$$M^{n+}_{(aq)} + n\,R^{0}_{(s)} \leftrightarrow M^{0}_{(s)} + n\,R^{+}_{(aq)} \tag{9}$$

Cementation processes are commonly implemented in acidic conditions to avoid metal hydrolysis and consider oxidation reactions which are located outside the water stability domain. The latter introduces in these systems parallel competitive reactions associated with the hydrogen evolution and others decreasing the efficiency of the cementation process. Other non-spontaneous electrochemical treatment is used such as electrodeposition, a clean technology claimed to avoid the generation of residues or electrocoagulation. The latter having an anode made of a polyvalent metal able to promote coagulation such as iron or aluminium. However, according to the elements to be removed, appropriate aqueous solution conditions must be reached. Relatively high cost associated with the application of voltage differences and mass transport limitations due to low conductivity and/or low element concentration make this wastewater treatment strategy expensive compared to other spontaneous processes.

Membrane-based separation treatments: This path includes ultrafiltration, nanofiltration and reverse osmosis. Ultrafiltration allows the removal of dissolved and colloidal materials. Isolated metal ions are difficult to remove directly due to their small sizes so the use of micelles and other complexing agents enhances its separation capabilities. Reverse osmosis removes a wide range of dissolved species from aqueous solutions; however, it has high power pumping requirements to restore the membranes. Membrane separation techniques have gone through significant progresses developing liquid membrane processes based on the formation of emulsions; however, there is still room for improvement with regard to their instability in salty and acidic conditions and fouling produced by other species present in the wastewater. Electrodialysis does not require the addition of external species to the wastewater, exhibits high selectivity and does not produce sludges, which makes it a promising technology. However, energy consumption and anodic/cationic membrane fouling is still a challenge.

Ion-exchange treatment: Ion-exchange resins either natural or synthetic are selected when high treatment capacity, high efficiency and fast kinetics are required. The stronger resins are built on sulphonic or carboxylic groups as in Eqs. (10) and (11). In both cases, the acidification of the aqueous media is commonly observed as it occurs in the case of solvent extraction using kerosene and implemented in countercurrent when using oximes. The only difference between ion exchange and solvent extraction would be that the organic phase would be in the dissolved phase in the organic phase rather than in the solid phase.

$$n\left(\mathrm{RSO_3H}\right)_{(s)} + M^{n+}_{(aq)} \leftrightarrow \left(\mathrm{RSO_3^-}\right)_n M^{n+}_{(s)} + n\,\mathrm{H}^+_{(aq)} \tag{10}$$

$$n\,(\mathrm{RCOOH})_{(s)} + M^{n+}_{(aq)} \leftrightarrow (\mathrm{RCOO^-})_n M^{n+}_{(s)} + n\,\mathrm{H}^+_{(aq)} \tag{11}$$

Froth flotation: This process can be used to treat wastewaters contaminated with heavy metals and can be implemented following different technologies such as dissolved air flotation (DAF), electroflotation, ion flotation or precipitation flotation. DAF processes are based on the gas oversaturation and decompression producing microbubbles that are able to separate small particles or agglomerates. The electroflotation process is associated with the water electrolysis producing the smallest bubble sizes of hydrogen and oxygen known at the industrial level. Ion flotation is based on ionic complex formation with surfactant molecules and subsequent frother-aided flotation. The precipitation flotation is a mixed technology between the precipitation followed by flotation operations (sulphide precipitation are commonly implemented).

Adsorption separation treatments: In heterogeneous systems, whenever two immiscible states of the matter, namely gas and liquid, gas and solid or liquid and solid, are set in contact they are separated by a surface layer having properties different to those of the two states forming it [9]. When one or more components present in one of the two phases (or in both) tend towards increasing its concentration in the surface layer, it is said that the adsorption process is taking place (**Figure 1**).

In **Figure 1**, the adsorbat is represented by the orange circles. The adsorbent is represented by the continuous marble-like colour. The adsorption process can be either physical or chemical in nature. The physical adsorption or physisorption requires energies in the order of some kcal/mol and is mainly associated with condensation processes of the species at the interface. Its spontaneity is enhanced by reducing the temperature of the system (Eq. (12)).

$$d\,G_{ads} = d\,H_{ads} - Td\,S_{ads} \tag{12}$$

where $d\,G_{ads}$, $d\,H_{ads}$, $d\,S_{ads}$ represent the adsorption Gibbs free energy, the adsorption enthalpy and the adsorption entropy, respectively, while T is the temperature of the system. The term $d\,H_{ads} < 0$ (heat of condensation). In consequence, the spontaneity $d\,G_{ads} < 0$ would be favoured by lower temperature conditions. In this case, the rate of the adsorption process is rapid, reversible between the adsorbed and desorbed states, and it can be either monomolecular or multimolecular with respect to the number of adsorbat molecules associated

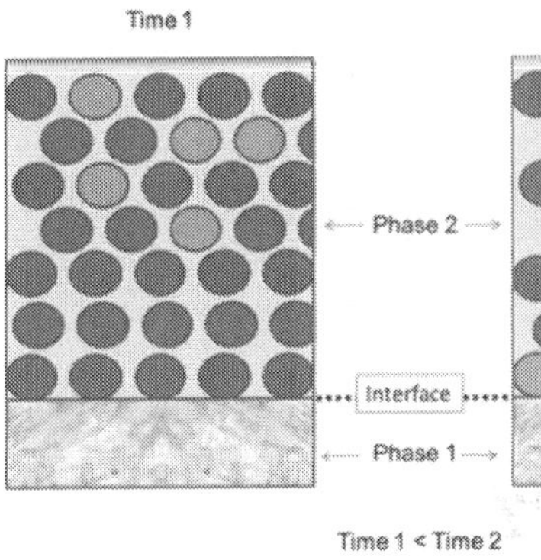

Figure 1. Schematics of an adsorption process.

with the number of available adsorption sites. On the other hand, chemical adsorption or chemisorption requires energies in the order of dozens of kcal/mol and involves the generation of bonding between the adsorbed specie and the interface. In general terms, its spontaneity is enhanced by increasing the temperature as $d\,H_{ads} > 0$ (activation energy needs to be reached). The adsorption process is slower and is highly dependent on the adsorbent nature and generally irreversible due to the strong forces between the adsorbat and the adsorbent taking place in the adsorption process. Adsorption operations have grown as a feasible way to treat wastewaters. The process is versatile in terms of the materials that can be used ranging from natural substrates or biosorbents such as stems, bark, leaves, among other; to synthetic sorbents like activated carbon, metal oxides, etc. The maximum adsorption capacity of the sorbent is commonly studied in thermodynamic experiments which are represented in terms of the adsorption isotherms. A list of the most used adsorption isotherms and the systems is presented in **Tables 1** and **2**. The $\frac{V}{V_{max}}$ ratio represents the adsorbed volume of adsorbat over the maximum volume of adsorbat able to be adsorbed. Otherwise named as a degree of coverage or q_i/q_{max} where q_i has units of mg of adsorbat/g of sorbent defined as θ. Please, note that q_{max} is referred to as the maximum adsorption associated with the monolayer. In all mathematical expressions, the variable P refers to the pressure of the adsorbat when a gas phase is set in contact with a condensed phase, namely, liquid or solid. In more general terms, the pressure P can be replaced by the activity or concentration of the adsorbat in aqueous phase.

Model No.	Name	Mathematical model	Specific features of the model and when to use it...
(1)	Langmuir [10]	$\theta = \frac{k_1 P}{1+k_1 P}$ (3) Or $k_1 P = \frac{\theta}{1-\theta}$ (3') Graphically: $\frac{1}{q_e} = \frac{1}{q_{max}} + \frac{1}{q_{max} k_1 P}$ (3.5) $\frac{P}{q_e} = \frac{1}{k_1 q_{max}} + \frac{P}{q_{max}}$ (3.6) $q_e = q_{max} - \frac{q_e}{k_1 P}$ (3.7) $\frac{q_e}{P} = k_1 q_{max} - k_1 q_e$ (3.8) $\frac{1}{P} = \frac{k_1 q_{max}}{q_e} - k_1$ (3.9)	This two-parameter model is widely used in cases where ideal localised monolayer is obtained by either chemical or physical adsorption. This is perhaps the most relevant model in the field of adsorption and has been derived from kinetic premises, thermodynamics conceptualisation and also statistically [11] The particular conditions under which the model is valid are as follows: (1) Once adsorbed the adsorbat is fixed onto one surface site and there is no migration throughout the sorbent surface (or interface). Therefore, the adsorbent exhibits a limited capacity for adsorption (2) One adsorbate molecule is associated with one surface site (3) The surface energy is identical for every surface site and therefore the same applies to the adsorption energy. In other words, the following equation holds $\Delta Ga = \Delta Ha - T\Delta Sa = cte$(3.3), where the subscript a refers to the adsorption process (4) A homogeneous surface without lateral interaction and steric hindrance between adsorbed species is required even if they are adsorbed in adjacent sorbent surface sites The strength of intermolecular attractive forces decreases dramatically with the distance to the surface sorbent
(2)	Freundlich (also known by the name Halsey and Taylor) (Freundlich, 1906) [12]	$q_e = k_1 P^{\frac{1}{n}}$ (11) Graphically: $\log q_e = \log k_1 + \frac{1}{n} \log P$ (11.3)	This two-parameter isotherm is applicable to chemical and physical adsorption without lateral interactions between adsorbed molecules. It is an empirical two parameter model which describes a multilayer adsorption process where the surface sites do not follow any uniform distribution in adsorption heat of adsorption or affinities between the sorbent and adsorbent. It is based on the fact that the concentration of the adsorbat on the surface increases with the concentration in the other phase The conditions are 1. The heat of adsorption can be represented by $-\Delta H_a = -\Delta H_0 \text{Log}\theta$ (11.1). 2. The model cannot be used at high and low degree of coverage. At very low adsorbat concentration, the Freundlich equation does not provide Henry's law; however, it is valid for ion adsorption at low solute concentrations If $\frac{1}{n}$ is close to zero, the surface heterogeneity of the sorbent is more significant. Such heterogeneity is characterised by the exponential decaying sorption site distribution (or the adsorption enthalpy changes logarithmically with the degree of coverage). Particularly, in this case, when n varies from 1 to 10, the adsorption is highly favourable. If , then, the partition between the two phases is independent of the concentration of the adsorbate and linear adsorption occurs. If $\frac{1}{n}$ is higher than 1, there is a symptom of cooperative adsorption

Model No.	Name	Mathematical model	Specific features of the model and when to use it...
(3)	Redlich-Peterson [13]	$q_e = \frac{k_r P}{1 + a_r P^{\varphi}}$ (24) Graphically, $\ln\left(k_r \frac{P}{q_e} - 1\right) = \varphi \ln P + \ln a_r$	This is an empirical thee parameter isotherm model that can be applied to sorbents exhibiting homogeneous and heterogeneous surface energy within a significant range of pressure or concentration of adsorbat. It is particularly useful for moderate pressures/concentrations. Its use requires evaluating three parameters (k_r, a_r *and* φ) combining aspects from Langmuir and Freundlich isotherms. The denominator consists of a hybrid Langmuir-Freundlich form The model consists of a linear dependence on the adsorbat concentration/pressure and an exponential behaviour with respect to the same variable in the denominator. At high adsorbat concentration/pressure the model is simplified to the Freundlich isotherm. In any case, if $\varphi = 1$ the Langmuir model is obtained. If $\varphi = 0$, the Henry's law is obtained

Table 1. Adsorption isotherms and their use.

No	Name	Mathematical model	Meaning of the parameters and when to use it...
(1)	Lagergren Pseudo-first order [14]	Derivative form: $\frac{dq_t}{dt} = k_{kin}(q_e - q_t)$ Integral form: $\ln(q_e - q_t) = \ln q_e - k_{kin} t$	k_{kin} is the specific rate of adsorption q_t is the adsorption parameter which evolves with time. This model is widely used for cases where the adsorbat is originally dissolved in aqueous phase and the adsorption follows a first order equation. The following assumptions are made: (1) The adsorption is localised and there is not any interaction between adsorbed molecules (2) The adsorption energy is independent of surface coverage (3) Maximum adsorption capacity is equivalent to the monolayer formation (4) The concentration of the adsorbat in the fluid phase is constant
(2)	Pseudo-second order	Derivative form: $\frac{dq_t}{dt} = k_{kin}(q_e - q_t)^2$ Or, $q_t = q_e\left[1 - \frac{1}{\beta_2 + k_2 t}\right]$ Integral form: $\frac{t}{q_t} = \frac{1}{k_{kin} q_e^2} + \frac{t}{q_e}$	k_{kin} is the specific rate of adsorption This model assumes that the chemisorption is the rate determining step in the process. These assumptions are similar to the pseudo-first order but in this case the adsorption follows a second-order kinetics
(4)	Intra-particle diffusion or Weber and Morris model [15]	$q_t = k_{id}\sqrt{t} + I$	k_{id} intraparticle diffusion rate constant I provides information about thickness of the boundary layer If the plot of q_t vs $\sqrt{t}$ passes through the origin. Otherwise, the overall adsorption mechanism is also controlled (to some degree) by a boundary layer

Table 2. Adsorption kinetics models.

2. Inorganic sorbents case study: Silica based sorbents - Nanostructured silicates

2.1. Introduction

The synthesis, characterisation and performance of sorbents based on nanostructured calcium silicates without modification or chemically modified with other atoms with the purpose to improve the adsorption of some specific contaminant species are reviewed. The nanostructured calcium silicates possess a very high capacity to remove contaminants from water even when these are present in higher concentrations, and also have the capability of a collective and simultaneous removal of cationic and anionic species from acidic aqueous solutions. Effectively, there is a strong need for adsorbent materials that have the capacity for removing simultaneously both anions and cations from industrial and mining solutions. The synthesized nanostructured calcium silicates exhibit a specific surface area, averaging values over 100–400 m^2/g, showing a good sorption capacities for removing high contents of different ionic species. The mechanisms governing the sorption process are related to the presence of silanol, calcium and hydroxyl groups in its structure, which would act as binding or nucleation sites on its surface [16]. Recent efforts for modifying nanostructured calcium silicates by introducing in their structures other metals that generate an improved sorbent addressed to enhance the sorption of other species present in solutions like the AMD, by using simple and low-cost processes, have been achieved. The efficiencies accomplished with nanostructured calcium silicates modified with iron, magnesium and aluminium introduced in its structure opens the possibility of tailoring a mix of modified silicates to suit the composition of various wastewaters requiring decontamination treatment. 'The more insoluble is the hydroxide precipitate or the salt formed during the reaction of the calcium silicate with the ionic species to be removed, the more efficient is the sorption process'.

The following modifications have been obtained successfully:

(a) Nanostructured calcium silicate derivatives containing Fe atoms have been prepared to selectively promote and enhance the adsorption of arsenic species through the formation of highly insoluble and stable double-salt of calcium and iron.

(b) The introduction of Mg atoms of *n*-calcium silicate by partial substitution by Ca atoms have permitted the preparation of a sorbent material that benefits the removal of phosphate and ammonium ions, frequently found in mining wastewaters, due the formation of a very insoluble calcium-containing double phosphate of Mg and ammonium.

(c) The partial replacement of Ca atoms in *n*-calcium silicate by Al atoms generates a stronger sorbent that allow and improve the sulphate removal by forming a high insoluble aluminium and calcium double basic salt.

(d) Nanostructured calcium silicate by forming a composite with magnetite in order to provide to this nanosorbent of magnetic properties and facilitate its separation from the resulting solution.

This monography addresses the base case and these four case studies.

2.2. Experimental procedures – Nanostructured silicate synthesis procedure

The synthesis of nanostructured calcium silicate can be conducted using several routes of preparation. In this study, calcium silicate was synthesized by precipitation as a result of the room temperature chemical reaction between liquid Na_2SiO_3 and Ca $(OH)_2$ at a pH value of 12.3, following a process in two steps. Firstly, a suspension of Ca $(OH)_2$ in water is treated with HCl 33% w/w to give a slurry with a pH value of 12–12.5, varying the stirring velocity in a range between 425 and 1000 min^{-1}. Secondly, a sodium silicate solution containing 28.5% w/w SiO_2 is diluted with water resulting in a 0.32% w/w SiO_2 solution. Then, this solution is vigorously mixed with the previously prepared Ca $(OH)_2$ suspension in a reactor, varying the stirring velocity between 1000 and 6000 min^{-1}, immediately forming a precipitate, nanostructured calcium silicates containing an average value of 35 g SiO_2/kg of formed solid. The formed slurry is aged for 20 min, allowed to settle for 13 h, before the obtained solid is recovered by vacuum filtration. Afterwards, the filter cake is washed with water and ethanol in order to disperse the nanostructured particles and lower the surface tension of the still wet solid. The solid is dried at 383 K for 2 days. Treatment with ethanol is necessary because during the drying of silicate, which presents a high surface area, is generated a great liquid-air interface that causes a high surface tension collapsing the structure of the material reducing its pore volume and surface area. The synthesised calcium silicate is an amorphous substance which does not present a defined chemical structure as is shown in **Figure 2**.

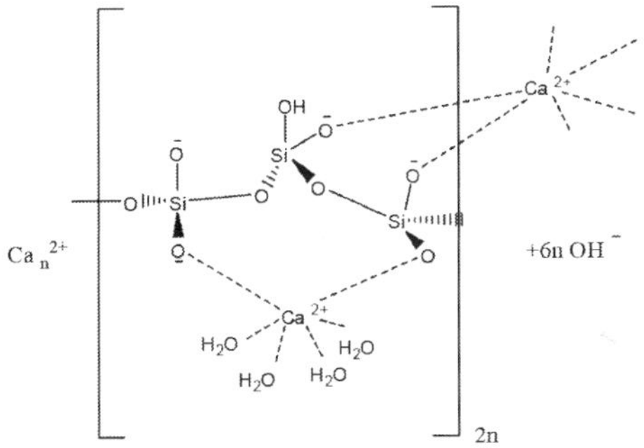

Figure 2. Probable structure of nanostructured calcium silicate.

Figure 2 shows the presence of OH^- groups and adsorbed water on the surface of the material. **Figure 3** shows scanning electron microscope (SEM) micrographs of the synthesised nanostructured calcium silicates. Particles exhibit a mean particle size between 0.2 and 1.0 μm. The particles appear in the SEM pictures forming larger agglomerates. This structure would correspond to a silicate backbone with plates of a thickness around of 10 nm and would consist of a tetrahedral silicate with Ca(II) ions and silanol groups on the surface forming a wollastonite-like structure, $CaSiO_3$, which contains SiO_3^{2-} species in which Ca(II) ions and the silanol groups would act as probable binding sites for the species being removed. The modifications done to the adsorbent by introducing other atoms that replace part of the calcium in the silicate, have not change significantly the mentioned structure, making possible the preparation of strong sorbent composites. Because of their low-cost and their disposable character, the regeneration of the modified calcium silicate is not considered. However, given their silicate-base composition similar to cement, later uses can be outlined.

Figure 3. SEM micrographs of the non-modified nanostructured silicate adsorbent.

2.3. Modifications to the nanostructured calcium silicate

Variants of this methodology to generate sorbent modifications are described below:

(a) Mg -substituted calcium silicate: add Mg during the synthesis of calcium silicate, a variable mole % magnesium is added as $Mg\,(OH)_2$ to replace part of the Ca atoms by Mg.

(b) Fe -calcium silicate: are prepared by replacing part of the calcium atoms of the silicate by Fe atoms using as iron source $FeCl_3$ or $Fe\,(OH)_3$

(c) Al -substituted calcium silicate, aluminium can be added as $NaAlO_2$, $Al\,(OH)_3$ or poly-aluminium chloride (PAC) at the start of the synthesis with the $Ca\,(OH)_2$.

(d) Magnetic properties of the calcium silicate that allows its easy separation from the treated waters, during the synthesis are added a suitable proportion of magnetite (Fe_3O_4), what avoid an important degrading of the accessible surface area of the silicate.

In all cases, there is a maximum element to calcium replacement ratio without compromising the nanostructure of silicate. Over certain replacement proportions, the nanostructure suffers a deleterious effect.

2.4. Characterization of nanostructured calcium silicate

The characterisation consisted of:

- Observing by scanning electron microscopy (SEM) JEOL JSM-25SII instrument.
- Measuring the mean particle size using a Malvern Mastersizer Hydro 2000 MU apparatus.

- Carrying out porosimetry analyses including the determination of the specific surface area are conducted using a N_2 sorptometer in a Micromeritics ASAP 2010 porosimeter at 20°C and 1 atm.
- Determining whether the nanostructured calcium silicates present a crystalline or amorphous structure, samples of the prepared solids were analysed using a Bruker D8 advance X-ray powder diffractometer which poses a LynxEye lineal detector.
- Determining the presence of free water molecules, O - H groups associated with water and silanol groups and Si - O bonds, infrared spectra were obtained in a Bruker-FTIR IFS 55.
- Performing differential scanning calorimetry (DSC) analysis using a DSC Perkin Elmer 6000 equipment to check the presence of water and of silanol groups in the silicate.
- Finding the elemental composition of the adsorbent using X-ray fluorescence spectrometry method and chemical analysis. In addition, the content of Ca and Fe in the adsorbent was also measured by flame atomic absorption analysis using a Perkin Elmer PinAAcle 900F apparatus, technique that was also used to check the chemical stability of the sorbent when contacted with acidic-aqueous solutions; these are necessary tests taking into account the acidic nature of most mine wastewaters.

2.5. Adsorption experiments

Batch equilibrium sorption tests are conducted at 303 K in a batch-type reactor by mixing a variable amount of nanostructured calcium silicate and different volumes of a copper mine aqueous solution having the following main composition:

(a) Main components: pH: 2–6;SO_4^{2-} :2–10 g/L; Cu(II): 20–120 mg/L; Fe (Fe(II) + Fe(III)): 120–330 mg/L; Zn(II): 15–120 mg/L; Mg(II): 140–250 mg/L; Ca(II): 250–400 mg/L; Mn(II):80–140 mg/L; P: 20–130 mg/L (in phosphate).

(b) Minor and toxic components: Cd(II): 5–20 mg/L; Pb(II): 8–30 mg/L; As: 40–200 mg/L; Cr(VI):0–20 mg/L; Ni(II): 4–5 mg/L; NO_3^-: 10–30 mg/L; Cl−: 20–40 mg/L; MoO_4^{2-}: 2–10 mg/L; TSS: 60–250 mg/L; pH: 2.1–4.6.

Batch experiments are conducted over sufficient time to reach equilibrium conditions. During the experiments, samples of the solution are collected at defined intervals and filtered using a 0.45 μm nitrocellulose Millipore membrane. The latter was performed before measuring the pH value and the concentration of metallic ions by atomic absorption spectrophotometry on a Perkin Elmer PinAAcle 900F instrument. The quantity of metal adsorbed is determined by the difference between the concentrations of metal in the initial aqueous feed phase and that in the raffinate solutions. At the end of the tests, the nanostructured calcium silicates were separated from the resulting aqueous solution by filtration. The SO_4^{2-} ion content is determined using a standard barium sulphate method [17] and PO_4^{3-} ion concentration is measured using the vanadate-molybdate-phosphoric UV-spectroscopy method [18].

2.6. Results

2.6.1. Synthesis and characterization

The chemical reaction between $Ca(OH)_2$ and Na_2SiO_3 is fast generating a thixotropic precipitate that produces particle agglomeration. The reaction of Ca(II) ions with soluble silicate as Na_2SiO_3 resulted in the precipitation of calcium silicate, an insoluble solid, where Ca(II) appeared to be strongly bound to a silicate backbone. The extent of the reaction and the particle size of the solid formed depend strongly on the pH of the reaction mixture, the proportions and concentration of calcium and silicate ions and the intensity of the stirring employed during the process. Intensive stirring in basic medium produces a quite homogeneous but colloidal and amorphous insoluble calcium silicate. The higher the stirring velocity, the smaller the particle size of the resultant calcium silicate and the larger and more accessible the substrate surface area comprising of micro-pores and meso-pores.

The modification of the adsorbent consists of a first step where $Ca(OH)_2$ and the source of the replacing elements ($Mg(OH)_2$or $FeCl_3$ or the Al compound) are mixed with the HCl solution. Afterwards, the resulting phase is mixed with the Na_2SiO_3 solution at high-stirring velocity, ideally over 2000 min^{-1}. It is not easy to establish a single stoichiometry of the chemical reaction due to the wide variety of silicate species possible to be formed and to the proportion of hydroxyl and silanol groups that silicates would contain. Even the synthesized calcium silicate as an amorphous substance does not present a defined chemical structure as shown in the scheme of **Figure 1**. Notwithstanding, in all cases, the prepared substrate corresponds effectively to an amorphous material without a defined structure. Particles present a mean particle size averaging 0.5–1.0 μm forming larger agglomerates. Details of nanostructured calcium silicate have been described in former communications by Cairns et al. using ^{29}Si-NMR spectroscopy and X-ray photoelectron spectroscopy (XPS) analysis [19]. They suggest the structure consists of a silicate backbone with plates of thickness around 10 nm. It would comprise silicate tetrahedral sites with Ca(II) ions and silanol groups on the surface forming a wollastonite-like structure, $CaSiO_3$, where Ca(II) ions and silanol groups would act as probable binding sites. Cairns et al. propose that the calcium silicate would contain 1.5% of hydroxyl groups meaning that approximately 8% of silicon atoms in the structure are silanol groups allowing metallic ions such as Cu(II) and Zn(II) to form the corresponding hydroxides on the surface of the silicates acting as nucleation sites [20]. Porosimetry analyses using the N_2 sorptometer, indicate a variable surface area ranging between 80 and 300 m^2/g, a value much higher than the surface area shown by other sorbents [21]. The mean pore diameter varies between 11 and 25 nm and the pore volume between 0.200 and 0.400 cm^3/g for calcium silicates varying with the stirring velocity used during the synthesis.

X-ray diffraction analysis confirms that the prepared compounds are basically amorphous or at least polycrystalline. However, elements of patterns associated with wollastonite, $CaSiO_3$, and larnyta-syn, Ca_2SiO_4, were observed confirming the synthesis of a calcium silicate. Although natural silicates are crystalline, like natural wollastonite, the solids prepared in this study appeared amorphous, probably because they were prepared by precipitation from aqueous solutions. Soluble silicates found in solutions comprise silicate ions of different size, most of them polymerized, which are not able to organise themselves in a crystal form resulting in a colloidal solid.

2.6.2. *Adsorption experiments*

Sorption experiments are carried out to measure the capability of these nanostructured calcium silicates to uptake diverse metallic ions and some anions found in many residual industrial and mining wastewaters. Most of the experiments were conducted using the aqueous solutions whose initial metals concentrations have been described before. However, in order to establish the maximum metal sorption capacity of the nanostructured calcium silicates, some experiments were carried out using a feed-aqueous solution with higher content of metals (up to 5 g/L) (**Figure 4**).

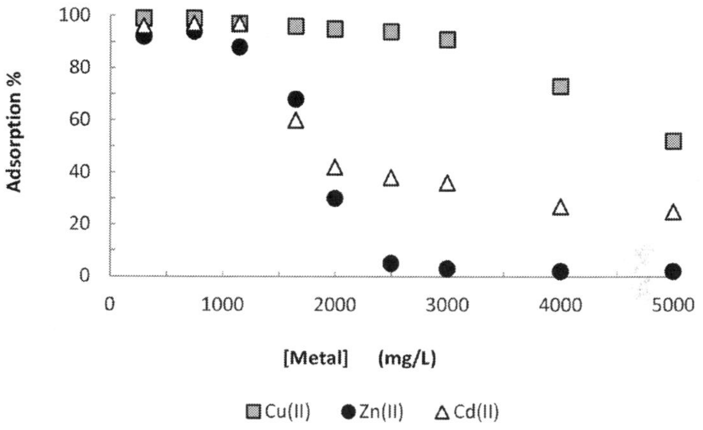

Figure 4. Adsorption of some metallic ions onto nanostructured calcium silicate.

Metallic ions such as Cu(II), Zn(II) and Cd(II) readily form insoluble hydroxides under basic conditions. However, their removal by precipitation only with lime is far from an ideal solution when their content in solution is high enough to consider them pollutants. In the presence of nanostructured calcium silicate the extent of precipitation reached is significantly higher than that observed with only lime being used. The degree of removal of Cu(II), Zn(II) and Cd(II) ions is higher and it produces more stable structures using the nanostructured calcium silicates prepared in this study. In fact, it is reasonable to think that silicates kept the pH value in a more basic region acting as a buffer. Thereby, they ensure a good precipitation of metallic ions probably onto the surface of the formed silicate, rather than in the bulk solution. This way, the precipitates generated in this case are of more granular nature, meaning that they could settle or get filtrated more easily than the precipitates obtained using lime or NaOH. This fact suggested that the metal removal from aqueous solutions would be a cation exchange between the metals to remove and the calcium ions associated with the silicate structure.

Figure 5 presents the results observed for the sorption of the anions PO_4^{3-}, SO_4^{2-} and CrO_4^{2-} introduced in the test solutions employing as a sorbent in the same sample of nanostructured calcium silicate without modification. The adsorption of PO_4^{3-} is fairly significant, $SO_4^{=}$ is also adsorbed, until a maximum uptake point. Cr(VI) is not adsorbed at all probably because this ion is present in aqueous solution as anionic species such CrO_4^{2-}, $HCrO_4^-$ or $Cr_2O_7^{2-}$ that do not tend to form insoluble salts with Ca(II) ions. The adsorption of PO_4^{3-} was higher than that of SO_4^{2-} in agreement with the lower solubility of $Ca_3(PO_4)_2$ compared to $CaSO_4$. The combined use of the nanostructured calcium silicate with Al(III) using poly-aluminium chloride improves the sulphate adsorption, even starting from initial solutions that contain over 2–5 g/L, obtaining a raffinate solution containing only a few mg/L accomplishing this way with the Chilean environmental regulations of discharge. The formation of ettringite, a double calcium and aluminium basic salt, during the adsorption would explain these results. Then, it is possible to achieve an almost complete sulphate removal.

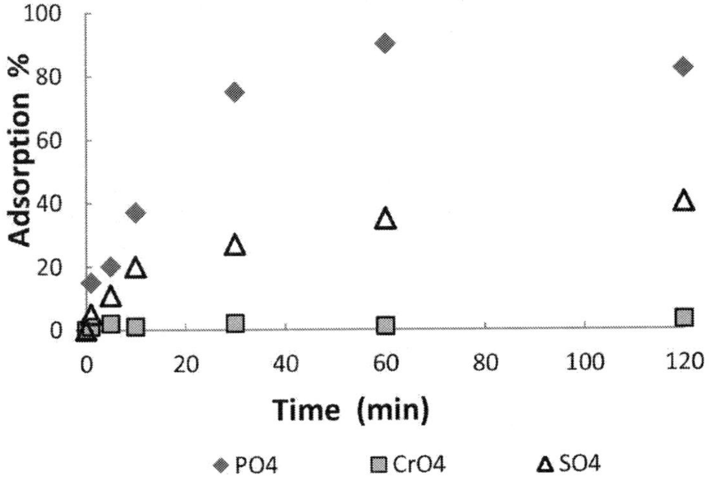

Figure 5. Adsorption of some anionic species onto nanostructured calcium silicate.

The dissolved species diffuse from the solution to the surface of the adsorbent, and then to the internal structure. The rate of adsorption is usually limited by mass transfer and depends on the properties of the sorbate and sorbent. The equilibrium adsorption results of ionic species adsorption have been explained using conventional equilibrium isotherms including the Langmuir model, the empirical Freundlich model and the Redlich-Peterson isotherm, which is a hybrid sorption model employed to analyse experimental data that do not fit well with other mentioned models. Normally, Langmuir isotherm only can explain the experimental results when synthetic and quite pure and ideal aqueous solutions are used. Freundlich and Redlich-Peterson used to fit equilibrium experimental results when real mining or more complex chemical matrices are used as aqueous solutions. With respect to kinetics experiments, normally experimental results are satisfactorily well explained by applying a pseudo-second-order kinetics model which is based on the sorption capacity of the sorbent.

http://dx.doi.org/10.5772/67097

2.7. Conclusions

In general, the modification with other atoms like Fe, Al and Mg of the nanostructured calcium silicate hydrate does not affect its high potential as adsorbent for treating polluted acidic solutions. The modified materials are superior to the unmodified material in terms of stability and sorption capacity by forming highly insoluble and stable double salts. The silicates would act as seeding material of insoluble hydroxides and salt species.

3. Natural sorbents case study: biosorbents - pine bark

3.1. Introduction

Several natural and low-cost products have been assessed as sorbents or as removal agents for heavy metal ions and other pollutants present in industry-discarded aqueous solutions [22–25]. Some examples of biosorbents are sugar maple [26]; mulch [27]; papaya wood [28]; tea industry wastes [29]; sawdust [30] and pine bark [31]. Particularly, the primary use of bark (and other by-products of the wood industry) is mainly associated with fuel with little added value. Therefore, their use as a sorbent is expected to produce an increase of the overall industry sustainability. Moreover, its use as biosorbent may become especially interesting when the pollutant loaded material can undertake further steps leading to its reutilisation as a sorbent (elution, pyrolysis, etc.) or carry on with other uses such as the production of activated carbon.

In this case, the adsorption efficiency, evaluated in terms of the adsorption capacity, varies from only a few micrograms up to 200 mg per grams of dry bark. The efficiency of the overall wastewater treatment process depends on the type of natural sorbent, the nature of the metal or pollutant to be adsorbed, the initial pollutant concentration, pH, temperature, pulp density, the cell design (batch, column, etc.) and the contacting time. In fact, many scientific reports have revealed that this material can adsorb and act simultaneously as biofilter for several types of pollutants, such as heavy metals and organic products.

This chapter focuses on the treatment of wastewaters containing heavy metals commonly observed at large scale operations related to industrial activities such as mining and metallurgical processes [32, 33]. The use of adsorption techniques overcomes this problem leaving aqueous solutions with heavy metal concentrations typically in the order of parts per billion [28].

It has been shown that the governing mechanism for the adsorption of cationic heavy metals using biosorbents is based on ionic exchange reactions [34], which occurs throughout the removal of protons present in the different bark molecular structures (Eq. (13)),

$$R - OH + M^{n+} \rightarrow R - O\,M^{(n-1)+} + H^{+} \tag{13}$$

where R represents the organic structure of the sorbent, M^{n+} is the dissolved heavy metal ion in solution. As a consequence of the adsorption, the acidity of the aqueous phase increases. The adsorption may coordinate one or more adsorption sites and the charge of the metallic ion can be totally or partially compensated by the surface charge of the biosorbent.

3.2. Substrate pre-treatment

One problem associated with the use of pine bark in the removal of heavy metals is its content of natural organic complexation agents such as tannins, which can be released in significant concentrations and stabilise contaminants in solution rather than removing them. There are a number of biosorbent pre-treatments reported in the literature aiming at removing or fixing such soluble compounds and simultaneously improving the subsequent pollutant adsorption efficiency/capacity. For instance, it has been shown that best adsorption results are obtained throughout carrying out a prior 'bark chemical activation'. This treatment consists of immersing the pine bark into diluted acid media at slightly high temperatures such as 50°C–70°C [34–36] before performing the adsorption tests. The acidic solution needs to be diluted to secure the integrity of the sorbent during the pre-treatment. The pH range in which the impact of chemical activation is the highest is narrow and it depends on the heavy metal nature and on the pulp density used [37]. Another pre-treatment consists of just washing the pine bark with water facilitating the removal of tannins and other soluble species which, if not removed, could again, irreversibly pollute even more the wastewater.

3.3. Experimental procedures

This experimental part gathers a number of findings concerning the bark structure, its response to chemical activation, and experiments generating adsorption equilibrium and kinetic data.

3.3.1. Pine bark characterisation and preparation

Pine bark samples are reduced in size to particle size distributions below 1 mm diameter (16# Tyler). The chemical activation is commonly performed using sulphuric acid 0.1–0.2 M and up to 1 M for 2 h at a temperature ranging from 20°C to 50°C with a solid to liquid ratio equal to 1:10.

Pine bark washing procedures are carried out with distilled water in a Soxhlet apparatus. Scanning electron microscopy (FEI Quanta 250 FEG SEM) is used to analyse the pine bark unloaded and loaded with the pollutant. Semi-quantitative elemental analysis was performed by EDX analysis of fluorescence intensities. FT-IR spectroscopy experiments are carried out on an IFS 55 spectrometer (Bruker) using diffuse reflectance mode (Harrick Attachment). The detector was of an MCT type and cooled at liquid nitrogen temperature (77 K).

XPS analysis is performed using a Kratos Axis Ultra instrument with a monochromatic Al K_α source. Samples are fixed on the sample holder using double-sided tape.

3.3.2. Adsorption studies

In all cases, there is an initial concentration of the pollutant (heavy metal) commonly in the dissolved state in the aqueous phase and introduced as a sulphate or a nitrate salt. The pH condition is close to the point where the hydrolysis starts producing insoluble or neutral species. All tests are carried out at 20°C and the pulp density used varies from 1 to 20 g sorbent/L aqueous solution unless stated otherwise. The adsorbed amount of pollutant is obtained by computing the difference between the initial amount of pollutant and the residual concentration after adsorption. The metal concentration in solution is obtained using atomic absorption spectrophotometry (AAS) Perkin Elmer Model 2380.

Equilibrium data are obtained at least after 48 h contact between the sorbent and the polluted aqueous phase. Kinetic data are generated following similar experimental protocols

In all cases, high purity reagents were used. Further details or changes in this procedure are pointed out when appropriate.

3.4. Results

3.4.1. *Characterisation and modelling of the substrate*

Table 3 shows the chemical composition of both raw and chemically (acid) activated bark resulting from the EDX elemental analysis of original washed bark and after the sulphuric acid treatment. As expected, most of the soluble species decrease their content due to the activation. The acid treatment reduces the concentrations of K, Mg, Mn and Ca significantly, whereas other metal concentrations (Na, Al, Si, Ti, Fe, Cu) remain almost constant.

	Chemical element									
Bark	Na	K	Mg	Ca	Al	Si	Ti	Mn	Fe	Cu
Washed	0.98	2.03	2.40	12.80	6.40	17.10	0.47	0.57	4.80	0.025
Activated	0.94	1.08	0.79	5.70	7.00	21.30	0.46	0.11	4.30	0.025

Table 3. Semi-quantitative EDX analysis of pine bark (values are in mg per g of dry bark).

Figure 6 shows the DRIFT analysis obtained for a section of the finger print region. A broad band is observed from 3600 to 3000 cm^{-1}, indicating the presence of O-H and C-H stretching vibrations. Between 2000 and 1500 cm^{-1} there is a strong band at 1610 cm^{-1} representing the carbonyl C = O stretch. When Pb(II) is adsorbed onto bark, a broadening of the band occurs and a simultaneous modifications at wavenumbers of approximately 1512 cm^{-1} related to aromatic skeletal vibrations. These changes are in agreement with the structures which are easier to ionise (**Figure 7**).

A high concentration of the metallic ions in solution may then activate the adsorption sites having lower ionization constant ultimately increasing the specific adsorption.

Figure 8 shows the SEM images after sorption using low pulp densities conditions. EDX elemental analysis at the spots vmarked in the figures revealed a high heterogeneity in the Pb distribution across the bark. The Pb/O weight ratio varied from 2:1 to 3:1 regardless the pulp density value. Similar observations of varying metal content were previously seen for untreated bark [39].

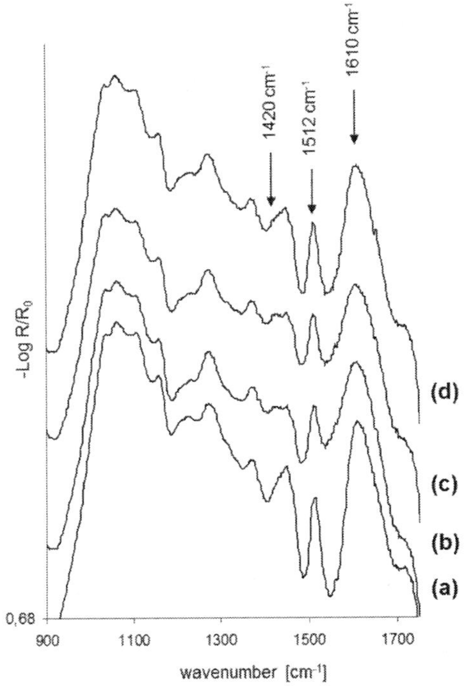

Figure 6. DRIFTS spectra for activated bark (a) and bark charged with solutions containing 1 (b), 1.5 (c) and 3 (d) g Pb L^{-1}. Pulp density: 1.5 g/L.

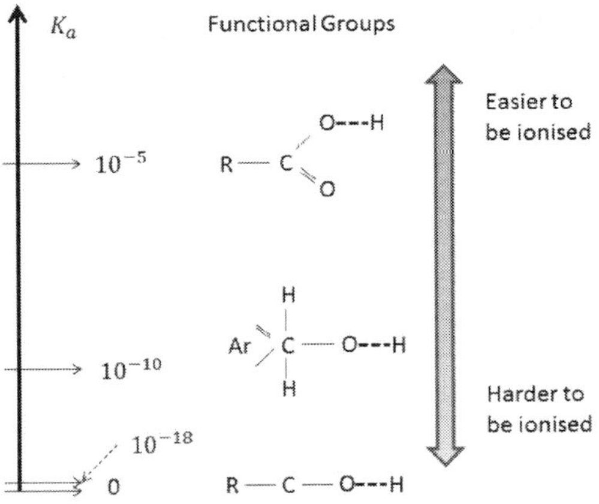

Figure 7. Schematics of the organic groups, its relative proportion in the structure of bark, and radicals through which the ionization could lead to heavy metals adsorption onto bark. *Ka* represents the acidity constant, C: carbon, O: oxygen, Ar: aromatic structure (from Ref. [38]).

http://dx.doi.org/10.5772/67097

Figure 8. SEM images of activated pine bark loaded with adsorbed Pb(II). Left: pulp density 1.0 g/L. EDX analysis indicates Pb/O weight ratios of 2:1 and 3:1 in spots (a) and (b), respectively. Right: pulp density 1.5 g/L. EDX indicates Pb/O weight ratios decreasing from (a) to (c) ranging from approximately 2:1 to 3:1.

3.4.2. *Adsorption tests: thermodynamics and kinetics*

As an example, **Figure 9** shows the adsorption percentage of Cu(II) from batch experiments. Langmuir and Freundlich isotherms fit well with the experimental data. The adsorption capacity varies from one chemical element to another and as a function of the experimental conditions used. Pb(II) ions reach similar adsorption capacity at 10 g/L pulp density but it can reach 90 mg/g at 1 g/L. Concerning adsorption competitive reactions, when comparing Cu(II) and Zn(II) adsorption, Cu(II) reaches higher adsorption capacities than Zn(II). Binary adsorption reveals both metallic ions which do not interfere in the adsorption of each other reaching similar adsorption capacities when mixed and separated; however, this is not the case for all metals. Al-Asheh and Duvnjak (1997) have proved that Ni^{2+} and Cu^{2+} interfere slightly whereas the pairs (Cu^{2+}, Cd^{2+}) and (Cd^{2+}, Ni^{2+}) show a more significant interaction [22].

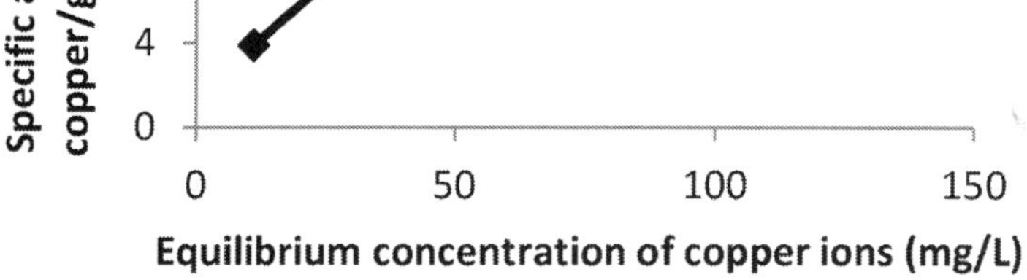

Figure 9. Adsorption isotherm of copper (II) ions onto washed pine bark at pH 5, 10 g/L pulp density [40].

If the heavy metal needs to be recovered, elution stages may allow removing and concentrating at least 90 and 71% or Cu(II) and Zn(II), respectively. The resulting solutions are 20 and 9 times more concentrated than the original polluted solution [41]. In consequence, the zinc surface complexes are more stable than the ones of copper. **Figure 10** shows the adsorption percentage and the specific adsorption results of copper obtained at different pulp densities with washed and activated pine bark.

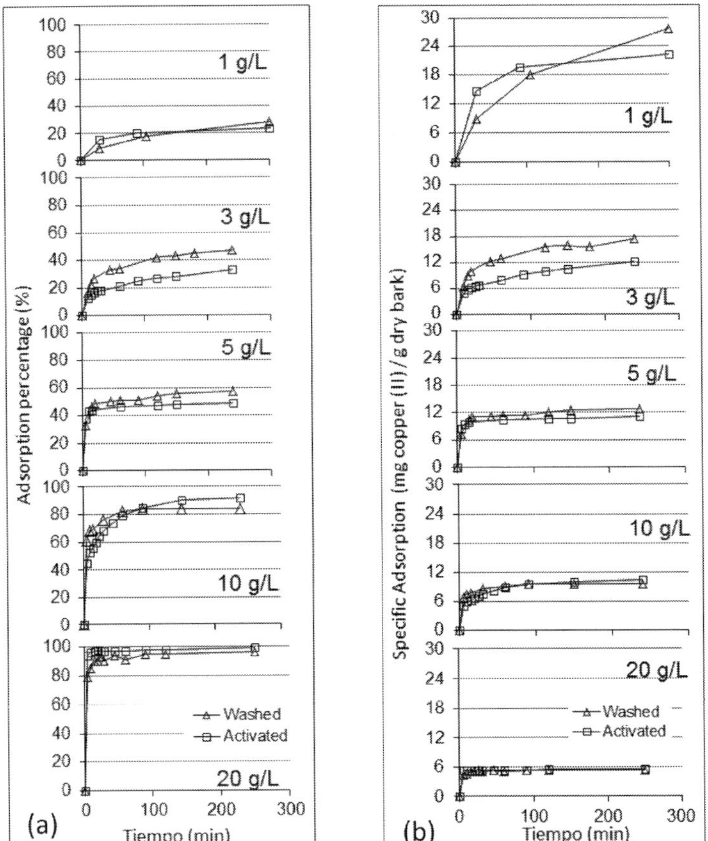

Figure 10. (a) Kinetics of adsorption percentage (%) and (b) specific adsorption (mg copper (II) ions/g dry bark) of copper (II) ions onto washed and activated bark [42].

The higher the pulp density (number of sorbent surface sites available), the higher the amount of metal removed from solution. Depending on the pulp density, the concentration of the heavy metal in solution and the contact time of the solution with the bark, the adsorption percentage may reach a *plateau* or maximum value. Such maximum value is usually referred to as 'saturation' which might be misleading. It can also be observed that the higher the pulp density, the lower the specific adsorption. At constant initial metal concentration, the increase of pulp density favours the adsorption process on bark surface sites that are easier to reach and to be replaced by a metal ion (high energy surface sites). In other words, the material is not being used at its maximum capacity.

Moving from the high to low pulp density values, the specific adsorption increases. This indicates that higher concentrations of copper (II) ions activate sites. This is expected, given that the increase of copper (II) ions will shift the adsorption equilibrium constants of secondary reactions towards increasing the metal concentration in its adsorbed state. Indeed, the ion exchange mechanism (the main responsible for metal adsorption) occurs throughout many organic radicals with different acidity constants.

3.4.3. *Surface analysis of the sorbent*

Several scientific reports have been devoted only to analyse and to interpret the results from X-ray photoelectron spectra of wood and compounds derived from it [43, 44]. **Figure 11** shows the O1s emission lines obtained for the untreated sample of pine bark and charged with Pb(II) ions. The energy scale was fixed and referred to 285 eV, in agreement with most studies reported [44].

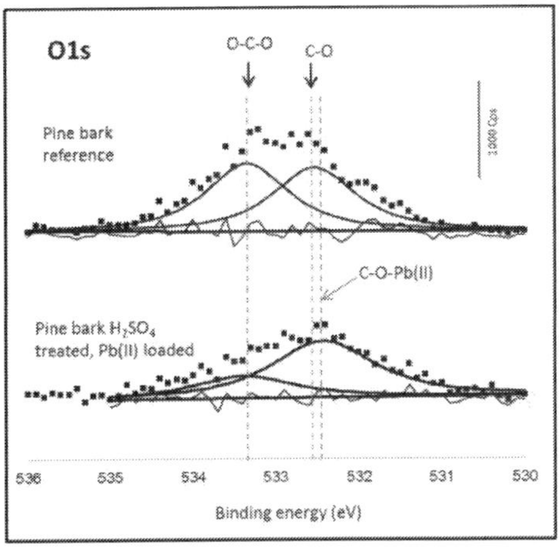

Figure 11. XPS spectra of pine bark samples. Sample 1: **Original** pine bark; sample 2 loaded for 24 h with Pb(II) in an acid aqueous solution (pH 5) at 5 g/L pulp density [45].

The O1s line is strongly perturbed by Pb(II) adsorption. The peak at the lowest position shifts to lower binding energy values, confirming a reductive environment due to Pb(II) adsorption. The peak associated with O-C-O is similarly shifted towards lower binding energies but in smaller magnitude compared to that representing C-O- type structures. Additionally, the intensity of C-O peak increases relative to the other peak. This suggests that not all sites present similar active sites. Most of them have C-O sites for adsorption while the amount of phenolic sites varies locally through bark structure. The mechanism, therefore, should take place in well distributed C-O groups and as a second type of sites in significance it would be the phenolic groups. The latter is in agreement with DRIFT results.

3.5. Conclusions

Results have shown that the adsorption efficiency depends not only on the pollutant nature but also on the presence of other species which may interact with the pollutant interfering positively or negatively on the adsorption performance. There are cases where the simultaneous presence of different pollutants results in influencing the overall adsorption capacity (for example Cd(II) and other divalent ions) and other cases where pollutants adsorb regardless the presence of others (for example Cu(II) and Zn(II)). The elution process carried out with 1N nitric acid solution proved to be promising to concentrate both ions. Cu(II) ions were more likely to be destabilised than Zn(II). The tests performed with a pulp density of 6 mg/g dry bark resulted in solutions 20 and 9 times more concentrated in the case of Cu(II) and Zn(II), respectively. The chemical activation of pine bark material increases the rate at which copper (II) ions are adsorbed and it simultaneously reduces the maximum achievable adsorption equilibrium. The chemical activation is useful primarily at high pulp density values (above 10 g/L). The adsorption capacity for Pb(II) at very low pulp density is about 93.7 mg Pb(II)/g dry bark. Mono-hydroxylated species and free lead ions were presented as the major responsible for adsorption. DRIFTS analysis revealed that the adsorption mechanism is complex, mainly driven by bark surface sites involving C-O groups and this was confirmed by XPS analysis. Phenolic and cellulose oxygen sites are also relevant.

Acknowledgements

The authors wish to thank The National Fund for Development of Science and Technology of Chile, FONDECYT, for all the financial support throughout the Grant N° 1140331. The authors would like to dedicate this manuscript to Professor Sergio Montes Sotomayor for all his contributions to the field of wastewater treatment. We thank Mrs. Ana Maria Rojo MSc. for all her comments that greatly improved the manuscript.

Author details

Gonzalo Montes-Atenas[1,*] and Fernando Valenzuela[2]

*Address all correspondence to: gmontes@ing.uchile.cl

1 Minerals and Metals Characterisation and Separation (M^{2}CS) Research Group, Departamento de Ingeniería de Minas, Facultad de Ciencias Físicas y Matemáticas, Universidad de Chile, Santiago, Chile

2 Laboratorio de Operaciones Unitarias e Hidrometalurgia, Facultad de Ciencias Químicas y Farmacéuticas, Universidad de Chile, Independencia, Santiago, Chile

References

[1] Johnson D., Hallberg K., Acid mine drainage remediation options: a review, *Sci. Total Environment*. 2005; 338: 3–14. DOI 10.1016/j.scitotenv.2004.09.002

[2] Ziemkiewicz P., Skousen J., Brant D., Sterner P., Lovett R., Acid mine drainage treatment with armoured-limestone in open limestone channels, *Journal of Environmental Quality*. 1997; 26: 1017–1024. DOI: 10.2134/jeq1997.00472425002600040013x

[3] Norm, D.S. MOP N° 601/2004, Maximum allowable limits for discharge of liquid wastes to continental and marine surface waters, Chile Government.

[4] Hubbard C., Black S., Coleman M., Aqueous geochemistry and oxygen isotope compositions of acid mine drainage from the Río Tinto, SW Spain, highlight inconsistencies in current models, *Chemical Geology*. 2009; 265: 321–334. DOI: 10.1016/j.chemgeo.2009.04.009

[5] Wisskirchen C., Dold B., Friese K., Spangenberg J., Morgenstern P., Glaesser W., Geochemistry of highly acidic mine water disposal into a natural lake with carbonate bedrock, *Applied Geochemistry*. 2010; 25: 1107–1119. DOI: 10.1016/j.apgeochem.2010.04.015

[6] Valenzuela F., Cabrera J., Basualto C., Sapag J., Separation of zinc ions from an acidic mine drainage using a stirred transfer cell–type emulsion liquid membrane contactor, *Separation Science and Technology*. 2007; 42: 363–377. DOI: 10.1080/01496390601069887

[7] Kalin M., Fyson A., Wheeler W., The chemistry of conventional and alternative treatment systems for the neutralization of acid mine drainage, *Science of the Total Environment*. 2006; 366: 395–408. DOI: 10.1016/j.scitotenv.2005.11.015.

[8] Martell A.E., Hancock R.D. Metal Complexes in Aqueous Solutions. 1st ed. 1996 Springer Science+Business Media New York, Plenum Press, New York, US. 253 p. DOI: 10.1007/978-1-4899-1486-6

[9] Defay R. and Prigogine I. Surface tension and adsorption (Treaty of Thermodynamics according to the methods of Gibbs and De Donder, Volume III, Tome III), Desoer Dunod Edition, 1951. 295 p. ISBN: 0582462843 9780582462847

[10] Langmuir I., The adsorption of gases on plane surfaces of glass, mica, and platinum, *Journal of American Chemical Society*. 1918; 40: 1361–1403 DOI: 10.1021/ja02242a004

[11] Brunauer S., The adsorption of gases and vapours. Princeton, USA: Princeton University Press, 1945, 528 p.

[12] Freundlich, H.M.F. Concerning adsorption in solutions. Zeitschrift für Physikalische Chemie-Stochiometrie und Verwandtschaftslehre, 1906, 57: 385–470. ISSN: 2196-7156

[13] Redlich O., Peterson D.L., A useful adsorption isotherm, *Journal of Physical Chemistry*. 1958; 63: 1024–1024 DOI: 10.1021/j150576a611

[14] Lagergren S. About the theory of so-called adsorption ofsoluble substances. 1898, Kungliga Svenska Vetenskapsakademiens. Handlingar. 1898; 24: 1-39. DOI: 10.1007@bf01501332

[15] Weber W.J.Jr, Morris J.C., Kinetics of adsorption on carbón from solution, *Journal of Sanitary Engineering Division of the American Society of Civil Engineers*. 1963; 89: 31–60. DOI:

[16] Valenzuela F., Basualto C., Sapag J., Ide V., Luis N., Narváez N., Yañez S., Borrmann T., Adsorption of pollutant ions from residual aqueous solutions onto nano-structured calcium silicate, *Journal of Chilean Chemical Society*. 2013; 58: 1744–1749. DOI: 10.4067/S0717-97072013000200023

[17] Chen R, Tanaka H, Kawamoto T, Asai M, Fukushima C, Na H, Kurihara M, Watanabe M, Arisaka M, Nankawa T, Selective removal of cesium ions from wastewater using copper hexacyanoferrate nanofilms in an electrochemical system, *Electrochimica Acta*. 2013; 87: 119–125. DOI:10.1016/j.electacta.2012.08.124

[18] Sirés I., Brillas E., Remediation of water pollution caused by pharmaceutical residues based on electrochemical separation and degradation technologies: a review, *Environment International*. 2012; 40: 212–229. DOI: 10.1016/j.envint.2011.07.012.

[19] Cairns M, Borrmann T, Höll W, Johnston J, A study of the uptake of copper ions by nanostructured calcium silicate, *Microporous and Mesoporous Materials*. 2006; 95: 126–134. DOI: 10.1016/j.micromeso.2006.05.009

[20] Qian Q, Chen Q, Machida M, Tatsumoto H, Mochidzuki K, Sakoda A, Removal of organic contaminants from aqueous solution by cattle manure compost (CMC) derived activated carbons, *Applied Surface Science*. 2009; 255: 6107–6114. DOI: 10.1016/j.apsusc.2009.01.060

[21] Güell R, Fontàs C, Anticó E, Salvadó V, Crespo J, Velizarov S, Transport and separation of arsenate and arsenite from aqueous media by supported liquid and anion-exchange membranes, *Separation and Purification Technology*. 2011; **80**:428.DOI: 10.1016/j.seppur.2011.05.015

[22] Al-Asheh S, Duvnjak Z, Sorption of cadmium and other heavy metals by pine bark, *Journal of. Hazardous Material*. 1997; 56: 35–51. DOI: 10.1016/S0304-3894(97)00040-X

[23] Tiwari D, Mishra SP, Mishra M, Dubey RS, Biosorptive behaviour of mango and neem bark for Hg+2, Cr +3 and Cd+2 toxic ions from aqueous solutions: a radiotracer study, *Applied Radiation and Isotopes*. 1999; 50: 631–642. DOI: 10.1016/S0969-8043(98)00104-3

[24] Vazquez G, Antorrena G, Gonzalez J, Doval MD, Adsorption of heavy metal ions by chemically modified Pinus pinaster Bark, Bioresource Technology. 1994; 48:251–255. DOI: 10.1016/0960-8524(94)90154-6

[25] Martin-Lara MA, Blazquez G, Ronda A, Pérez A, Calero M, Development and characterisation of biosorbents to remove heavy metals from aqueous solutions by chemical treatment of Olive Stone, *Industrial and Engineering Chemistry Research*. 2013; 52:10809–10819. DOI: 10.1021/ie401246c

[26] Watmough SA, Hutchinson TC Uptake of 207Pb and 111Cd through bark of mature sugar maple, white ash and white pine: a field experiment, *Environmental Pollution*. 2003; 121:39–48. DOI: 10.1016/S0269-7491(02)00208-7

[27] Jang A, Seo Y, Bishop PL, The removal of heavy metals in urban runoff by sorption of mulch, *Environmental Pollution*. 2005; 133:117–127. DOI: 10.1016/j.envpol.2004.05.020

[28] Saeed A, Akhter MW, Iqbal M Removal and recovery of heavy metals from aqueous solution using papaya wood as a new biosorbent, *Separation and Purification Technology*. 2005; 45:25–31. DOI: 10.1016/j.seppur.2005.02.004

[29] Cay S, Uyanik A, Ozasik A, Single and binary component adsorption of copper(II) and cadmium(II) from aqueous solution using tea-industry waste, *Separation and Purification Technology*. 2004; 38:273–280 DOI: 10.1016/j.seppur.2003.12.003

[30] Taty-Costodes VC, Fauduer H, Porte C, Delacroix A, Removal of Cd(II) and Pb(II) ions from aqueous solutions, by adsorption onto sawdust of Pinus sylvestris, *Journal of Hazardous Material*. 2003; 105:121–142 DOI: 10.1016/j.jhazmat.2003.07.009

[31] Khokhotva AP, Adsorption of heavy metals by a sorbent based on pine bark, *Journal of Water Chemistry and Technology*. 2010; 32: 336–339. DOI: 10.3103/S1063455X10060044

[32] Martins M, Faleiro ML, Barros RJ, Verissimo AR, Barreiros MA, Costa MC Characterization and activity studies of highly heavy metal resistant sulphate-reducing bacteria to be used in acid mine drainage decontamination, *Journal of Hazardous Material*. 2009; 166:706–713. DOI: 10.1016/j.jhazmat.2008.11.088

[33] Huang S-C, Chang F-C, Lo S-L, Lee M-Y, Wang C-F, Lin J-D, Production of lightweight aggregates from mining residues, heavy metal sludge, and incinerator fly ash, *Journal of Hazardous Material*. 2007; 144:52–58 DOI: 10.1016/j.jhazmat.2006.09.094

[34] Martin-Dupont F, Gloagen V, Guilloton M, Granet R, Krausz P, Study of chemical interaction between barks and heavy metal cations in the adsorption process, *Journal of Environmental Science Health, Part A*. 2006; 41:149–160. DOI: 10.1080/10934520500349250

[35] Al-Asheh S, Banat F, Al-Omari R, Duvnjak Z, Predictions of binary sorption isotherms for the sorption of heavy metals by pine bark using single isotherm data, *Chemosphere*. 2000; 41:659–665. DOI: 10.1016/S0045-6535(99)00497-X

[36] Teles de Vasconcelos LA, Gonzalez Beca CG, Chemical activation of pine bark to improve its adsorption capacity of heavy metal ions. Part 2: by conversion to a salt form, *European Water Pollution Control: Official Publication of the European Water Pollution Control Association* (EWPCA) 1997; 7:47–55. ISSN: 0925-5060

[37] Celik A, Demirbas A, Removal of heavy metal ions from aqueous solutions via adsorption onto modified lignin from pulping wastes, *Energy Source*. 2005; 27:1167–1177. DOI: 10.1080/00908310490479583

[38] Brown WH, Foote CS, Iverson BL, Anslyn E, Organic chemistry, 5 edition. 20 Davis Drive, Belmont, CA: Brooks Cole Cengage Learning, United States of America 2009, 1312 p. ISBN-10: 084005498X

[39] Montes S, Montes-Atenas G, Vilches A, Study of removing copper (II) and chromium (VI) from aqueous solution using bark pine. Proceedings of the TMS Fall 2002 Extraction and Processing Division Meeting; September 2002. Lulea, Sweden: pp. 1–5.

[40] Montes S, Montes-Atenas G, Salomo F, Valero E, Diaz O, On the adsorption mechanisms of copper ions over modified biomass, *Bulletin of Environmental Contamination and Toxicology*. 2006; 76: 171–178. DOI: 10.1007/s00128-005-0904-8

[41] Montes S, Montes-Atenas G, García-García F, Valenzuela M, Valero E, Díaz O, Evaluation of an adsorption-desorption process for concentrating heavy metal ions from acidic wastewaters, *Adsorption Science and Technology*, 2009; 27: 513–521. DOI: 10.1260/0263-6174.27.5.513

[42] Montes-Atenas G, Valenzuela F, Montes S, The application of diffusion-reaction mixed model to assess the best experimental conditions for bark chemical activation to improve copper(II) ions adsorption, *Environmental Earth Sciences*. 2014, 72: 1625–1631. DOI: 10.1007/s12665-014-3066-3

[43] Nzokou P, Kamdem DP, You have free access to this content X-ray photoelectron spectroscopy study of red oak-(*Quercus rubra*), black cherry- (Prunus serotina) and red pine-(*Pinus resinosa*) extracted wood surfaces, *Surface and Interface Analysis*, 2005; 37: 689–694. DOI: 10.1002/sia.2064

[44] Shchukarev A, Sundberg B, Mellerowicz E, Persson P, XPS study of living tree, *Surface and Interface analysis*. 2002; 34: 284–288. DOI: 10.1002/sia.1301

[45] Montes-Atenas G, Schroeder SLM, Sustainable natural adsorbents for heavy metal removal from wastewater: lead sorption on pine bark (Pinus radiate D.Don). *Surface and Interface Analysis*. 2015; 47: 996–1000. DOI: 10.1002/sia.5807

Chapter 11

Integral use of *Nejayote*: Characterization, New Strategies for Physicochemical Treatment and Recovery of Valuable By-Products

Eduardo Alberto López-Maldonado,
Mercedes Teresita Oropeza-Guzmán and
Karla Alejandra Suárez-Meraz

Additional information is available at the end of the chapter

http://dx.doi.org/10.5772/66223

Abstract

In this research, an innovative physicochemical strategy is presented to address the problem of *nejayote*, from two perspectives: the first focusing on sanitation and reuse of *nejayote* using waste from shrimp shells, thereby adding value to the recovered solids of *nejayote*. Zeta potential measurements are a proactive electrochemical tool to define the strategy to allow integral use of *nejayote* in the industry nixtamalization. The treated water can be discharged from the municipal sewer system using a process of coagulation-flocculation, with an optimal dose of 1250 mg/L chitosan at pH 5, achieving removal of up to 80% of total suspended solids and turbidity. Moreover, zeta potential measurements show that the anionic biopolyelectrolyte obtained from *nejayote* has potential to be applied in the area of water treatment as a green chelating agent.

Keywords: nejayote, nixtamalization, biopolyelectrolytes, zeta potential, coagulation-flocculation

1. Introduction

Nixtamalized products such as maize tortillas originated in Mexico, are the main sources of energy, protein, calcium and other important nutrients and are considered the national breads and consumed with other fillings such as beans, meats, eggs and vegetables [1–3].

The ancient, laborious or traditional process (nixtamalization) to obtain tortillas is a process widely used by indigenous people in Mexico (41%), the Southern United States, Central America, Asia and parts of Europe, that consumes significant amounts of water, energy and time [2]. Traditional maize is lime-cooked in clay pots over a fire, followed by steeping for 8–16 h (generally overnight), the supernatant called maize wastewater or commonly known as "*nejayote*", derived from the Nahuatl word meaning "lime broth ashes" is discarded and then the nixtamal is hand-washed. Nixtamal is ground into a fine masa with a stone grinder called metate and then hand-molded, patted or pressed into disks, which are baked on both sides on a hot griddle [4–10].

Nixtamalization causes a loss of about 5% by weight dry basis of corn; 3% is suspended and the remaining 2% is dissolved. The suspended matter can be separated easily and inexpensively by sedimentation and the dissolved substance should "precipitate" to separate solids which is also done by sedimentation [2, 11–16].

A typical maize nixtamalization facility, processing 50 kg of maize everyday, uses over 75 L of water per day and generates nearly the equivalent amount of alkaline wastewater on a daily basis [4]. The estimated monthly volume of *nejayote* generated in Mexico is about 1.2 m^3 [17].

Nejayote is considered an environmental pollutant because it is an alkaline wastewater, with high chemical and biological oxygen demand [2, 9]. Due to the presence of lime in the process, the pH of the wastewater is very high (12–14), with a high temperature between 40 and 70°C), containing suspended solids (corn husks and broken grains) and a very high portion of dissolved material from the alkaline hydrolysis of corn components [14]. The *nejayote* with these physicochemical characteristics is thrown, often without treatment, into drainage systems and even directly to the soil and groundwater. Thus, alternatives for sanitation of *nejayote* and utilization are needed [14]. Among the solutions that have been reported, they are from biological treatment processes [9], membrane filtration, nixtamalization methods that minimize water use and the use of *nejayote* as a supplement in animal feed [17].

In this research, an innovative physicochemical strategy is presented to address the problem of *nejayote* from two perspectives: the first focused on remediation of *nejayote* and the second is on water reuse using biopolyelectrolyte (BPE) from waste shrimp shells. With the use of effluents generated by 20 tons of corn nixtamalized equivalent to one ton of corn or sorghum protein is obtained [2, 16]. Another benefit, both economic and social, which could have *nejayote* recovery is that wastewater could be recycled, either in nixtamalization industry itself or for any other use. The second is based on the use of *nejayote* for obtaining anionic BPE (maize gum) for treating wastewater from electroplating industry. In both cases, zeta potential (ζ) measurements as electrochemical parameters were used to develop the process of sanitation and water reuse and for the extraction and application of anionic BPE in the separation of heavy metals.

2. Experimental

It is shown in **Figure 1** that zeta potential measurements were used to interconnect the physicochemical characteristics of *nejayote* and chitosan flocculant capacity to achieve sanitation and water reuse in nixtamalization industry. In the first stage, plots of ζ vs pH of *nejayote,* chitosan and maize gum were constructed to determine the behavior of surface charge and isoelectric point (IEP). Then the strategic dosage of chitosan was done in the process of coagulation-flocculation of *nejayote*. The coagulation-flocculation window was constructed by measuring the water-quality parameters of environmental interest (turbidity and total suspended solids) and zeta potential. Moreover, zeta potential measurements were used to to exploring the interaction capacity of maize gum obtained from *nejayote* with metal ions, frequently contained in wastewater from the electroplating industry.

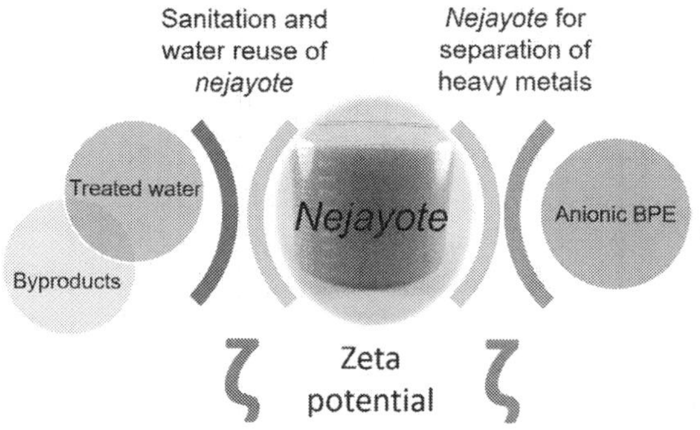

Figure 1. Using zeta potential measurements for *nejayote* sanitation and water reuse, and its use for obtaining a green flocculant for the separation of heavy metals.

2.1. Materials

Commercial testing water-quality reagents from HACH® were used. Milli-Q grade water was used in all the experiments. All other reagents were of analytical grade and were used without further purification.

2.1.1. Wastewater sampling in the nixtamalization industry

Nejayote was provided by a local tortilla-making industry. The wastewater sampling protocol was followed as recommended by Mexican sampling standard (NMX-AA-003-1980).

2.1.2. Chitosan extraction from waste shrimp shells

Chitosan is obtained from waste shrimp shells using the method proposed by the authors Goycoolea et al. [15].

2.1.3. Maize gum extraction of nejayote

Maize gum was obtained by fractional separation, using hexane, ethanol and hydrochloric acid, isopropanol, acetone, methanol formed by the steps of desalmidonado, deproteinization, delipidation, delignification which are proposed by the authors of [8, 18, 19].

2.2. Methods

2.2.1. Physicochemical characterization of nejayote

The main parameters of quality wastewater used in this research were performed following the Mexican standard procedures to determine the biochemical oxygen demand (BOD_5), chemical oxygen demand (COD), total nitrogen (TN), the solids content, total organic carbon (TOC), total phosphorus (TP), and other parameter fields such as pH, electrical conductivity (EC) and temperature were carried out based on the Hach methods.

2.2.2. Profiles of $\zeta = f(pH)$ of nejayote, maize gum and chitosan

The charge density, isoelectric point and chitosan-dosing strategy for treating *nejayote* were determined in a $\zeta = f(\text{pH})$ plot. The zeta potential measurement was performed using the SZ-100 of Horiba Scientific equipment based on studies by López-Maldonado et al. [20, 21].

2.2.3. Nejayote treatability tests by coagulation-flocculation using chitosan

A sample of 20 mL of *nejayote* was taken in a vial and the pH was adjusted to 5. The chitosan dosage tests were performed in 20-mL-vials. Progressive additions of 0.1 g/L chitosan solution were done and after each one, the vials were shaken for 2 min at 250 rpm and 5 min at 50 rpm and allowed to settle for 5 more min. Finally the supernatant to a height of 2 cm from the vial was suctioned to determine the parameters of water quality in the supernatant.

2.2.4. Evaluation of the capacity of polyelectrolyte maize gum for decontaminating wastewater

The anionic BPE obtained from *nejayote* is characterized by Fourier Transform Infra-Red spectroscopy (FTIR), scanning electron microscopy (SEM) and measurements of zeta potential (ζ). FTIR spectra of maize gum were recorded using a Nicolet FT-IR spectrometer. The samples of maize gum were analyzed by SEM and X-ray microanalysis. The analysis was performed on SEM (ZEISS EVO-MA15), equipped with an EDS (energy dispersive spectroscopy) BRUKER detector microscope to observe the composition. The zeta potential measurement was performed using the SZ-100 of Horiba Scientific equipment based on studies by López-Maldonado et al. [20]. This was developed with the maize gum dispersion in a 0.1% solution, which took different levels of acidity and alkalinity in the range of 2–12 and injected into a cell with electrode graphite.

http://dx.doi.org/10.5772/66223

3. Results and discussion

3.1. Physicochemical characterization of *nejayote*

In this investigation the *nejayote* generated by a tortilla factory in Mexico was taken as the object of study. A typical maize nixtamalization facility, processing 500 kg of maize every day, uses over 750 L of water per day and generates nearly the equivalent amount of alkaline wastewater on a daily basis. **Figure 2** shows the stages of the nixtamalization process used for the manufacture of nixtamal mass and the generation of *nejayote*.

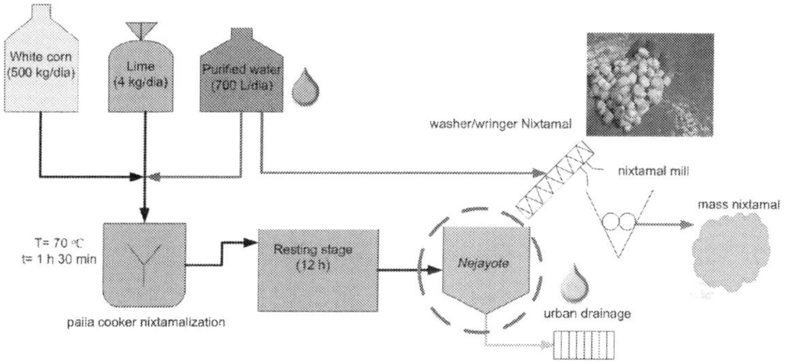

Figure 2. Diagram of the nixtamalization process and the point of generation *nejayote*.

As shown in **Table 1**, the physicochemical characteristics of *nejayote* concerning the content of organic matter determined by the parameters COD, TP, BOD_5 and TOC normed indicate that najeyote exceeds the maximum permissible limits of NOM-002-SEMARNAT-1996. The *nejayote* has a pH of 11.6 as already well known, is a wastewater alkaline by the use of lime in the nixtamalization.

Parameter	*Nejayote*	Maximum permissible limit
Suspended Solids, SS (mL/L)	800–900	1[b]
Total Solids, TS (mg/L)	46,523.00	200[b]
Total Dissolved Solids, TDS (mg/L)	46,339.70	NI
Total Suspended Solids, TSS (mg/L)	2000.00	NI
Turbidity (FAU)	690–1500	NI
Alkalinity (mg/L $CaCO_3$)	1020–1050	NI
Electric conductivity, EC (mS/cm)	4.29–6.42	NI
ζ (mV)	−10.5	NI
Particle size of the dissolved part (nm)	100–600	NI

Parameter	*Nejayote*	**Maximum permissible limit**
Temperature (°C)	30–39	40[b]
Color (Pt-Co)	5653–8580	NI
pH	11.61–12.1	5.5–10[a]
Chemical Oxygen Demand, COD (mg O_2/L)	9800–28,450	NI
Total Organic Carbon, TOC (mg C/L)	7337–9836	NI
Inorganic Carbon, IC (mg/L)	23–28	NI
Total Carbon, TC (mg C/L)	7360–9864	NI
Biochemical Oxygen Demand, BOD_5 (mg O_2/L)	2700	200[b]
Total Phosphorus, TP (mg P/L)	905–1321	30[b]
Total Nitrogen, TN (mg N/L)	303–418	60[b]
Biodegradability (BOD_5/COD)	0.27	NI

NI= Not included in the standard.
[a]NOM-001-SEMARNAT-1996.
[b]NOM-002-SEMARNAT-1996.

Table 1. Maize industry wastewater physicochemical composition.

For this research the measurement of other nonregulatory parameters was performed, and they are key to evaluate the performance of coagulation-flocculation process and determine the best operating conditions. $\zeta = -10$ mV (pH = 12) and particle size of the dissolved part of *nejayote* (100–600 nm), which indicates containing dispersed particles very stable. Considering the surface charge of the *nejayote* colloids and particle size to be separated by coagulation-flocculation, it requires the addition of a cationic BPE.

3.2. Profiles of $\zeta = f$(pH) of *nejayote* and chitosan

The zeta potential is a parameter by electrochemical nature that allows to study and predict the interactions occurring at the molecular level between the colloidal particles *nejayote* and the different ionic species of the medium, also it indicates the degree of stability of dispersion in an aqueous medium from the point electrically. The aim is to employ ζ measurements to know and understand the behavior of the BPE type chitosan in this kind of wastewater treatment (see **Figure 3**).

Surface charge of chitosan and *nejayote* colloids are pH-dependent and their behavior has great influence on coagulation-flocculation performance [22]. In addition, ζ measurements are required to characterize the colloidal system to understand repulsion and aggregation between colloidal particles.

In **Figure 4**, chitosan shows an amphoteric behavior, in the region of pH = 2–5.5 has a positive surface charge ($\zeta = 51.1$ mV) due to protonation of amine groups, at pH = 6–10 its surface charge remains neutral, this is due to the insolubilization phenomenon occuring at pH > IEP (pH = 5–6) of chitosan and increases their hydrophobicity.

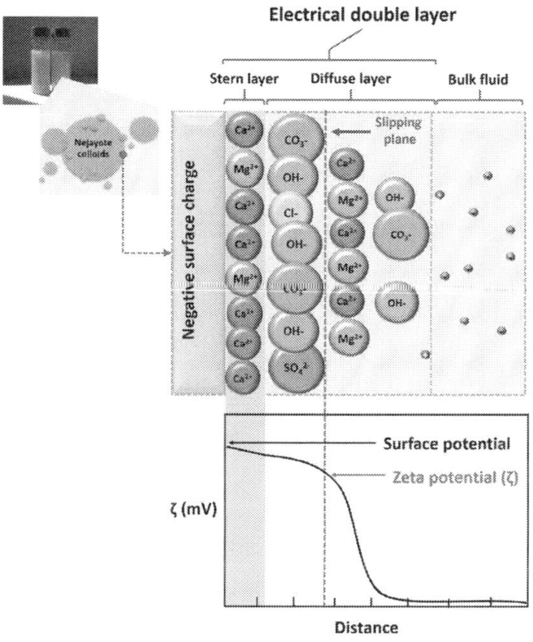

Figure 3. Model of the electrical double layer and zeta potential concept adopted for sanitation of *nejayote*.

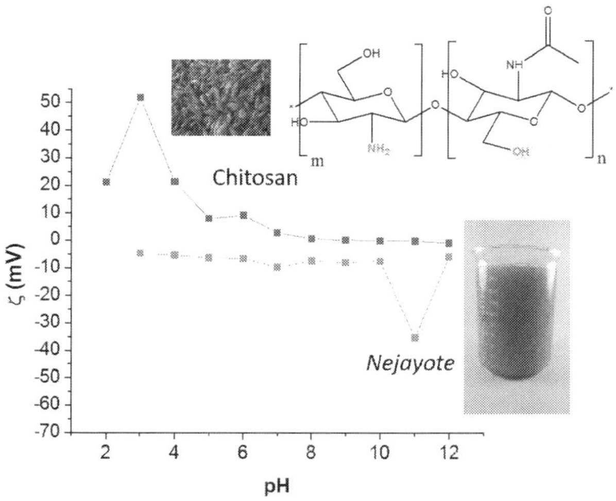

Figure 4. Electrokinetic properties of *nejayote* and biopolyelectrolyte type chitosan.

Moreover, the *nejayote* (ζ = 0) has a negative surface charge throughout the pH range. From the electrical viewpoint, at pH = 5, the interaction between oppositely charged species chitosan-*nejayote* is ensured and therefore the strategic dosage cationic BPE was performed.

3.3. *Nejayote* treatability tests by coagulation-flocculation using chitosan

Dosing strategy for chitosan was determined by ζ of *nejayote* colloids and chitosan, and also by observing critical pH value of the IEP. In this study, charged chitosan purpose is to reduce the repulsion forces between particles by neutralizing the negatively charged molecules. In general, the electroneutrality zone for chitosan-*nejayote* system is below pH = 6 (see **Figure 4**), this has a practical application since higher charge density with less BPE concentration can be achieved.

Since the best wastewater clarification was at pH = 5.5 for chitosan, a turbidity-dosage profile was performed near the same pH to determine the optimal quantity of chitosan needed to flocculate *nejayote* colloids.

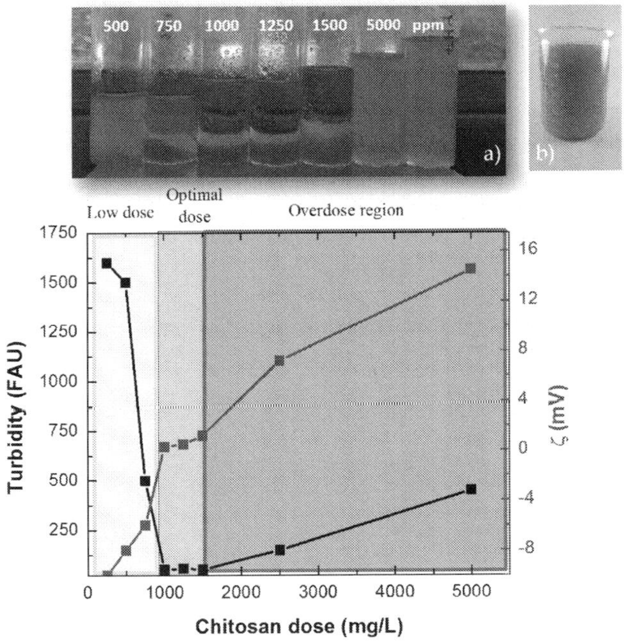

Figure 5. Turbidity and ζ of the supernatant in the coagulation-flocculation of nejayote at pH= 5 with chitosan: a) Nejayote and b) Nejayote visual appearance treated with chitosan.

The coagulation-flocculation window of *nejayote* using chitosan at pH = 5 was constructed based on the methodology reported by López Maldonado et al. [23].

Figure 5 shows the behavior of the zeta potential and turbidity with respect to the concentration of chitosan. In the region of low doses (250 and 750 mg/L), a decrease in turbidity (1590–500 FAU) is achieved, and the variation of zeta potential $\zeta = -10$ mV to more positive values ($\zeta = -5.4$ mV) shows that the mechanism of destabilization of *nejayote* colloids occurs by charge neutralization [24]. At a dose of 1250 mg/L chitosan, point of zero charge was reached and the better quality of treated water (turbidity = 22 FAU, color = 315 Pt-Co, TSS = 12 mg/L) was obtained. At higher concentration (>1250 mg/L) the best dose, the restabilization processes occur due to excess chitosan adsorbed on the colloids of *nejayote*. In this region of overdose, the addition of chitosan had an adverse effect on the quality of wastewater, increasing turbidity and stability of the dispersed particles ($\zeta = 15$ mV and turbidity = 450 FAU).

The coagulation-flocculation window was obtained from 1000 to 1500 mg/L chitosan with optimal dosage of 1250 mg/L chitosan, obtaining with this removal turbidity and suspended solids of about 80% (see **Figure 6**). At this dose, the surface charges of *nejayote* colloids were neutralized by chitosan molecules, resulting in a ζ value very close to zero.

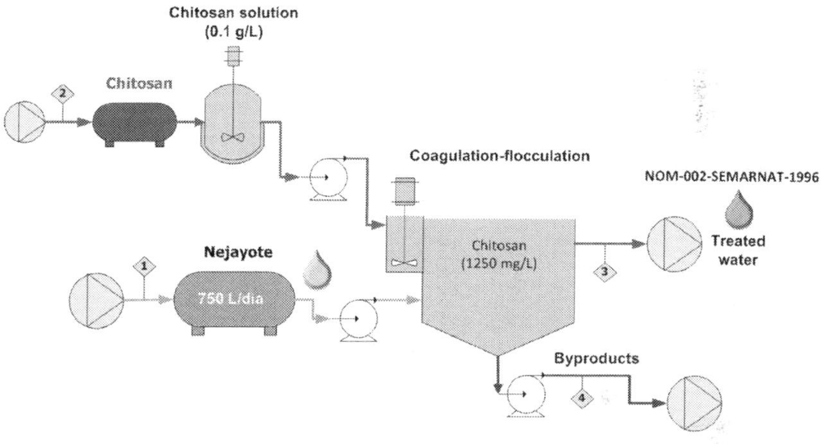

Figure 6. Diagram of the engineering for the sanitation process of *nejayote* using chitosan.

3.4. Evaluation of the polyelectrolyte capacity of maize gum for decontaminating wastewater

The behavior of zeta potential vs pH of anionic BPE obtained from *nejayote* is shown in **Figure 7**, which has a high negative charge density (−35 mV) in the pH range 6–12, having the isoelectric point close to pH = 2.

This negative surface charge is very interesting for the treatment of wastewater containing high concentration of heavy metal. In the FTIR spectrum (see **Figure 8**) shows that the BPE has the characteristic functional groups of a polysaccharide (3400 cm^{-1} corresponding to stretching of the OH groups and 2900 cm^{-1} corresponding to the CH_2 groups) which give the negative surface charge and that can interact with oppositely charged species, such as heavy metal ions [25].

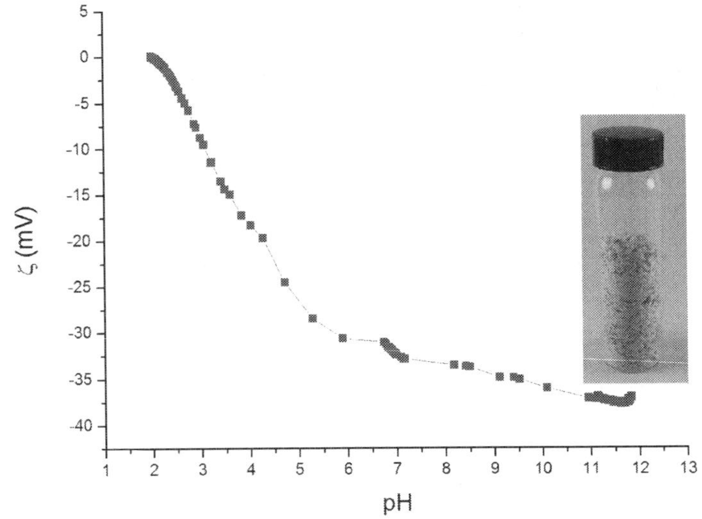

Figure 7. Zeta potential vs pH profiles of anionic BPE.

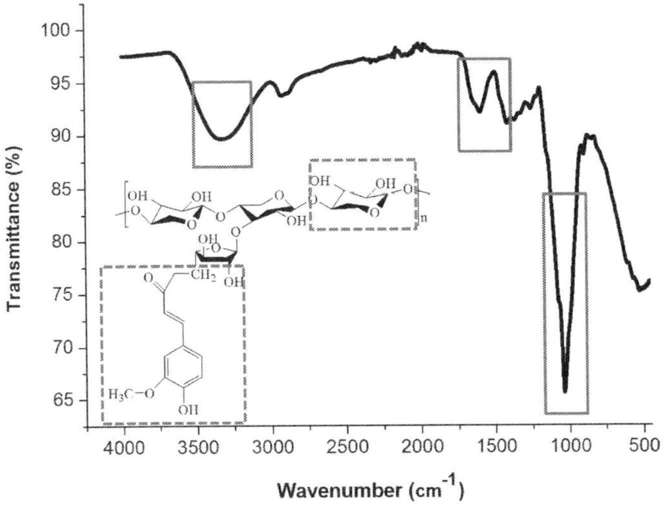

Figure 8. FT-IR spectrum of anionic BPE.

Figure 9 shows the morphology of anionic BPE and analysis of chemical composition, indicating that its content is primarily carbon, oxygen and calcium, because lime is used in the nixtamalization.

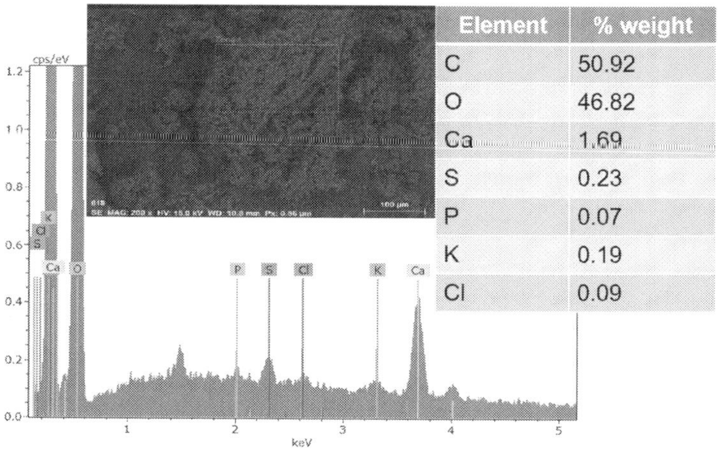

Figure 9. SEM micrograph and EDS spectrum of the anionic BPE: Inset Table shows the composition analysis.

4. Conclusion

Zeta potential measurements are a proactive electrochemical tool to define the strategy of chitosan dosage that allows sanitation and water reuse industry nixtamalization. The use of chitosan allows the use and reuse of byproducts recovered from *nejayote* and it serves as a source of protein for animal feed. The treated water can be discharged into the municipal sewer system using an optimal dose of 1250 mg/L chitosan at pH = 5, achieving removal of up to 80% in the removal of total suspended solids and turbidity. This work evidenced the potential use of *nejayote* as a raw material for obtaining anionic biopolyelectrolyte in the treatment of wastewater with heavy metals of the electroplating industry.

Acknowledgements

The authors gratefully acknowledge support from Consejo Nacional de Ciencia y Tecnologia, México (CONACyT) Project Ciencia Básica 2015 No. 237032, Project Problemas Nacionales No. 247236.

Author details

Eduardo Alberto López-Maldonado*, Mercedes Teresita Oropeza-Guzmán and Karla Alejandra Suárez-Meraz

*Address all correspondence to: elopez@cideteq.mx

Faculty of Chemical Sciences and Engineering, Autonomous University of Baja California, Tijuana, Baja California, Mexico

References

[1] Serna Saldivar SO, Chuck-Hernandez C. Tortillas. Encyclopedia of Food and Health. 2016; 1: 319–325. DOI: 10.1016/B978-0-12-384947-2.00697-8.

[2] Rosentrater KA. Review of corn masa processing residues: generation, properties, and potential utilization. Waste Management 2006; 26: 284–292. DOI: 10.1016/j.wasman.2005.03.010.

[3] Valderrama C, Rojas A, Gutiérrez E, Rojas I, Oaxaca A, Rosa E, Rodríguez ME. Mechanism of calcium uptake in corn kernels during the traditional nixtamalization process: diffusion, accumulation and percolation. Journal of Food Engineering 2010; 98: 126–132. DOI: 10.1016/j.jfoodeng.2009.12.018.

[4] Fernandez JL, Acosta AA, Gruintal MA, Zelaya O. Kinetics of water diffusion in corn grain during the alkaline cooking at different temperatures and calcium hydroxide concentration. Journal of Food Engineering 2011; 106: 60–64.

[5] Gutiérrez E, Rojas I, Rojas A, Arjona JL, Cornejo MA, Zepeda Y, Velázquez R, Ibarra C, Rodríguez ME. Microstructural changes in the maize kernel pericarp during cooking stage in nixtamalization process. Journal of Cereal Science 2010; 51: 81–88. DOI: 10.1016/j.jcs.2009.09.008.

[6] Fernández JL, Martín ES, Díaz JAI, Calderon A, Alvarado A, Ortiz H, Leal M. Steeping time and cooking temperature dependence of calcium ion diffusion during microwave nixtamalization of corn. Journal of Food Engineering 2006; 76: 568–572. DOI: 10.1016/j.jfoodeng.2005.06.004.

[7] Osorio P, Agama E, Bello LA, Islas JJ, Gomez NO, Paredes O. Effect of endosperm type on texture and in vitro starch digestibility of maize tortillas, LWT. Food Science and Technology 2011; 44: 611–615.

[8] Niño-Medina G, Carvajal-Millán E, Lizardi J, Rascon-Chu A, Marquez-Escalante J, Gardea A, Martínez-López A, Guerrero V. Maize processing wastewater arabinoxylans:

gelling capability and cross-linking content. Food Chemistry 2009; 115: 1286–1290. DOI: 10.1016/j.foodchem.2009.01.046.

[9] Salmeron-Alcocer A, Rodriguez-Mendoza N, Pineda-Santiago S, Cristiani-Urbina E, Juarez Ramirez C, Ruiz-Ordaz N, Galindez-Mayer J. Aerobic treatment of maize processing wastewater (*nejayote*) in a single stream multistage reactor. Journal of Environmental Engineering and Science 2003; 2: 401–406. DOI: 10.1139/s03-046.

[10] Bressani R, Paz y Paz R, Scrimshaw NS. Corn nutrient losses. Chemical changes in corn during preparation of tortillas. Journal of Agricultural and Food Chemistry 1958; 6: 770. DOI: 10.1021/jf60092a009.

[11] Sefa S, Cornelius B, Sakyi E, Ohene E. Effect of nixtamalization on the chemical and functional properties of maize. Food Chemistry 2004; 86: 317–324. DOI: 10.1016/j.foodchem.2003.08.033.

[12] Pflugfelder RL, Rooney LW, Waniska RD. Dry matter losses in commercial corn masa production. Cereal Chemistry 1988; 65: 127–132.

[13] Jackson DS, Rooney LW, Kunze OR, Waniska RD. Alkaline processing properties of stress cracked and broken corn (*Zea mays* L.). Cereal Chemistry 1988; 65: 133–137.

[14] Gutiérrez-Uribe J, Rojas-García C, García-Lara S, Serna-Saldivar S. Phytochemical analysis of wastewater (nejayote) obtained after lime-cooking of different types of maize processed into masa for tortillas. Journal of Cereal Science 2010; 52: 410–416. DOI: 10.1016/j.jcs.2010.07.003.

[15] Goycoolea F, Agullo E, Mata R, Sources and processes of obtaining. In: Pastor de Abram A, editor. Chitin and chitosan: obtaining, characterization and applications.1ra ed. Fondo editorial de la Pontificia Universidad Catolica del Perú; 2004. p. 103–154.

[16] Ochoa A, Vinieagra G, editores. Selected topics in the maize-tortilla chain. A multidisciplinary approach.1ra ed. Autonomous Metropolitan University of Iztapalapa; 2004. 333 p.

[17] Campechano Carrera EM, de Dios Figueroa Cárdenas J, Arámbula Villa G, Martínez Flores HE, Jiménez Sandoval SJ, Luna Bárcenas JG. New ecological nixtamalization process for tortilla production and its impact on the chemical properties of whole corn flour and wastewater effluents. International Journal of Food Science & Technology. 2012; 564–571. DOI: 10.1111/j.1365-2621.2011.02878.x.

[18] Simpson J. A survey of microorganisms for the production of enzymes that attack the pentosans of wheat flour. Canadian Journal of Microbiology 1954; 131–139. DOI: 10.1139/m55-017.

[19] Zhang Z. Extraction and modification technology of arabinoxylans from cereal by-products: a critical review. Food Research International 2014; 423–436. DOI: 10.1016/j.foodres.2014.05.068.

[20] López-Maldonado EA, Oropeza-Guzman MT, Jurado-Baizaval JL, Ochoa-Terán A. Coagulation–flocculation mechanisms in wastewater treatment plants through zeta potential measurements. Journal of Hazardous Materials 2014; 279: 1–10. DOI: 10.1016/j.jhazmat.2014.06.025.

[21] López-Maldonado EA, Ochoa-Terán A, Oropeza-Guzmán MT. A multiparameter colloidal titrations for the determination of cationic polyelectrolytes. Journal of Enviromental Protection. 2012; 3: 1559–1570. DOI: 10.4236/jep.2012.311172.

[22] Suarez Meraz KA, Ponce Vargas SM, Lopez Maldonado JT, Cornejo Bravo JM, Oropeza-Guzmán MT, López-Maldonado EA. Eco-friendly innovation for nejayote coagulation–flocculation process using chitosan: evaluation through zeta potential measurements. Chemical Engineering Journal 2015; 284: 536–542. DOI: 10.1016/j.cej.2015.09.026.

[23] López Maldonado EA, Oropeza Guzmán MT, Ochoa Terán A. Improving the efficiency of a coagulation-flocculation wastewater treatment of the semiconductor industry through zeta potential measurements. In: Monsalvo, VM, editor. Ecological Technologies for Industrial. Wastewater Management. Oakville, Canada: Apple Academic Press Inc.; 2015. p. 167–195.

[24] Bolto B, Gregory J. Organic polyelectrolytes in water treatment. Water Research 2007; 41: 2301–2324. DOI: 10.1016/j.watres.2007.03.012.

[25] Egüés I, Stepan AM, Eceiza A, Toriz G, Gatenholm P, Labidi J. Corncob arabinoxylan for new materials. Carbohydrate Polymers 2014; 102: 12–20. DOI: 10.1016/j.carbpol.2013.11.011.

Chapter 12

Social Perspectives on the Effective Management of Wastewater

Dalia Saad, Deirdre Byrne and Pay Drechsel

Additional information is available at the end of the chapter

http://dx.doi.org/10.5772/67312

Abstract

The chapter discusses how adopting a holistic methodology that acknowledges sociological factors, including community participation, public involvement, social perception, attitudes, gender roles and public acceptance, would lead to improvements in wastewater management practice. It highlights the social dimension as a tool, a lens through which wastewater management and reuse can take on new dimensions. In this way, this chapter aims to shift the focus from perceiving wastewater as a nuisance that needs disposal, toward a resource not to be wasted, which can contribute to food security, human and environmental health, access to energy as well as water security.

Keywords: wastewater reuse, social dimension, community participation, public involvement, public acceptance, gender

1. Introduction

The global water crisis, the shortage of fresh water, contamination of water and increasing volumes of wastewater being produced have eventually necessitated the use of wastewater. A paradigm shift is therefore required not only to prevent further damage to the ecosystems, but also to emphasize that wastewater is a resource whose effective management is essential for future water security [1, 2].

Wastewater can be recycled and reused for a variety of water demanding activities such as agriculture, firefighting, flushing of toilets, industrial cooling, park watering, formation of wetlands for wildlife habitats, etc. [3]. Treated wastewater reuse can be seen as a sustainable way of addressing long-term imbalances between water dem and and supply, which makes sound economic sense also in view of increased imbalances due to climate change [4]. The focus of most wastewater research has been on the technical aspects and improvements in terms of water quality and on minimizing the environmental and health impacts, without paying sufficient attention to their basic social and sustainability dimensions. Recent research has shown that ignoring broader social issues that impact the adoption of sustainable solutions prolongs global environmental problems as well as unjust public health and social conditions [5]. Thus, more attention is needed to the social aspects of wastewater management strategies.

2. The Global Demand for Water

The development of human societies is heavily dependent on the availability of water of suitable quality in adequate quantities. However, the demand for water is ever increasing due to population growth, technological advancement, industrial expansion, pollution and urbanization, which put great stress on the natural water cycle [1]. These demands were met by constructing ever-larger dams, which in turn affect both water quality and quantity [6]. Moreover, the available freshwater supplies are not evenly distributed, and there is a growing competition for water from different sectors, including industry, agriculture, power generation, domestic use, etc. As a result, one-third of the world's population is currently experiencing water scarcity. In water-scarce regions and countries, inequity in access to water resources is increasing because of competition for limited resources, and this particularly affects poor people [7]. However, the focus on freshwater without enough attention to its end products (wastewater) will exacerbate the water quality problem. It is therefore very important to consider wastewater management as a critical component in achieving future water security through integrated water resources management [8]. This is particularly true as wastewater is the only source of additional water that actually increases in quantity as population and water consumption grow.

3. Wastewater crisis

With increasing urbanization and changing lifestyles, increasing amounts of wastewater is being generated and where these are not sufficiently treated, freshwater bodies are continuously threatened [9]. Achieving the Sustainable Development Goal 6, which targets improved wastewater management, thus puts immense economic pressure especially on poor countries [10]. Inadequate infrastructure and sustainable management systems for the increasing volume of produced wastewater are at the heart of wastewater crisis in developing countries. As a result of inadequate infrastructure of wastewater treatment in most of the big cities where half of the world's population lives, the majority of wastewaters are discharged into the environment without any form of treatment, harming both the ecosystem and humans [9, 11]. Over half of the world's hospital beds are occupied by people suffering from diseases caused by contaminated water, and more people die as a result of polluted water than are killed by all forms of war. In many developing countries, an estimated 1.8 million children under 5 years old die every year due to water-related disease [12].

Wastewater damages the ecosystem in many ways: For example, wastewater may contain high levels of nutrients such as nitrates and phosphates. When water bodies receive excess amounts of these nutrients, it may stimulate excessive plant growth, which may release toxins into the water bodies, leading to oxygen depletion and causing what is known as de-oxygenated dead zones. This phenomenon decreases biodiversity and changes species composition and dominance, as well as decreasing water quality for reuse [11, 13]. Another example is the impact of wastewater on the climate around the globe: Wastewater treatment-related emissions of methane and nitrous oxide (powerful global warming gases) could rise by 50% and 25%, respectively, between 1990 and 2020 [1].

According to the fourth World Water Development Report by UNESCO [14], only 20% of globally produced wastewater receives proper treatment. Treatment capacity typically depends on the income level of the country; thus, in high-income countries, the treatment capacity reaches up to 70% of the generated wastewater compared to 8% in low-income countries [15]. Meeting the wastewater treatment challenge is thus not a luxury but a prudent, practical and transformative act, able to maintain public health and secure ecosystem health.

While so far, wastewater has mostly been seen as a treatment challenge, a paradigm shift toward its recognition as a resource for sustainable development is emerging. In this sense, wastewater can be reframed from being a problem to be disposed of to being a resource with social and economic value [5]. This shift offers wastewater to become part of an integrated, full life cycle, ecosystem-based management system that operates across the three main dimensions of sustainable development, that is, its social, economic and environmental pillars [1, 16].

4. Wastewater as a resource

Depending on the treatment or lack of it, as well as the degree of dilution, wastewater can be rich in resources such as nutrients, inorganic and organic compounds as well as energy, making it worthwhile for recovery and reuse. On the other hand, it can also be rich in chemical and microbial contaminants, and the improper use of untreated wastewater can have adverse effect on both human health and environment [13]. Wise wastewater management can therefore be a positive addition to the environment with significant returns in terms of enhancing food security, creating livelihood opportunities, climate change adaptation and sustainable ecosystem [1].

Successful examples of this paradigm shift can be found around the globe. There have been dramatic successes in using treated water for drinking purposes; for example, in Namibia, 35% of all drinking water is treated wastewater, and in Singapore, 30% of all water used is reclaimed water (and this percentage is increasing) [17]. The United States of America has also been seen several successes in treating wastewater for drinking purposes.

Wastewater can also be treated to provide energy. Various forms of energy can be recovered from wastewater and its biosolids, with biogas being the most prominent. It can be combusted on-site for heat or electricity generation, cleaned and sold to local natural gas providers or as fuel for vehicles [18–20]. Wastewater treatment plants are increasingly generating their own energy, which is an important achievement because energy consumption is a major cost in treatment plants. Another example from low-income countries is the transformation of fecal sludge (and other organic waste) into dry fuel like briquettes [21, 22]. The most common materials, however, that are recovered from wastewater are the water itself, which can be used for irrigation and its crop nutrients and biosolids as fertilizer. The use of fecal sludge as fertilizer is a well-known practice, especially from septic treatment plants given the low contamination within household-based on-site sanitation systems, compared to biosolids recovered from wastewater treatment plants. Some treatment processes recover nutrients, such as the N- and P-rich struvite, from wastewater during treatment rather than from the final products of the treatment [23–25].

5. The overlooked social dimension

The focus of most wastewater-related research has been on the technical aspects of the problem and improvements in terms of water quality and in minimizing environmental and health impacts, with very limited attention to its basic social and cultural sustainability dimensions [5, 9]. While, with increasing urbanization, wastewater treatment has moved further away from the household and its social roots, three types of campaigns (i) against open defecation, (ii) for the promotion of water-saving dry toilets and (iii) for using reclaimed water for drinking made it clear that sanitation depends strongly on social habits and acceptance. Where treatment is not keeping pace with population growth, and environmental pollution is threatening public health, the social dimension of wastewater management becomes obvious. Recognizing the role of the social base for wastewater management from risk reduction to reuse can have major implications, for example, on the choice and effectiveness of the technologies employed. Yet, usually, only limited information is available on the social perspective [5].

Wastewater management strategies have been traditionally driven by considerations of efficiency, safety, and cost-effectiveness. Even technology choices are often made by finance institutions outside the country, especially in low-income countries, often favoring "Northern" technology options. The emphasis on costs and benefits in this context would be acceptable if, in addition, other relevant factors could be included in the decision-making process by adopting a holistic methodology that includes the voices of all stakeholders and an analysis of sociological factors. Unless a holistic methodology is adopted, even cutting-edge technology might impede progress toward sustainable development, as the example of Toowoomba shows (Box 1). Likewise, the Singaporean success story would have had a very different outcome if public buy-in for wastewater reuse for drinking purposes had not been secured [5].

http://dx.doi.org/10.5772/67312

Queensland's Toowoomba in Australia is an often cited case illustrating the strength of public opinion regarding wastewater use. A plan to turn wastewater into drinking water failed in Toowoomba at a referendum in 2006, although water scarcity in the community was severe, to the point that water use for gardening was completely prohibited in the "Garden City." With no major river nearby, the community water supply had to be pumped uphill. During several years of drought, the 140,000 residents of Toowoomba and surrounding areas endured tough water restrictions. Local officials considered that the city had no choice but to treat and use parts of its wastewater for drinking water, and given the water crisis, they expected the program would be acceptable. However, the proposal met with fierce opposition from the community. In 2006, the residents of Toowoomba voted strongly against treating and using 25% of the city's wastewater. They relied instead on water piped from Brisbane's Wivenhoe Dam, at a cost to ratepayers of nearly $100 million more than the reuse program would have cost.

The Toowoomba proposal was an indirect wastewater use program, in which highly treated wastewater would be passed through an environmental buffer before being treated again, as part of the drinking water system. The public poll was accompanied by two dynamic campaigns building on the "yuck" and "fear" factors on one side, and social and financial arguments on the other. In the end, 62% of those polled opposed the project.

Sources: Ref. [5].

Box 1: Community resistance to wastewater reuse

A primary shortcoming in wastewater reuse is the lack of a combined sociotechnological planning and design methodology to identify and deploy the most sustainable solution in a given geographic and cultural context. The best practice, once a treatment or reuse technology has been developed, is to get early stakeholder buy-in and identify the best way to implement the technology in a participatory manner that is socially acceptable from the local perspective [26–28]. Stakeholders can be included in the decision-making process in different ways, including facilitating positive social learning processes, minimizing and resolving conflicts and, most importantly, using local knowledge and community participation [4].

5.1. Community participation and public involvement

Successful employment of appropriate technologies requires deep understanding of the social dynamics of the community in which they are applied [29–31]. This is only achieved through effective public involvement and community participation. Public involvement is best achieved through participation and involvement of users in all parts of the project cycle, from planning and design to implementation and decision-making, which produces more efficient and sustainable projects/outcomes [32]. In a sense, when communities have influence and control over decisions that affect them, they have a greater stake in the outcomes and are more committed to ensuring success.

Public involvement is of particular relevance when it comes to wastewater reuse, which is associated with major social concerns, including impacts on public health and safety, impacts on environmental quality as well as the benefits and risks of reuse. Thus, having an effective public involvement strategy from the planning phase to full implementation leads to greater acceptance and facilitates the implementation process of the wastewater reuse scheme. In other words, community participation can assure the social viability of the wastewater reuse practices [33–35]. Effective public involvement begins with early contact with potential users through the actual inclusion of all stakeholders and can involve educational and public awareness programs, the formation of advisory committees, and holding public workshops to discuss the benefits and risks of reuse [5, 36]. According to Ashley et al. [30], publicity, including advertisement in the media, education and inclusion of all stakeholders (politicians, experts and general public) in the decision-making process are the key elements for successful design and implementation of wastewater schemes. Gibson and Apostolidis [37] argue that the best way to involve the general public and to gain its support and acceptance is through successful demonstration projects.

For community participation to be as inclusive and effective as possible, the diversity of people within the same community should be acknowledged and dealt with. Communities are made up of individuals of different genders and groups of people who command different levels of power and wealth. Within each community, there are always competing interest groups. For example, there are rich and the poor, the farmers who have fields and livestock to water and the landless farmhands with children to care for, marginalized groups and members of socioeconomic minorities, housewives who need water for drinking and household and businessmen who own industries that require water. Thus, perception studies are a key component of any social analysis [26, 38].

5.2. Social perception and public acceptance

Even when wastewater is treated using advanced technologies and health risks are carefully addressed and controlled, irrespective of all scientific evidence, social perception remains the driver of the success or failure of wastewater reuse schemes. Depending on public perceptions, impressions and attitudes, the development of a wastewater scheme can be supported or constrained. Negative public perception can prevent well-planned projects from moving forward. On the other hand, positive public perception, which leads to greater acceptance, is the key element for successful implementation of wastewater recycling [5, 39]. Experience shows that the local communities have rejected a number of wastewater recycling projects by the governments and water boards around the world as a result of inadequate community consultation which led to negative public perception [40].

The degree of acceptance of wastewater reuse varies widely depending on the reuse purposes and is influenced by many factors, such as the degree of contact; expressions of disgust; education; risk awareness; the degree of water scarcity or availability of alternative water sources; calculated costs and benefits; trust and knowledge; issues of choice; attitudes toward the environment; economic considerations; involvement in decision-making; the source of water to be recycled; and experience with treated wastewater. Other factors that depend on the region and case include cultural, religious, educational and/or socioeconomic factors [5, 27, 35, 41–43].

Education and the level of physical contact (potable/no potable reuse) are the most influential factors that have been frequently associated with levels of acceptance of treated wastewater. In

Kuwait and Greece, for example, the willingness to accept recycled water increased with educational levels [44, 45]. However, as much as education and knowledge support public acceptance, nevertheless, direct exposure to the recycled water strongly influences its acceptance [46, 47]. For example, potable use is usually rejected due to health concerns. Wastewater use in agriculture generally is preferred to potable use, but more distant uses, such as landscape irrigation, are the most preferred [48, 49].

Several authors have investigated the association of sociodemographic descriptors with the acceptance of treated wastewater. The D'Angelo report [50] indicated that the acceptability of using recycled water in agriculture is higher for nonedible crops than for edible crops. For edible crops, the preference is for crops that must be peeled prior to human consumption, such as oranges and sweet corn. A relevant study [51] reported that the public's acceptance of reuse increases as the degree of human contact with the recycled water increases, with 97% and 96% of the public supporting wastewater reuse for irrigation and for toilet flushing, respectively, whereas only 20%–30% support potable reuse. Another study conducted by Friedler and Lahav [39] to determine the attitudes of the Israeli urban public toward possible urban reuse revealed that the majority of participants supported options perceived as low contact, such as irrigation of public parks (96%), sidewalk landscaping (95%) and use in the construction industry (94%), while higher contact reuse options, such as commercial launderettes (60%), found less support. According to Bruvold [52], the degree of human contact has a greater effect when people were asked about general use options, whereas when the specific use scheme was used, other factors such as health, environment, treatment, distribution and conservation had greater impact on people's perceptions. Therefore, he argues that it is essential to weigh the different objectives of the recycling options in coordination with people's/users' acceptability and preference and select the recycling projects which are most likely to be accepted by the community and therefore make the project implementation successful.

5.3. Gender roles and implications

As mentioned above, successful community participation is better achieved by acknowledging the diversity of people within the community. This includes gender, age, education level, power, wealth and so on [53]. In this context, it is very important to acknowledge the differences of interests and roles between men and women as different stakeholders. There are a number of gender aspects which influence how both genders are involved in and benefit from improvements to the water. In many developing countries, women have limited access to education and other resources and services, have heavier workloads, are more constrained by poor health, have a lower social status, and are poorly represented in decision-making at both household and community levels [54]. Thus, balanced attention is needed in the form of distinctions between what women and men know, do and decide and what the effects are for them, their families and communities (Box 2).

In general, women are most vulnerable to water-related disasters, including water scarcity and bad water quality. Many infectious diseases are associated with poor water quality, and these are reported as being among the fifth biggest killer of women worldwide, causing more deaths than AIDS, diabetes or breast cancer [55]. Dirty water and poor sanitation are also at the root of problems such as maternal and child mortality and sexual violence. Many women in developing countries give birth at home without access to clean water, exposing themselves and their babies to infections. More than 50 million primary-school-aged girls in developing countries are not in school because they are required to fetch water and firewood [56]. Thus, it becomes a necessity to bring women frequently on the scene for consultation and allow their full participation in wastewater management. Implementing a gender-sensitive approach produces more effective, efficient and affordable outcomes. Including women in water and wastewater management planning often makes for fewer oversights in technical planning and improves resource and financial management, as well as allowing for greater transparency [54].

Acknowledging gender roles and differences not only contributes to the success of a project, but offers planning options to optimize the overall social and economic development and reduces competition and conflicts over water resources. In most societies, the provision of water for the fulfillment of fundamental human needs has always been women's responsibility, yet their participation in decision-making is very limited if there at all. To bring about constructive change, more efforts are needed to better understand the gender implications in water sector [57, 58].

Thoughtful safety interventions must be gender sensitive. In many cultures, women carry the main responsibility for hygiene and health, also vis-à-vis greywater or wastewater use as reported, for example, from Jordan, Vietnam and Tunisia. The strong connection between women and water use at household level offers significant potential for innovative training approaches to improve the social acceptance of safe water reuse, as recently demonstrated in Jordan. Also the use of protective clothing should be gender-specific. In Vietnam, women were observed wearing protective gloves and boots more consistently than men. The differences were attributed to the gendered work division on the farm, with men walking around the farms much more than women, and where protective clothing constrained men's movements. Sources [5, 59–61].

Box 2: Gender roles

6. Key improvement areas

There is little known about public perceptions of wastewater reuse in the literature, and it is mostly documented in a limited number of locations, that is, the United States, Australia and Western Europe. Still a lot of more studies are required at national and subnational context in order to avoid outcomes being transferred from one country to another, which is always inappropriate due to the range of factors that influence public acceptance from country to another, including culture, religion, economy, climate and water availability [40].

In general, public acceptance of reuse is not straightforward, but it is always easier when water scarcity is already affecting the public, so that they perceive wastewater reuse as a solution rather than a problem [5, 62]. However, for greater acceptance, public and private concerns and benefits must be aligned. Public concerns about risks are to be weighed against the benefits of using treated water. The dialogue should be built on mutual trust to provide the right climate for negotiation and conflict resolution [5].

Certain social factors have always been associated with poor acceptance of wastewater reuse, including the lack of coordination between the authorities involved in planning; inadequate community consultation; lack of trust in technology; social pressure and fear of social backlash; and fear of losing markets in case of wastewater reuse in irrigation [41, 63, 64]. Another factor is overlooking the gender dimension [65].

In order to fill the gap in knowledge regarding the social dimension of wastewater reuse, extensive social research into public perceptions of wastewater reuse is needed. Some of the priority areas to focus on are as follows (1) to understand judgment strategies that shape public decisions to support or reject wastewater reuse; (2) to identify factors influencing people's risk perceptions; (3) to investigate the role of trust in the authorities and the limits in scientific knowledge in people's decision-making processes to either accept or reject the reuse; (4) to examine how factors such as health, environment, treatment, distribution and conservation can affect people's willingness to use recycled water; (5) to examine people's sensitivity with regard to the disgust emotion or "yuck" factor; (6) to understand the impact of the source of wastewater on people's decisions; (7) to understand how the economic advantages in using recycled water can facilitate public acceptance; and (8) to identify possible environmental justice issues that may affect public acceptance [66, 67].

With regard to gender implications on community participation and public acceptance, greater women's participation is needed through effective gender mainstreaming strategies. Obstacles to women's participation generally include lack of confidence, family commitment including child care, heavy workload and time constraints, traditional values and stereotypes, fears of men and husbands who prevent women from participating (many women said that their husbands do not support their participation in public life) [68, 69].

Apart from the social and cultural issues, another reality with regard to gender mainstreaming is the lack of general awareness of the significance of gender factors in water and wastewater management, which applies to both leaders and decision-makers who work in water management programs. Another shortcoming is the lack of gender-disaggregated data, which is the only way to move forward from principles to practice in gender mainstreaming [70, 71].

In order to fill the gender gap in wastewater reuse, investing time and effort in awareness raising on the different needs and impacts for women and men at all levels is part of the necessary training for all professionals in the wastewater sector. Nevertheless, gender mainstreaming is a continuous process and a holistic approach, which cannot be achieved by a single training session.

Some of the key issues to focus on are as follows: (1) to acknowledge both men's and women's roles and responsibilities, energy, experience and knowledge in contributing to the effectiveness of wastewater reuse programs as well as identifying their different needs and priorities; (2) to mainstream gender throughout all projects' cycle from planning and design to implementation as well as related policy; (3) to ensure women's participation in consultation committees and educational workshops in terms of timing and allocation of these meetings by taking into consideration their family responsibilities (e.g., domestic work), otherwise, women may choose not to participate to avoid conflict with their responsibilities; and (4) "at institutional level" to train the technical staff working in research and development to integrate gender dimensions into the socioeconomic aspects of research work, in order to address the differential impacts of structural interventions and the appropriation of new technologies [53, 54, 72].

7. Conclusions

With increasing pressures on water resources, wastewater recycling and reuse have rapidly become an imperative for integrated water management strategies. However, along with the technology advancement in wastewater treatment, societal factors such as public perception, public acceptance and the dimension of gender have great implications on the success of wastewater reuse.

Adopting a sociotechnological approach by means of considering all social factors together with technology in wastewater recycling results in great improvements in terms of effectiveness and efficiency as the infrastructure will be more widely used and optimally sustained by all user groups including women and men. It will also contribute to the overall development of the society by increasing consumption, production, income, environmental security, health and overall family welfare, along with securing water resources when addressing the societal issues of the service delivered. Another gain of the sociotechnological approach in water sector is the sustainability of the service, in the sense that equal participation of all stakeholders in research and project implementation can increase the potential, flexibility and creative innovation in responding to water insecurity.

Acknowledgements

The authors would like to acknowledge the World Academy of Science (TWAS) for making this work possible through TWAS-UNESCO Scheme, as well as the support of the CGIAR Research Program on Water, Land and Ecosystems.

Author details

Dalia Saad[1,2*], Deirdre Byrne[1] and Pay Drechsel[2]

*Address all correspondence to: dalianono@gmail.com

1 International Water Management Institute, Colombo, Sri Lanka

2 Institute for Gender Studies, University of South Africa, Pretoria, South Africa

References

[1] Corcoran E, Nellemann C, Baker E, Bos R, Osborn D, Savelli H. (eds.). Sick water? The central role of wastewater management in sustainable development. A rapid response assessment. United Nations Environment Programme, UN-HABITAT, GRID-Arendal; 2010. www.grida.no. ISBN: 978-82-7701-075-5.

[2] Qadir M, Wichelns D, Raschid-Sally L, McCornick PG, Drechsel P, Bahri A, Minhas PS. The challenges of waste-water irrigation in developing countries. Agricultural Water Management. 2010; **97**: 561–568.

[3] Yang H, Abbaspour K. Analysis of wastewater reuse potential in Beijing. Desalination. 2007; **212**: 238–250.

[4] Drechsel P, Qadir M., Wichelns D. (eds.). Wastewater: economic asset in an urbanizing world. Dordrecht, Netherlands: Springer; 2015, pp. 3–14.

[5] Drechsel P, Mahjoub O, Keraita B. Social and cultural dimensions in wastewater use. In: Drechsel P, Qadir M, Wichelns, D. (eds.). Wastewater: Economic Asset in an Urbanizing World. Dordrecht, The Netherlands: Springer; 2015, pp. 75–92.

[6] Vigneswaran S, Sundaravadivel M. Recycle and reuse of domestic wastewater. Saravanamuthu (Vigi) Vigneswaran, in Encyclopedia of Life Support Systems (EOLSS). Developed under the Auspices of the UNESCO, Eolss Publishers, Oxford, UK. http://www. eolss. net 2004.

[7] UNEP. Africa: Atlas of our Changing Environment. United Nations Environment Programme. Division of Early Warning, Nairobi; 2008.

[8] OECD. Water Quality and Agriculture—Meeting the Policy Challenge. OECD Studies on Water. OECD Publishing; 2012.

[9] Mekala G, Davidson B, Samad M, Boland A. A Framework for Efficient Wastewater Treatment and Recycling Systems. Colombo, Sri Lanka: International Water Management Institute; 2008, p. 23 (IWMI Working Paper 129).

[10] Hutton G. Editorial: Can we meet the costs of achieving safely managed drinking-water, sanitation and hygiene services under the new sustainable development goals? Water, Sanitation and Hygiene for Development. 2016; **6**: 191–194.

[11] World Water Council. 6th World Water Forum, Marseille, 12–17 March 2012. http://www.solutionsforwater.org/objectifs/1-2-8-operator-efficiency-and-effectiveness-in-urban-wastewater-collection-and-treatment. Accessed 31/3/14.

[12] WHO. The Global Burden of Disease: 2004 update. Geneva: World Health Organization; 2008.

[13] UN-Water. Wastewater Management. A UN-Water Analytical Brief; 2015.

[14] UNESCO. Managing water under uncertainty and risk. The United Nations World Water Development Report 4. Paris: United Nations Educational, Scientific and Cultural Organization; 2012.

[15] Sato T, Qadir M, Yamamoto S, Endo T, Zahoor A. Global, regional, a country level need for data on wastewater generation, treatment, and use. Agricultural Water Management. 2013; **130**: 1–13.

[16] Özerol G, Günther D. The role of socio-economic indicators for the assessment of wastewater reuse in the Mediterranean region. In: Hamdy A, El Gamal F, Lamaddalena N, Bogliotti C, Guelloubi R. (eds.). Non-Conventional Water Use: WASAMED Project. Bari: CIHEAM/EU DG Research; 2005, pp. 169–178 (Options Méditerranéennes: Série B. Etudes et Recherches; n. 53.)

[17] Lazarova V, Asano T, Bahri A, Anderson J. Milestones in Water Reuse: The Best Success Stories. IWA; 2013, p. 408.

[18] Oki T, Kanae S. Global hydrological cycle and world water resources. Science. 2006; **313**: 1068–1072.

[19] Zimmerman J, Mihelcic, Smith J. Global stressors on water quality and quantity. Environmental Science Technology. 2008; **42**: 4247–4254.

[20] Conley D, Paerl H, Howarth R, Boesch D, Seitzinger S, Havens K, Lancelot C, Likens G. Controlling eutrophication: nitrogen and phosphorus. Science. 2009; **323**: 1014–1015.

[21] Funamizu N, Iida M, Sakakura Y, Takakuwa T. Reuse of heat energy in wastewater: implementation examples in Japan. Water Science and Technolgy. 2001; **43**: 277–285.

[22] Logan B, Hamelers B, Rozendal R, Schröder U, Keller J, Freguia S, Aelterman P, Verstraete W, Rabaey K. Microbial fuel cells: methodology and technology. Environmental Science and Technolgy. 2006; **40**: 5181–5192.

[23] De-Bashan L, Bashan Y. Recent advances in removing phosphorus from wastewater and its future use as a fertilizer (1997–2003). Water Research. 2004; **38**: 4222–4246.

[24] Guest et al. A new planning and design paradigm to achieve sustainable resource recovery from wastewater. Environmental Science and Technology. 2009; **43**: 6126–6130.

[25] Larsen T, Alder A, Eggen R, Maurer M, Lienert J. Source separation: will we see a paradigm shift in wastewater handling? Environmental Science and Technology. 2009; **43**. doi:10.1021/es803001r

[26] Hartley TW. Public perception and participation in water reuse. Desalination. 2006; **187**: 115–126.

[27] Po M, Kaercher J, Nancarrow B. Literature review of factors influencing public perceptions of water reuse. Australian Water Conservation and Reuse Research Program, CSIRO Land and Water; 2004.

[28] Macpherson L, Slovic P. Stigma and fear—changing mental models about reuse. WE&T Magazine. July 2008; **20**(7).

[29] Jeffrey P, Temple C. Sustainable water management: some technological and social dimensions of water recycling. Sustainable Development International. 1999; **1**: 63–66.

[30] Ashley R, Blackwood D, Butler D, Jowitt P, Davies J, Smith H, Gilmour D, Oltean-Dumbrava C. Making asset investment decisions for wastewater systems that include sustainability. Environmental Engineering. 2008; **134**: 200–209.

[31] Marks J, Martin B, Zadoroznyj M. How Australians order acceptance of recycled water—national baseline data. Sociology. 2008; **44**: 83–99.

[32] Maharaj N, Athukorala K, Vargas M, Richardson G. Mainstreaming Gender in Water Resources Management: Why and How. Background Paper for the World Vision Process. Paris, France: World Water Vision Unit; 1999. Available at: http://www.worldwatercouncil.org/fileadmin/wwc/Library/Publications_and_reports/Visions/GenderMainstreaming.pdf

[33] Davis R, Hirji R. Wastewater reuse. Working Paper—Report No: 26325, Water Resources and Environment. Technical Note no. F.3. Washington, DC: The World Bank; 2003.

[34] WHO. Guidelines of the Safe Use of Wastewater, Excreta and Grey Water; Vol. 2: Wastewater Use in Agriculture. Geneva, Switzerland: World Health Organization; 2006.

[35] U.S. Environmental Protection Agency (USEPA). Guidelines for Water Reuse. Chapter 8. EPA. EPA/600/R-12/618. Washington, DC: Environmental Protection Agency; 2012.

[36] Friedler E, Lahav O. Centralised urban wastewater reuse: what is the public attitude? Water Science & Technology. 2006; **54**: 423–430.

[37] Gibson H, Apostolidis N. Demonstration, the solution to successful community acceptance of water recycling. Water Science Technology. 2001; **43**: 259–266.

[38] Abu-Madi M, Al-Sa'ed R, Braadbaart O, Alaerts G. Perceptions of farmers and public towards irrigation with reclaimed wastewater in Jordan and Tunisia. Arab Water Council Journal. 2008; **1**. Available at: http://www.academia.edu/1152959/Public_perceptions_towards_wastewater_reuse_in_Jordan_and_Tunisia

[39] Friedler E, Lahav O, Jizhaki H. Study of urban population attitudes towards various wastewater reuse options: Israel as a case study. Environmental Management. 2006; **81**: 360–370.

[40] PMSEIC (Prime Minister's Science, Engineering and Innovation Council). Recycling Water for Our Cities. Paper Prepared by an Independent Working Group for PMSEIC, 28 November 2003, pp. 1–45.

[41] Wilson Z, Pfaff B. Religious, philosophical and environmentalist perspectives on potable wastewater reuse in Durban, South Africa. Desalination. 2008; **228**: 1–9.

[42] Mara D. The production of microbiologically safe effluents for wastewater reuse in the Middle East and North Africa. Water, Air, and Soil Pollution. 2000; **123**: 595–603.

[43] Kaercher J, Po M, Nancarrow B. Water Recycling Community Discussion Meeting I. Perth: Australian Research Centre for Water in Society (ARCWIS); 2003.

[44] Tsagarakis K, Georgantzís N. The role of information on farmers' willingness to use recycled water for irrigation. Water Science and Technology: Water Supply. 2003; **3**: 105–113.

[45] Alhumoud J, Madzikanda D. Public perceptions on water reuse options: the case of sulaibiya wastewater treatment plant in Kuwait. International Business and Economics Research Journal. 2010; **9**: 141–158.

[46] Po M, Nancarrow BE, Leviston Z, Poter NB, Syme GJ, Kaercher JD. Predicting Community Behaviour in Relation to Wastewater Reuse: What Drives Decisions to Accept or Reject? Perth: CSIRO; 2005.

[47] Hamilton A, Stagnitti F, Xiong X, Kreidl S, Benke K, Maher P. Wastewater irrigation the state of play. Vadose Zone. 2007; **6**: 823–840.

[48] Higgins J, Warnken J, Sherman PP, Teasdale PR. Surveys of users and providers of recycled water: quality concerns and directions for applied research. Water Research. 2003; **36**: 5045–5056.

[49] Dolnicar S, Saunders C. Recycled water for consumer markets—a marketing research review and agenda. Desalination. 2006; **187**: 203–214.

[50] D'Angelo Report. Using Reclaimed Water to Augment Potable Water Resources. Public Information Outreach Programs (Special Publication, Salvatore D'Angelo, Chairperson). Water Environment Federation & American Water works Association; 1998.

[51] Denlay J, Dowsett B. Water Reuse the Most Reliable Water Supply Available. Sydney, Australia: Friends of the Earth Inc.; 1994, p. 57. Report prepared as part of the Sydney Water Project.

[52] Bruvold W. Public opinion on water use options. Journal WPCF. 1988; **60**: 45–49.

[53] McConville J, Mihelcic J. Adapting life-cycle thinking toolsW to evaluate project sustainability in international water and sanitation development work. Environmental Engineering Science. 2007; **24**: 937–948.

[54] Mahaarcha W, Kittisuksathit S. Gender disparities in access to safe drinking water: evidence from Kanchanaburi, Thailand. Proceedings: Regional Conference on Urban Water and Sanitation in Southeast Asian Cities; 2006.

[55] Fletcher A, Schonewille R. Overview of Resources on Gender-Sensitive Data Related to Water. Gender and Water Series. WWAP. Paris: UNESCO; 2015.

[56] United Nations Children's Fund. Harnessing the Power of Data for Girls: Taking Stock and Looking Ahead to 2030. New York: UNICEF; 2016.

[57] CAP-NET, GWA. Why Gender Matters: A Tutorial for Water Managers. Multimedia CD and Booklet. Delft: CAP-NET International network for Capacity Building in Integrated Water Resources Management; 2006.

[58] Carolyn H, Andersson I. Gender Perspectives on Ecological Sanitation. Stockholm: Ecosanres; 2002.

[59] Boufaroua M, Albalawneh A, Oweis T. Assessing the efficiency of grey-water reuse at household level and its suitability for sustainable rural and human development. British. Applied Science and Technology. 2013; **3**: 962–972.

[60] Knudsen LG, Phuc PD, Hiep NT, Samuelsen H, Jensen PK, Dalsgaard A, Raschid-Sally L, Konradsen F. The fear of awful smell: risk perceptions among farmers in Vietnam using waster and huma excreta in agriculture. The Southeast Asian Journal of Tropical Medicine and Public Health. 2008; **39**: 341–352.

[61] Mahjoub O. Ateliers de sensibilisation au profit des agriculteurs et des femmes rurales aux risques liés à la réutilisation des eaux usées en agriculture: Application à la région de Oued Souhil, Nabeul, Tunisie. In: UN-Water. Proceedings of the Safe Use of Wastewater in Agriculture, International wrap-up event, Teheran, Iran; 2013.

[62] Fawell J, Fewtrell L, Hydes O, Watkins J, Wyn-Jones P. A Protocol for Developing Water Reuse Criteria with Reference to Drinking Water Supplies (UKWIR Report Ref No. 05/ WR/29/1). London, UK: Water Industry Research Limited; 2005.

[63] Sydney Water. Community Views on Recycled Water. Sydney; 1999.

[64] Boland A. The Use of Recycled Water in Australian Horticulture Keynote Address at Irrigation 2005—Irrigation Association of Australia Conference. Townsville; 2005.

[65] Mainstreaming Gender in Water Management, Resource Guide, UNDP; 2006.

[66] Po M, Kaercher J, Nancarrow B. Literature Review of Factors Influencing Public Perceptions of Water Reuse. Technical Report 54/03. 2003.

[67] Dolnicar S, Schafer A. Desalinated versus recycled water: public perceptions and profiles of the accepters. Environmental Management. 2009; **90**: 888–900.

[68] Gordon S. Non-Formal Education and Training of Women in Development Projects. Washington, DC: World Bank; 1982.

[69] Tanzania, Ministry of Water and Energy; DANIDA, and Institute of Resource Assessment and Centre for Development Studies. Water master plans for Iringa, Ruvuma and Mbeya regions, socio-economic studies: village participation. Water and health. 1983; Vol. 13. Dar es Salaam, Tanzania, Ministry of Water and Energy.

[70] IFAD (International Fund for Agricultural Development). Gender and Water. Securing Water for Improved Rural Livelihoods: The Multiple-Uses System Approach. Rome: IFAD; 2007. Available at: http://www.ifad.org/gender/thematic/water/gender_water.pdf

[71] Seager J. Gender and water: Good rhetoric, but it doesn't "count". Geoforum. 2010; **41**: 1–3.

[72] Ray I. Women, water and development. Annual Review of Environment and Resources. 2007; **32**: 421–449.